*Valim Levitin*

**Interatomic Bonding in Solids**

## *Related Titles*

Har, J.J., Tamma, K.K.

**Advances in Computational Dynamics of Particles, Materials and Structures**

2012
ISBN: 978-0-470-74980-7 (Also available in digital formats)

Horstemeyer, M.F.

**Integrated Computational Materials Engineering (ICME) for Metals**
**Using Multiscale Modeling to Invigorate Engineering Design with Science**

2012
ISBN: 978-1-118-02252-8 (Also available in digital formats)

Schmitz, G.J., Prahl, U. (eds.)

**Integrative Computational Materials Engineering**
**Concepts and Applications of a Modular Simulation Platform**

2012
ISBN: 978-3-527-33081-2 (Also available in digital formats)

Nikrityuk, P.A.

**Computational Thermo-Fluid Dynamics**
**In Materials Science and Engineering**

2011
ISBN: 978-3-527-33101-7 (Also available in digital formats)

Oganov, A.R. (ed.)

**Modern Methods of Crystal Structure Prediction**

2011
ISBN: 978-3-527-40939-6 (Also available in digital formats)

Vaz Junior, M., de Souza Neto, E.A., Munoz-Rojas, P.A. (eds.)

**Advanced Computational Materials Modeling**
**From Classical to Multi-Scale Techniques**

2011
ISBN: 978-3-527-32479-8 (Also available in digital formats)

Roters, F., Eisenlohr, P., Bieler, T.R., Raabe, D.

**Crystal Plasticity Finite Element Methods**
**in Materials Science and Engineering**

2010
ISBN: 978-3-527-32447-7 (Also available in digital formats)

*Valim Levitin*

# Interatomic Bonding in Solids

## Fundamentals, Simulation, and Applications

**Author**

***Prof. Dr. Valim Levitin***
Friedrich-Ebert-Str. 66
35039 Marburg
Germany



**Library of Congress Card No.:**
applied for

**British Library Cataloguing-in-Publication Data:**
A catalogue record for this book is available from the British Library.

**Bibliographic information published by the Deutsche Nationalbibliothek**
The Deutsche Nationalbibliothek lists this publication in the Deutsche Nationalbibliografie; detailed bibliographic data are available on the Internet at http://dnb.d-nb.de.



**Print ISBN** 978-3-527-33507-7
**ePDF ISBN** 978-3-527-67158-8
**ePub ISBN** 978-3-527-67157-1
**Mobi ISBN** 978-3-527-67156-4
**obook ISBN** 978-3-527-67155-7

**Cover Design** Adam Design, Weinheim
**Typesetting** le-tex publishing services GmbH, Leipzig
**Printing and Binding** Markono Print Media Pte Ltd, Singapore

Printed in Singapore
Printed on acid-free paper

# Contents

*This book is dedicated to my wife Lydia and my son Viktor with sincere gratitude for understanding, encouragement and help.*

# Preface

The progress in many areas of industry and technology depends increasingly on the development of new materials and processing techniques. Materials science explores fundamental properties of solids. The knowledge acquired in this area is of great importance for production and everyday life.

The aim of materials science is to explain fundamentals of processes in solids and to predict the material behavior in real conditions of operation. Achievements of condensed matter physics and chemistry combined with various experimental data make it possible to solve these problems.

A variety of solids is related to the distinction of a type and strength of the interatomic bonding.

The minute particles, which a solid consists of, have the extraordinary quantum features. However, there is a gap between quantum theory on the one hand and engineering on the other hand. Even the principal notions and terms are different. The quantum physics operates with such notions as electron, nucleus, atom, energy, the electronic band structure, wave vector, wave function, Fermi surface, phonon, and so on. The objects in the engineering material science are: crystal lattice, microstructure, grain size, alloy, strength, strain, wear properties, robustness, creep, fatigue, and so on.

It is important to understand a connection between a quantum behavior of the structure elements of a substance and parameters that determine the macroscopic properties of materials.

Mechanical, electrical, magnetic properties had, for many years, a basic meaning in order to study and predict the behavior of a material in designs, components, circuits, and engines. The technique of determining the properties by the trial and error method is still widely used.

Many properties of materials can now be determined directly from the fundamental equations for systems nuclei–electrons providing new insights into critical problems in materials science, physics, and chemistry. "Increasingly, electronic structure calculations are becoming tools used by both experimentalists and theorists to understand characteristic properties of matter and to make predictions for real materials and experimentally observable phenomena [1]."

The trend of the knowledge development is as follows: one goes from the known electronic structure of atoms to features of molecules and solids. In other words, one tries to derive the peculiarities of interatomic bonding in solids from the peculiarities of electrons in free atoms. Such research in condensed matter physics and chemistry has been performed a lot over the last 20 years mainly caused by advances in calculation techniques owing to the design of numerical algorithms.

There are also applications of quantum theory for instance in the onset of a failure in a material. The failure starts on the atomic scale when an interatomic bonding is stressed beyond its yield-stress threshold and breaks. The initiation and diffusion of point defects in crystal lattice turn out to be a starting point of many failures. These events occur in a stress field at certain temperatures. The phenomena of strain, fatigue crack initiation and propagation, wear, and high-temperature creep are of particular interest. The processes of nucleation and diffusion of vacancies in the crystal lattice determines the material behavior at many operation conditions.

The subject of this book is of an interdisciplinary nature. The theme is at a junction of physics, chemistry, materials science, computer simulation, physical metallurgy, and crystallography. The goal of the book is particularly to demonstrate how the properties of materials can be derived and predicted proceeding from the quantum properties of their structural elements. We would like to show the methods of the model construction and simulation.

Thanks to the research of Walter Kohn and his collaborators the first principle simulation of structure and properties of solids is available. This simulation using contemporary density functional theory has proved to be a reliable and computationally tractable tool in condensed matter physics and chemistry.

One feels, however, that we are now on the threshold with regard to the applications in materials science. This is because, as a result of the progress in the area, we now deal with much more complex structures of real materials. Also, the practitioners (engineers and technologists) are not sufficiently acquaint enough with up-to-date methods of structural investigations, simulation, and computations.

This book covers the important topics of quantum physics, chemistry, simulation, and modeling in solid state theory, an application of electronic and atomic properties to service performance of materials. The book is not an exhaustive survey of the applications. Other authors would have chosen different topics. Nevertheless, I hope that the book will introduce the reader into this vast area of solid-state physics and chemistry and its applications.

Science as well as engineering is very differentiated now. We need an interdisciplinary approach to processes in science, technology, and engineering. In addition, another view of events, phenomena and laws can be very useful in our own specialty. On the other hand, a presentation of theories and laws has to be done at an accessible level.

I hope that this book can be used for courses such as:

- solid-state physics and chemistry as engineering fundamentals;
- the computer simulation of solids;
- condensed matter fundamentals in modern technology;
- quantum physics as basis of materials behavior.

This book is appropriate for final-year undergraduates and first-year graduate students. First of all I mean students in the area of materials science, solid-state physics, chemistry, engineering, materials technology, and machine building. The book may also be used as a preparatory material for students starting a doctorate in condensed matter physics or for recent graduates starting research in these fields of industry. It should be used as a textbook in an upper-level graduate engineering course. I believe the book will be useful for researches and practitioners from industry sectors including metallurgical, mechanical, chemical, and structural engineers.

I would like to suggest that the reader would take up the simulation of a crystal structure, which is of interest for him or her. The study of solids by means of the first-principle simulation based on the density functional theory allows one to obtain new and useful data. The definite advantage of the method is a possibility in a variation of a composition, a type of the crystal structure and the unit cell. The computer simulation is a promising addition to experimental results. Besides, it is a creative and fascinating occupation for a researcher.

# 1
# Introduction

The experiment is the supreme judge of any physical theory.

*Lev Landau*

The behavior of a solid in the force, electric, or magnetic field depends on the type and energy of interatomic bonding.

As a first approximation, we can consider processes, which determine properties and structures of solids, at five levels, namely: the electronic, the atomic, the microscopic, the mesoscopic, and the macroscopic levels. Table 1.1 presents these conventional levels. At each level, the processes take place in a space dimension given by a characteristic length. The characteristic length is a linear dimension, where a corresponding process occurs. It goes without saying that there are no clear boundaries between characteristics defining these levels. Some physical phenomena have significant manifestations on more than one level of length. The corresponding experimental techniques and some methods used for the theoretical study and the simulation of the phenomena are also shown in Table 1.1.

The smallest length scale of interest is about tenths and hundredths of a nanometer. On this scale one deals in a system directly with the electrons, which are governed by the Schrödinger equation of quantum physics. The techniques that have been developed for solving this equation are extremely computationally intensive. These calculations are theoretically the most rigorous; their data are also used for developing and validating more approximate but computationally more facilitated descriptions.

The electronic level of properties of solids is of primary importance. This is no mere chance that an important subfield of condensed matter physics and chemistry is focused on the electrons in solids. Basic sciences are fundamentally concerned with understanding and exploiting the properties of interacting electrons and atomic nuclei. With this comes the recognition that, at least in principal, almost all problems of materials can be and should be addressed within quantum theory. An understanding of the behavior of electrons in solids is essential for explanation and prediction of solid state properties.

In a sense, electrons form the glue holding solids as whole, and are central in determining structural, mechanical, electrical, and vibrational properties. The un-

*Interatomic Bonding in Solids: Fundamentals,Simulation,andApplications*, First Edition. Valim Levitin.

**Table 1.1** The levels of properties in solids.

| Level | Characteristic length (m) | Experimental technique | Methods of theoretical study |
|---|---|---|---|
| Electronic | $10^{-11}$–$10^{-9}$ | Emission and absorption spectra; photoelectric effect; electric, magnetic, thermal properties of solids | The Schrödinger equation; pseudopotentials; density functional theory; tight-binding approximation; embedded atom method |
| Atomic | $10^{-9}$–$10^{-6}$ | X-ray diffraction; transmission electron microscopy; atomic-force microscopy | Molecular dynamics; Monte-Karlo method |
| Microscopic | $10^{-6}$–$10^{-4}$ | Light microscopy; nondestructive testing | Finite-element analysis; solution of differential equations |
| Mesoscopic | $10^{-4}$–$10^{-2}$ | Mechanical testing | Classical mechanics |
| Macroscopic | $> 10^{-2}$ | Mechanical testing; strain measurement | Theory of elasticity; elastic fracture mechanics |

derstanding of strength, plasticity, electric properties, magnetism, superconductivity, and most properties of solids requires a detailed knowledge of the "electronic structure," which is the term associated with the study of the electronic energy levels. The concept of the energy is central in physics. Nearly all physical properties are related to total energies or to differences between total energies.

The atomic level spans from nanometers to micrometers. Here, theoretical and experimental techniques are well developed, requiring the specification with parameters fitted to electronic-structure calculations. The most important feature of atomic simulation is that one can study a system of a relatively large number of atoms.

Above the atomic level the relevant length scale is 1 μm.

The terms simulation, modeling, calculating, and computing all refer to formulating and solving various equations which describe, explain, and predict properties of materials. If we also want to study formation and breaking of bonds, optical properties, and chemical reactions, we have to use the principles of the quantum theory as the basis for our simulation.

Most simulations utilize idealized crystalline symmetry, thus diverging from accurate description of technologically important "real materials." This is to make models more tractable or solvable at all.

*Ab initio* quantum chemistry has now achieved such a level of maturity that it can satisfactorily predict most properties of isolated, relatively small molecules from a theoretical point of view. One now attempts to apply the theory to more complicated and expensive experimental observations. However, there is an even greater need for the computer simulation of solids to be equally predictive. Good correspondence with experiments is the criterion, and it is the accuracy of this correspondence which measures the worth of simulations rather than pure numerical precision of results.

Major advances in prediction of the structural and electronic properties of solids come from two sources: improved performance of hardware and development of new algorithms, and their software. Improved hardware follows technical advances in computer design and electronic components. Such advances are frequently characterized by the Moore law, which states that computer power doubles every 2 years or so. This law has held true for the past 20 or 30 years and one expects that it will hold for the next decade, suggesting that such technical advances can be predicted. In clear contrast to hardware, the development of new high performance algorithms did not show such rapid growth. Nonetheless, over the past half century, most advances in the theory of the electronic structure of matter have been made with new algorithms as opposed to better hardware. One may reasonably expect these advantage to continue. Physical concepts such as density functional theory and pseudopotentials coupled with numerical methods such as iterative methods have permitted one to examine much larger systems than one could handle solely by more and more power hardware.

This book consists of 18 chapters. This introduction is the first one.

A succinct description of the quantum physics fundamentals is presented in Chapter 2. I recall particle–wave dualism, uncertainty principle, concepts of wave motion. The wave function and the Schrödinger equation and an abstract notion of $k$-space are discussed.

Chapter 3 is devoted to atoms. One-electron atom and multi-electron atoms of chemical elements are considered. The probability density functions (orbitals) for electrons are illustrated. The Hartree theory is presented as a first method of approximation that has been proposed in order to calculate wave functions and energies of electrons in atoms. The covalently bonded diatomic molecules are subject of the consequent consideration.

Chapter 4 deals with the crystal lattice. Here, we discuss a basic concept of reciprocal lattice in detail and present the Wigner–Seitz cell and the Brillouin zone. These notions are commonly used in any description of the energy of electrons in solids.

We consider a homogenous gas that consists of free electrons in Chapter 5. Notions of exchange energy and correlation energy of electrons are introduced. The theory enables one to calculate some macroscopic properties of simple metals, which have the $ns^1$ external electronic shell. The calculated cohesive energy of simple metals turns out to fit the experimental values satisfactorily.

Chapter 6 is devoted to behavior of electrons in a crystal lattice. The obvious Kronig–Penny model demonstrates the influence of periodic potential of crystal lattice on the electronic structure. The Bloch waves in the crystal lattice are described. Description of conductors, insulators, and semiconductors follows consideration of a general structure of energetic bands.

Some criteria of strength of the interatomic bonding in solids are treated in Chapter 7. Especially, we consider elastic constants, amplitudes of atomic vibrations, melting temperature, and the energy of the vacancy formation.

A technique of the solid simulation starting from the first principles (*ab initio* theory) is the subject of Chapter 8. We study milestones in solution of the many-body problem. We describe the density functional theory as an essence of the technique. The Kohn–Sham approach, pseudopotential method, iterative technique of calculations are described here. These methods enable one to determine and calculate the equilibrium structure of a solid quantitatively and self-consistently.

The content of Chapter 9 sheds light on validity and application for different solids of the theory, which has been considered in previous chapters. The calculated values of cohesive energy and bulk modulus are compared with the experimental results. We present data on superconductivity, the embrittlement of metals, the electronic density of states, properties of intermetallic compounds, and the energy of vacancy formation.

The $Ni_3Al$-based solid solutions are the subject of Chapter 10. Experimental methods and also the computer simulating technique are used for these technologically important intermetallic compounds. An increase in elastic constants as a result of the replacement of aluminum atoms by the atoms of 3d and 4d transition elements is described and discussed. We demonstrate the electron density distributions that evidence delocalization of electrons in alloyed intermetallic compounds.

Chapter 11 deals with the tight-binding and the embedded-atom models of solid state. The method of the local combination of atomic orbitals is described. We present examples of the technique application. Description of atom systems in the embedded-atom method, embedding functions and applications are considered. In conclusion the reader will find the review of interatomic pair potentials.

The crystal lattice vibration and the force coefficients are the subject of Chapter 12. We describe the experimental dispersion curves and conclusions that follow from their examination. The interplanar force constants are introduced. Group velocity of lattice waves is computed and discussed. It allows one to make conclusions about the interatomic bonding strength. Energy of atomic displacements during lattice vibration (that is propagation of phonons) is related to electron structure of metals.

The transition metals are presented in Chapter 13. We describe their structure, physical models, cohesive energy, density of states for these metals.

Chapter 14 contains data on band structure, the covalent bond strength and properties of semiconductors. Here we describe the graphene, a material that "should not exist." Here, we also dwell on the nanomaterials.

Chapter 15 is devoted to the bonding nature in molecular and ionic crystals. We recall the dipole–dipole, dipole-induced and dispersion intermolecular forces. The van der Waals and hydrogen bonds are considered. We discuss intermolecular structure and strength of ice and the solid noble gases. The description of organic molecular crystals is presented. In conclusion we consider ionic crystals and calculate their interatomic bonding.

Fundamentals of the high-temperature creep in metals is described in Chapter 16.

A physical mechanism of fatigue is reported in Chapter 17.

Finally, in Chapter 18 we present instances of modeling of kinetic processes in solids by a system of differential equations.

# 2
# From Classical Bodies to Microscopic Particles

Let us first of all recall some peculiarities of the microscopic world which are studied by quantum physics. The laws of this world are different from those of our "usual" macroscopic world.

The interatomic bonding in solids is ensured by substances that have unusual properties compared with our daily experiences. The laws of the quantum physics deal with energy and motion of microscopic objects.

## 2.1
## Concepts of Quantum Physics

The fundamentals of quantum physics are as follows:

- the electromagnetic radiation that reveals as a particle flux;
- the particle–wave duality of microscopic objects;
- the discreteness of their energy;
- the uncertainty principle;
- the probabilistic nature of the space position of the microscopic particles.

We now briefly discuss these phenomena.

Thermal radiation is electromagnetic radiation emitting from the surface of a body as a result of its temperature. The radiation intensity (that is, the energy divided by wavelength) emitted by a heated body as a function of temperature and wavelength is shown in Figure 2.1. The curves have a maximum. Note that the maximum of the intensity shifts to shorter wavelengths as the temperature increases.

Laws of classical physics can be used to derive an equation which describes the intensity of blackbody radiation as a function of frequency for a fixed temperature – the result is known as the Rayleigh–Jeans law.[1] Although the Rayleigh–Jeans law agrees with experimental data for low frequencies (long wavelengths), it diverges

1) A blackbody is an ideal body that completely absorbs all radiant energy falling upon it with no reflection and that radiates at all wavelengths with a spectral energy distribution dependent on its absolute temperature.

*Interatomic Bonding in Solids: Fundamentals,Simulation,andApplications*, First Edition. Valim Levitin.

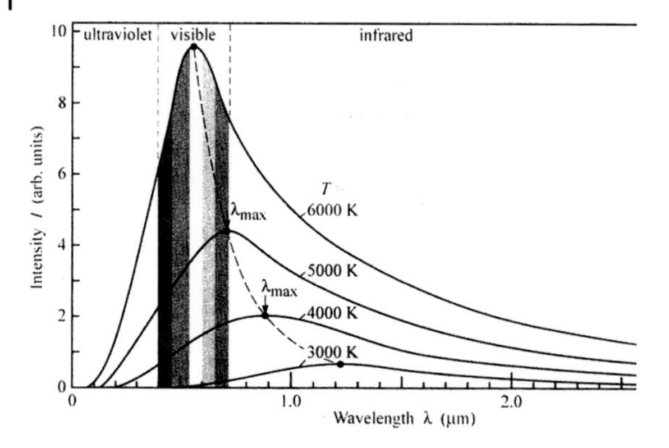


**Figure 2.1** Dependence of a radiation of the blackbody on the wavelength and temperature. Take note of three areas of the blackbody spectrum (infrared, visible, and ultraviolet).

as the frequency increases; physicists of the beginning of nineteenth century even named this discrepancy as an "ultraviolet catastrophe."

The electromagnetic radiation can diffract and interfere, that is it possesses certain wave properties. However, the electromagnetic radiation behaves also as a flow of particles in other phenomena (the Compton effect, photoelectric emission).

It turns out that the waves, which heated bodies radiate, are identical to the flow of particles that are called photons. According to the Planck equation the energy of a photon can be expressed as

$$E = h\nu \,, \tag{2.1}$$

where $h = 6.626 \times 10^{-34}$ J s is the Planck constant, $\nu$ is the frequency of the radiation. Energy of a free quantum particle is also expressed as

$$E = \hbar\omega \,, \tag{2.2}$$

where $\omega = 2\pi\nu$ is the angular frequency, $\hbar = h/2\pi$. Consequently, a heated body radiates energy by small but finite portions, that is by quants. As a result the energy spectrum is not continuous. The quantum of electromagnetic radiation is the photon.

There is an experimental evidence that microscopic particles move according to the laws of wave motion. On the one hand, the electron is a particle with a certain rest mass and a certain charge. On the other hand, flow of electrons can diffract, that is to say the electrons behave as waves when they interact with atoms of a solid. This experimental evidence forms the basis for considering a microscopic object as a dual particle–wave one. According to the de Broglie ideas any moving particle associates with a wave. The wavelength $\lambda$ of the moving particle is given by

$$\lambda = \frac{h}{p} \,, \tag{2.3}$$

where $h$ is the Planck constant, $p = mv$ is the momentum of the particle, $m$ and $v$ are the mass and the velocity of the particle, respectively. Not only the electrons but

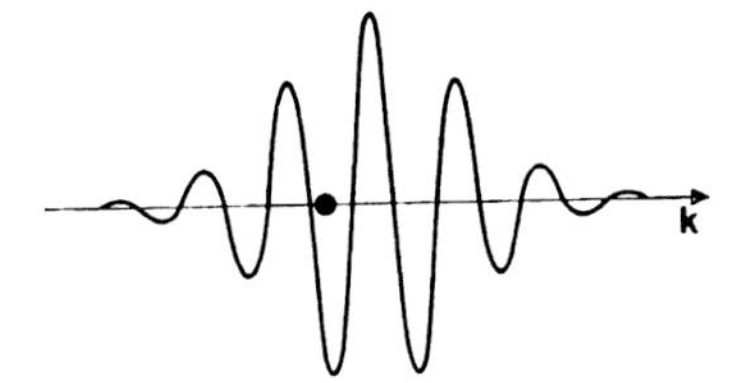


**Figure 2.2** The schematic picture of a particle and a traveling wave packet associated with it. $\boldsymbol{k}$ is the wave vector; $|\boldsymbol{k}| = 2\pi\lambda$.

all objects, charged or uncharged, show wavelike characteristics in their motion and interaction with solids. The neutron flow, molecular beams of helium or hydrogen atoms all exhibit diffraction patterns.

It can be seen from (2.3) that for macroscopic bodies having a relatively large mass the wavelength $\lambda$ is too little to be detected.

Figure 2.2 presents a scheme of a moving quantum particle and a wave packet that is associated with the particle. The motion of a quantum particle can be determined by the wave vector $\boldsymbol{k}$. The direction of the wave vector coincides with the wave propagation. The magnitude of this vector is expressed as

$$|\boldsymbol{k}| = \frac{2\pi}{\lambda} . \tag{2.4}$$

The classical expression of total energy $E$ of a body of a mass $m$ and velocity $v$ is the sum of the body's kinetic and potential energy

$$E = \frac{mv^2}{2} + V = \frac{p^2}{2m} + V , \tag{2.5}$$

where $p$ is the momentum. Equation (2.5) does not restrict the energy of a classical body, it may vary continuously. We shall see further that energy of a quantum object varies discretely if its motion is limited by a certain finite part of the space.

Combining (2.2)–(2.5) we obtain the total energy of a quantum particle

$$\hbar\omega = \frac{h^2 k^2}{2m} + V . \tag{2.6}$$

Let us compare how the motion parameters can be measured in the classic physics and in the quantum physics. For the sake of simplicity we consider one-dimensional motion of a body.

The basic equation of the classical mechanics asserts that the acceleration $a$ of the body is directly proportional to the sum of the applied forces $\sum_i F_i$ and inversely proportional to the mass $m$ of the body. An external resultant force is the cause of a change in the velocity. The Newton Second Law is given by[2)]

$$a = \frac{d^2 x}{dt^2} = \frac{\sum_i F_i}{m} . \tag{2.7}$$

2) In the vector form the Second Law is expressed as $d^2\boldsymbol{r}/dt^2 = \sum_i \boldsymbol{F}_i/m$. The acceleration of the body is collinear to the sum of the applied forces.

If $\sum F_i = \text{const.}$ one can find the velocity $v$ of the body and the path $x$ solving for

$$v = \frac{dx}{dt} = \int_0^t a dt = at + v_0 \tag{2.8}$$

and

$$x = \int_0^t (at + v_0) dt = \frac{at^2}{2} + v_0 t + x_0 , \tag{2.9}$$

where $v_0$ and $x_0$ are an initial velocity and an initial path, respectively.

Note that in the classical physics the solution of the differential equation (2.7) determines coordinate of the body depending on its velocity and the time. This coordinate is measured with an error $\pm\Delta x$, of course. The value of the error depends on the errors $\pm\Delta a$ and $\pm\Delta t$. The equation of motion of a body with given sum of forces can be solved to find out the position $x$, velocity $v$, and the momentum $p = mv$. We can therefore predict the location, momentum, and energy of the body at any given moment of time. We only need to know the initial velocity and time.

It is important that there are not any restrictions on decreasing the errors in measurement $\Delta a$ and $\Delta t$. One can decrease the errors by improving methods of measurements.

However, the situation becomes completely different when we deal with microscopic particles. The point is that a measurement of the particle position changes its state.

The Heisenberg uncertainty principle states that an experiment cannot simultaneously determine the exact values of the position and the momentum of a particle. The precision of measurement is inherently limited by the measurement process itself such that

$$\Delta p_x \Delta x \geq \frac{\hbar}{2} , \tag{2.10}$$

where the momentum $p_x$ is known to within an uncertainty (that is, an error) of $\Delta p_x$ and the position $x$ at the same time to within an uncertainty $\Delta x$. The Heisenberg principle has nothing to do with improvements in instrumentation leading to better simultaneous determinations of $p_x$ and $x$. The principle rather says that even with ideal instruments we can never measure better than $\Delta p_x \Delta x \geq \hbar/2$.

Let us illustrate the Heisenberg principle by considering motion of an electron in a hydrogen atom. The position of the electron and the momentum is assumed to be measured with an uncertainty equal to 0.01% – a commonly acceptable accuracy in engineering. The centripetal force equals to the force of electrostatic attraction between the electron and the nuclei. The velocity of the electron can be found from the equation

$$\frac{mv^2}{r_B} = \frac{1}{4\pi\varepsilon_0} \frac{e^2}{r_B^2} . \tag{2.11}$$

The Bohr atom radius $r_B = 5.29 \times 10^{-11}$ m, the electric constant $\varepsilon_0 = 8.85 \times 10^{-12}\,\mathrm{C\,V^{-1}\,m^{-1}}$, electron charge $e = -1.60 \times 10^{-19}$ C. From (2.11) we obtain velocity of the electron $v = 2.19 \times 10^6\,\mathrm{m\,s^{-1}}$. The momentum $p = mv$; $p = 9.11 \times 10^{-31} \cdot 2.19 \times 10^6\,\mathrm{kg\,m\,s^{-1}} = 2.00 \times 10^{-24}\,\mathrm{kg\,m\,s^{-1}}$.

Substituting $\Delta p = 0.0001p$ to (2.10) we find that $\Delta x \geq 5.175 \times 10^{-7}$ and, consequently, $\Delta x/x \geq 9.78 \times 10^3$. That is to say, the uncertainty in the electron location in the atom is by four orders of magnitude greater than the atomic radius.

Thus, the coordinate of the electron becomes undefinable even if we determine its momentum with an error of 0.01%. If we determine the coordinate of the electron then its momentum becomes undefinable.

## 2.2 Wave Motion

A wave motion can be characterized by several important parameters: amplitude, wavelength, and frequency. The equation of the wave propagation in a medium represents a dependence of the deviation $u$ of a chosen point relative to its equilibrium position on its coordinate $x$ at the time $t$. The general expression of this dependence for a plane harmonic wave has the following form:

$$u(x,t) = A \exp(i(k \cdot x - \omega t)) \,, \tag{2.12}$$

where $u(x,t)$ is the displacement from the equilibrium position, $A$ is the amplitude, $k$ is the magnitude of the wave vector that is determined by (2.4), $\omega$ is the angular frequency.

The function $\exp(i\varphi)$ is a periodic one with the period $2\pi$: $\exp(i\varphi) = \exp(i\varphi + 2\pi n)$ where $n$ is an integer. Therefore, we see that the wave displacement of a point is doubly periodic: with respect to time and coordinate, where the time period is $T = 2\pi/\omega$ and the coordinate period is the wave length $\lambda$.

Note that the function in (2.12) is complex. Generally speaking, one should equate only the real part of the expression as only it has the physical sense. One could also use functions sine or cosine for the wave equation as follows,

$$u = A \sin(kx - \omega t) \,; \quad u = A \cos(kx - \omega t) \,. \tag{2.13}$$

However, the exponential form (2.12) is more convenient, especially when summing waves with different amplitudes and phases. It is just convenient to sum waves with the same $\omega$ on the complex plane. The transition from one form of the equation to the other one can be done using the Euler formula

$$\exp(i\varphi) = \cos\varphi + i \sin\varphi \,. \tag{2.14}$$

The graph of the transverse traveling wave is shown in Figure 2.3.

When the traveling wave reflects from the interface of two media, a standing wave is formed as a result of interactions of the direct and inverse waves. The

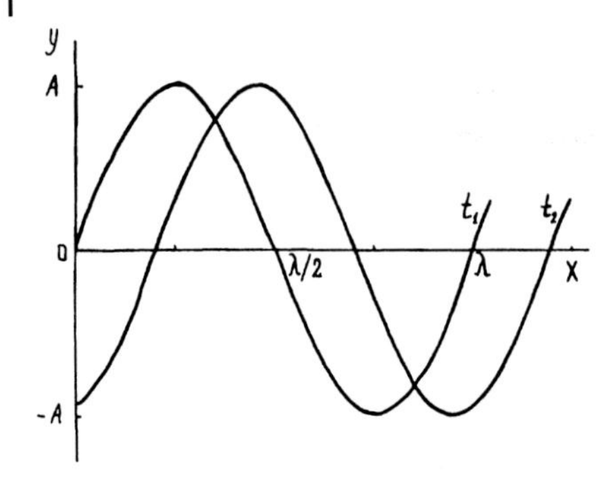


**Figure 2.3** The graph of the traveling wave moving from left to right. The two curves correspond to two successive retention intervals; $t_2 > t_1$.

specific features of the standing wave are appearance of both nodes and antinodes and absence of one-way transfer of the energy in the medium. The equation of the standing wave can be obtained by summing (2.13) with the opposite signs of the second term:

$$u = A\sin(kx - \omega t) + A\sin(kx + \omega t) = 2A(\sin kx)(\cos \omega t) . \quad (2.15)$$

The graph of the standing wave (Figure 2.4) shows the evolution of displacements over time.

Let us return to Figure 2.2. In Appendix B we expand the concepts of wave packet and group and phase velocities.

There are two velocities associated with the moving wave packet, the phase velocity and the group velocity. The phase velocity of a wave $v_\mathrm{p}$ is the rate at which the phase of the wave propagates in space. This is the speed at which the phase of

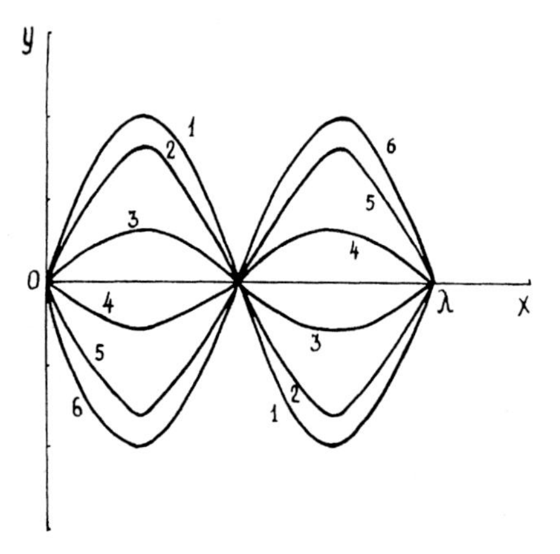


**Figure 2.4** The graph of the standing wave. 1, 2, 3, . . ., 6 are the positions in successive time intervals: 0; 0.1$T$; 0.2$T$; . . .; 0.5$T$.

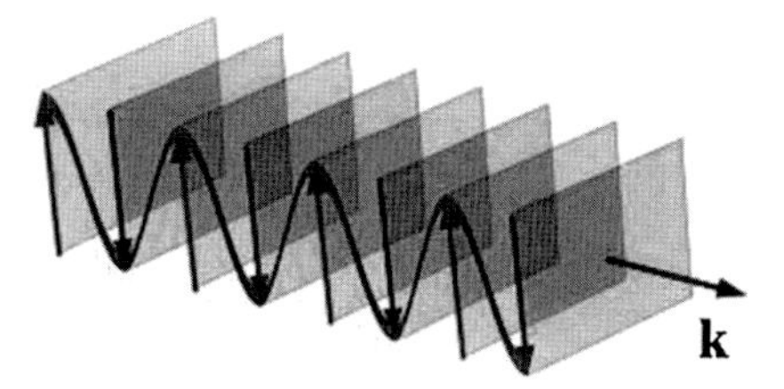


**Figure 2.5** Three-dimensional image of a plane wave.

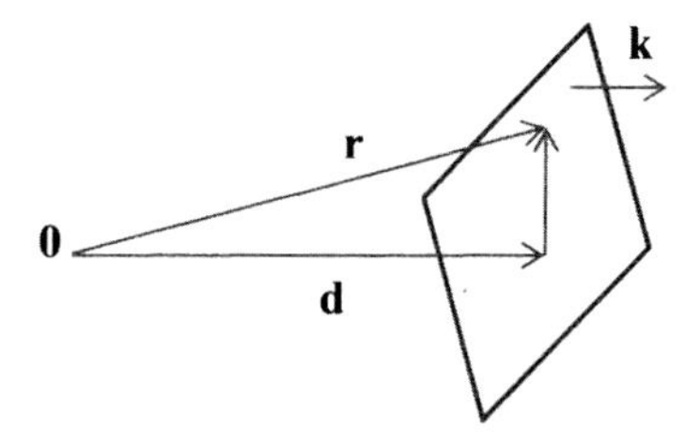


**Figure 2.6** Scheme of plane perpendicular to propagation of the plane wave. Product $\boldsymbol{k} \cdot \boldsymbol{r} = d$ at any point of the plane.

any one frequency component of the wave travels. It is expressed as

$$v_\text{p} = \frac{\omega}{k} = \frac{\lambda}{T} = \lambda \nu \; . \tag{2.16}$$

The group velocity of a wave is the velocity with which the overall shape of the amplitudes, which is known as the envelope of the wave, propagates through the space. The group velocity is given by

$$v_\text{g} = \frac{\partial \omega}{\partial k} \; . \tag{2.17}$$

Figures 2.5 and 2.6 show a plane wave. Equation of the plane wave is given by

$$\psi(\boldsymbol{r}) = u(\boldsymbol{k}, \boldsymbol{r}) \exp(i \boldsymbol{k} \boldsymbol{r}) \; . \tag{2.18}$$

Plane wave is constant in any plane perpendicular to the wave vector $\boldsymbol{k}$ (since such planes are defined by the equation $\boldsymbol{k} \cdot \boldsymbol{r} = \text{const}$) and it is periodic along lines parallel to $\boldsymbol{k}$, with wavelength $\lambda = 2\pi/k$.

## 2.3 Wave Function

We are forced to replace the deterministic notion of a state of a microscopic particle by a probabilistic determination of a state. Consequently, the deterministic equation of the state has to become a probabilistic one. By "state of a particle" we mean its location and its energy.

As the quantum particle possesses wave properties it is logical to describe its state by a corresponding function.

It is appropriate to introduce a wave function $\Psi$ for description of the state of the particle. In the one-dimensional case, the wave function depends upon coordinate $x$ and time $t$. The function $\Psi(x,t)$ is related to the probability that the particle is in a point $x$ at a moment $t$. The certain quantum equation that describes behavior of a quantum particle ought to be a differential one since it tells us about a force acting on the particle by specifying potential energy corresponding to the force. We are particularly interested in how particle state changes.

A mathematical expression for a wave function can be taken in a form of a traveling wave, such as

$$\Psi(x,t) = \exp i\left(\frac{2\pi}{\lambda}x - \omega t\right) . \tag{2.19}$$

The wave function is a complex-valued one. One should consider the wave function (2.19) as a part of the Schrödinger theory. The connection between the behavior of a quantum particle and properties of the wave function $\Psi(x,t)$ is expressed in terms of the probability density $p(x,t)$. This value determines the probability, per unit length of the $x$ axis of finding the particle at the coordinate $x$ at the time $t$. The probability density is given by

$$p(x,t) = \Psi(x,t)\Psi^*(x,t) , \tag{2.20}$$

where the symbol $\Psi^*(x,t)$ represents the complex conjugate of $\Psi(x,t)$. The probability $p(x,t)dx$ that the particle is situated at a coordinate between $x$ and $x + dx$ equals to $\Psi(x,t)\Psi^*(x,t)dx$. Thus, only square of the wave function has certain physical meaning.

The following property says that the particle is always somewhere in the space as the probability of such an event has to be one,

$$\int_{-\infty}^{+\infty} \Psi(x,t)\Psi^*(x,t)dx = 1 . \tag{2.21}$$

## 2.4 The Schrödinger Wave Equation

An equation, which describes states of a microscopic particle, must satisfy all assumptions concerning quantum properties of the particle.

The time-dependent differential equation for the wave function $\Psi(x,t)$ bears the name of Schrödinger. For a one-dimensional case it is given by

$$-\frac{\hbar^2}{2m}\frac{\partial^2\Psi(x,t)}{\partial x^2} + V(x,t)\Psi(x,t) = i\hbar\frac{\partial\Psi(x,t)}{\partial t} , \tag{2.22}$$

where $V(x, t)$ is the potential energy. There is an important special case where $V(x, t) = V_0$ is a constant. In such a case the force acting on the particle $F = 0$ since $F = -\partial V(x, t)/\partial x$.

Equation (2.22) is a general equation that describes peculiarities of the objects in the microscopic world. The differential equation has particular solutions,

$$\Psi(x, t) = A \exp(i(kx - \omega t)) \; ; \tag{2.23}$$

$$\Psi(x, t) = A \cos(kx - \omega t) + Ai \sin(kx - \omega t) \; . \tag{2.24}$$

It is easily seen that these formulas describe wave motion which we are already acquainted with. The wave function is a complex value. Let us check the solution (2.23) for $V(x, t) = V_0 = \text{const}$.

$$\begin{aligned} \frac{\partial \Psi(x, t)}{\partial x} &= ikA \exp(i(kx - \omega t)) \; ; \\ \frac{\partial^2 \Psi(x, t)}{\partial x^2} &= -k^2 A \exp(i(kx - \omega t)) \; ; \end{aligned} \tag{2.25}$$

$$\frac{\partial \Psi(x, t)}{\partial t} = -iA\omega \exp(i(kx - \omega t)) \; . \tag{2.26}$$

Substituting (2.25) and (2.26) into (2.22) and canceling we arrive at

$$\frac{\hbar^2 k^2}{2m} + V = \hbar \omega \; , \tag{2.27}$$

which coincides with (2.6) for total energy of the quantum particle.

It might seem that the Schrödinger equation is unfounded. As a matter of fact, it cannot be derived either from laws of classic physic or from any known dependences. However, the equation is substantiated by correctly predicting of a number of natural phenomena.

The wave function is linear in $\Psi(x, t)$. That is, if $\Psi_1(x, t)$ and $\Psi_2(x, t)$ are two different solutions to the Schrödinger wave equation then any of their linear combinations,

$$\Psi(x, t) = c_1 \Psi_1(x, t) + c_2 \Psi_2(x, t) \; , \tag{2.28}$$

where $c_1$ and $c_2$ are some constants, is also a solution.

For three-dimensional cases, the Schrödinger equation is expressed as

$$-\frac{\hbar^2}{2m} \nabla^2 \Psi(\mathbf{r}, t) + V(\mathbf{r}, t) \Psi(\mathbf{r}, t) = i\hbar \frac{\partial \Psi(\mathbf{r}, t)}{\partial t} \; , \tag{2.29}$$

where $\nabla$ is the Laplacian operator in rectangular coordinates,

$$\nabla^2 = \frac{\partial^2}{\partial x^2} + \frac{\partial^2}{\partial y^2} + \frac{\partial^2}{\partial z^2} \tag{2.30}$$

and $\boldsymbol{r}$ is the radius-vector of a point where the particle is found. This equation contains a space variable $\boldsymbol{r}$ and the time variable $t$. The solution of the time-dependent equation (2.29) may be separated into time-dependent and time-independent parts.

$$\Psi(\boldsymbol{r}, t) = \psi(\boldsymbol{r}) \cdot \varphi(t) . \tag{2.31}$$

Let us separate variables in (2.29). The time-dependent part can be expressed as

$$\varphi(t) = \exp\left(-\frac{iE}{\hbar}t\right) , \tag{2.32}$$

where $E$ is the total energy of the particle.

Substituting

$$\Psi(\boldsymbol{r}, t) = \psi(\boldsymbol{r}) \cdot \exp\left(-\frac{iE}{\hbar}t\right) \tag{2.33}$$

into (2.29) we arrive at the time-independent Schrödinger equation

$$-\frac{\hbar^2}{2m}\nabla^2\psi(\boldsymbol{r}) + V(\boldsymbol{r})\psi(\boldsymbol{r}) = E\psi(\boldsymbol{r}) . \tag{2.34}$$

The wave functions $\psi(\boldsymbol{r})$ are called the eigenfunctions of the Schrödinger equation. The term originates from the German adjective "eigen" (own). The time-independent Schrödinger equation (2.34) does not contain the imaginary unit $i$. In one-dimensional cases it involves only one independent variable $x$ and is an ordinary differential equation. Three-dimensional version involves more independent variables and is therefore a partial differential equation.

Differential equations may have a wide variety of possible solutions. Acceptable solutions to the time-independent Schrödinger equation must satisfy certain requirements. An eigenfunction $\psi$ and its derivative $d\psi/dx$ must be finite, single valued, and continuous.

Certain values of energy $E$ that correspond to solutions $\psi(\boldsymbol{r})$ are called the eigenvalues of the potential $V(\boldsymbol{r})$.

## 2.5
## An Electron in a Square Well: One-Dimensional Case

A free electron can have any value of energy. However, the state of an electron changes if its motion is limited by external interactions.

We consider first the one-dimensional case. The square well potential is used in quantum physics to represent a situation when a particle is confined to some region of space by forces holding it in the region. This model represents an infinite potential barrier placed at the ends of an interval $0 \le x \le L$. The infinite square well potential is written as

$$V(x) = \infty \quad \text{if} \quad x < 0 \quad \text{or} \quad x > L\,; \quad V(x) = 0 \quad \text{if} \quad 0 \le x \le L\,, \tag{2.35}$$

where $L$ is the length of the well. In other words, the electron cannot overstep the limits of the segment. This simple model is useful because the motion of the electron in the isolated atom is also confined to some area.

Two conditions have to be satisfied. The wave function must vanish at both ends of the interval $L$, that is $\psi^2(x) = 0$. Further, the localization of the electron somewhere inside the interval is the sure event, that is the probability of this event has to be 1,

$$\int_0^L \psi^2(x)dx = 1 . \qquad (2.36)$$

The states of the electron are described by the Schrödinger equation

$$-\frac{\hbar^2}{2m}\nabla^2\psi(x) = E\psi(x) . \qquad (2.37)$$

A possible solution of (2.37) is

$$\psi = A\sin kx , \qquad (2.38)$$

where $A$ is a proportionality constant. It follows from the boundary conditions (2.35) that

$$k = \frac{\pi n}{L} , \qquad (2.39)$$

where $n = 1, 2, 3, \ldots$ is an integer. The electron is necessarily inside the segment $0\text{–}L$. We find a value of $A$ from the condition (2.36).

Substituting (2.38) and (2.39) into (2.36) we arrive at

$$A = \left(\frac{2}{L}\right)^{\frac{1}{2}} \qquad (2.40)$$

and

$$\psi = \left(\frac{2}{L}\right)^{\frac{1}{2}} \sin\frac{\pi n}{L}x . \qquad (2.41)$$

A probability to find the electron within the segment $dx$ equals to $\psi^2 dx$,

$$\psi^2(x)dx = \frac{2}{L}\sin^2\left(\frac{\pi n}{L}x\right)dx , \qquad (2.42)$$

$n$ is called a quantum number. The distribution of probabilities of the electron location along the segment $0\text{–}L$ for $n = 1, 2, 3$ is presented in Figure 2.7.

Two basic results are as follows.

The probability distribution is not a uniform one, that is probabilities to find the electron in different places within the segment are different. For $n = 1$ the probability is maximal at $x = L/2$, it decreases towards ends of the segment and

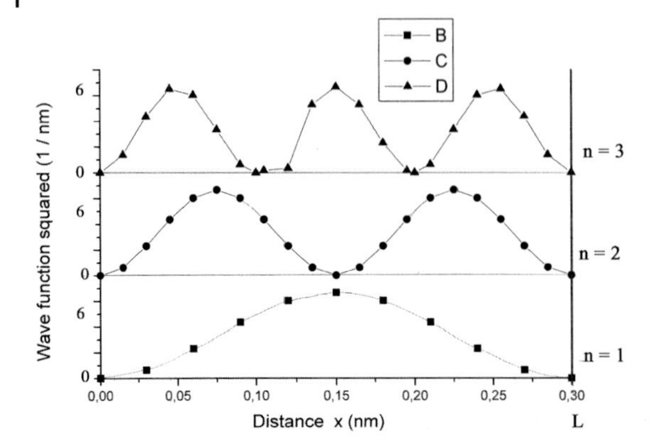


**Figure 2.7** The squared wave functions of an electron for a one-dimensional model with an interval of 0–$L$. The segment $L$ is taken to be 0.3 nm in length.

vanishes at the ends. For $n = 2$ it is zero at the point $x = L/2$. The maximal probability is observed at $x = L/4$ and $x = 3L/4$. If $n = 3$ then points of the greatest probability are $x/L = 1/6, 3/6, 5/6$. At $x/L = 1/3$ and $2/3$ the probability is zero. The graphs in Figure 2.7 look like standing waves. It is not surprising because the electron behaves like a wave. The direct wave interacts with the wave reflected from the opposite wand of the well.

Substituting solution (2.38) to (2.37) one obtains

$$E = \frac{\pi^2 \hbar^2}{2mL^2} n^2 . \tag{2.43}$$

The energy of the electron under consideration cannot change uninterruptedly. The equation (2.43) gives the allowed energy levels of the electron, the motion of which is limited by conditions (2.35). Under these circumstances energy can take discrete values only. The ratios of the values of allowed energies are $1 : 4 : 9 : 16 \ldots$ Energy of the electron cannot be zero. The quantum number $n$ defines the state of the electron. The greater $n$ the higher energy of the electron. The ground energy state corresponds to $n = 1$, higher values of $n$ correspond to excited states of the electron.

## 2.6
## Electron in a Potential Rectangular Box: *k*-Space

Turning to a three-dimensional case let us consider an electron in a rectangular box with sides $L_x$, $L_y$ and $L_z$ and the electrostatic potential $V(x, y, z) = 0$ in the box and $V(x, y, z) = \infty$ outside it.

We are looking for a solution of the three-dimensional time-independent Schrödinger equation,

$$-\frac{\hbar^2}{2m} \nabla^2 \psi(x, y, z) + V(x, y, z)\psi(x, y, z) = E\psi(x, y, z) . \tag{2.44}$$

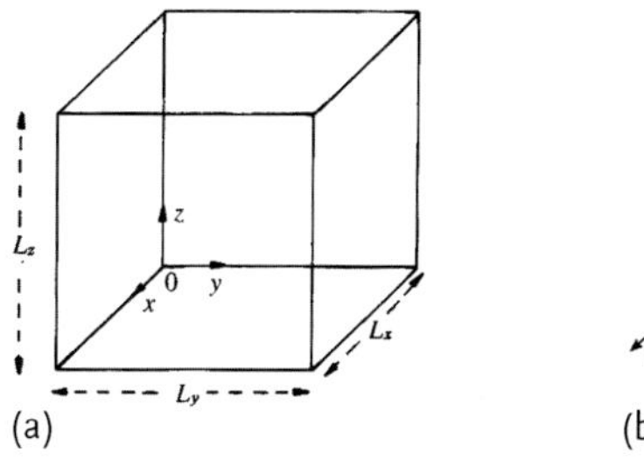


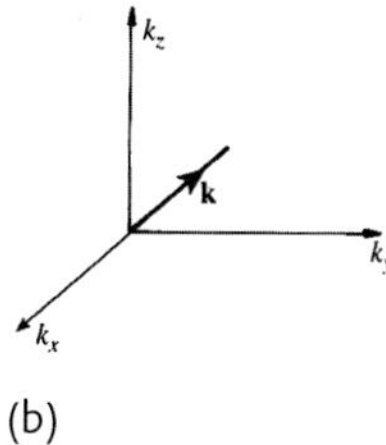


**Figure 2.8** (a) A three-dimensional crystal in the real-space; $L_x$, $L_y$, $L_z$ are dimensions of the crystal. (b) The electron wave vector $\boldsymbol{k}$ in $\boldsymbol{k}$-space. The components of $\boldsymbol{k}$ are $k_x = \pi n_x / L_x$; $k_y = \pi n_y / L_y$; $k_z = \pi n_z / L_z$.

On the surfaces of the box the wave function vanishes: $\psi(x, y, z) = 0$. Inside the box a solution of (2.44) has the form

$$\psi(x, y, z) = \left(\frac{8}{L_x L_y L_z}\right)^{\frac{1}{2}} \sin\left(\frac{\pi n_x}{L_x} x\right) \sin\left(\frac{\pi n_y}{L_y} y\right) \sin\left(\frac{\pi n_z}{L_z} z\right) . \tag{2.45}$$

Or

$$\psi(x, y, z) = \left(\frac{8}{L_x L_y L_z}\right)^{\frac{1}{2}} \sin(k_x x) \sin(k_y y) \sin(k_z z) , \tag{2.46}$$

where $k_x$, $k_y$, $k_z$ are the components of the wave vector $\boldsymbol{k}$:

$$\boldsymbol{k}^2 = k_x^2 + k_y^2 + k_z^2 , \tag{2.47}$$

$$k_x = \frac{\pi n_x}{L_x} ; \quad k_y = \frac{\pi n_y}{L_y} ; \quad k_z = \frac{\pi n_z}{L_z} \tag{2.48}$$

with $n_x$, $n_y$, $n_z = 1, 2, 3, 4, \ldots$

The energy of the electron can take only discrete values. They are given by

$$E = \frac{\pi^2 \hbar^2}{2 m L^2} \left(n_x^2 + n_y^2 + n_z^2\right) . \tag{2.49}$$

The $n$ values can again be interpreted as quantum numbers which define energetic states of the electron and therefore determine the allowed values of energy. It can be seen from (2.48) that the values of $k$ and $n$ are closely related.

Quantum numbers of the electron in the ground state are equal to 1, 1, 1; corresponding energy of the electron equals to $3\pi^2\hbar^2/2mL^2$. The first excited state of the electron is determined by one of three combinations of quantum numbers: 2, 1, 1; 1, 2, 1 and 1, 1, 2. Any of three sets corresponds to the same value of energy, namely $6\pi^2\hbar^2/(2mL^2)$. A state, to which several combinations of the quantum numbers correspond to, is called a degenerated state. In this case it is the triply degenerated state.

The three-dimensional time independent Schrödinger equation (2.44) has a family solutions of the (2.46) type. Every solution has corresponding wave vector $\boldsymbol{k}$ with components $k_x, k_y, k_z$ given by (2.48). The $\boldsymbol{k}$ vectors constitute a vector space called $\boldsymbol{k}$-space (reciprocal space and phase space are other terms used in the literature). The concept of $\boldsymbol{k}$-space is extremely useful by considering the interaction of a radiation with a matter.

For the real-space with a coordinate system $x, y, z$ that defines a three-dimensional crystal there is a $\boldsymbol{k}$-space. The $\boldsymbol{k}$-space is a mathematical image direct bounded with the real-space. The real-space and the $\boldsymbol{k}$-space are presented in Figure 2.8.

# 3
# Electrons in Atoms

This chapter is devoted to the electronic structure of atoms.

We examine at first an one-electron atom. A state of the electron is described by quantum numbers. We consider a more complex problem of a multi-electron atom further and turn our attention to theories that describe the many-body problem.

D.R. Hartree and V.A. Fock made an important contribution to the science of systems consisting of a large number of electrons and nuclei. We review the Hartree theory, which gives an approximation method for the determination of the ground-state eigenfunction and the corresponding eigenvalue of a many-body quantum system.

Finally, we will reintroduce the atomic structures of chemical elements.

## 3.1
## Atomic Units

Atomic units are used in quantum physics.

The Bohr radius $a_0 = 0.529\,177 \times 10^{-10}$ m is a unit of length. This unit is called Bohr (atom unit, a.u.).

The unit of energy is Rydberg; $1\,\text{Ry} = 13.605\,\text{eV} = 13.605\,\text{eV} \cdot 1.602\,176 \times 10^{-19}\,\text{C} = 2.179\,760 \times 10^{-18}\,\text{J}$.

The rest mass of the electron, which is equal to $m_e = 9.109\,534 \times 10^{-31}$ kg, is accepted as the unit of the mass.

Other atomic units are the electron charge $e = 1.602\,169 \times 10^{-19}$ C, the Planck constant (divided by $2\pi$) $\hbar = 1.054\,589 \times 10^{-34}$ J s, and the multiplier $1/(4\pi\varepsilon_0) = 8.987\,552 \times 10^9\,\text{N}\,\text{m}^2\,\text{C}^{-2}$.

In order to convert SI units into atomic units one has to put $m = \hbar = e = 1/(4\pi\varepsilon_0) = 1$.

Thus, to determine the value of distance in meters one should multiply the value expressed in Bohr by $a_0$. Conversely, in order to obtain the value of distance in atomic units (a.u.) one should divide the value expressed in meters by $a_0 = 0.529\,177 \times 10^{-10}$ and so on.

*Interatomic Bonding in Solids: Fundamentals,Simulation,andApplications*, First Edition. Valim Levitin.

## 3.2
## One-Electron Atom: Quantum Numbers

The theory of one-electron atom seems to be the simplest one but it is also the most important one.

An one-electron atom contains two particles, that is, a positively charged nucleus and the negatively charged electron. The two particles are bound together by the Coulomb attraction force. In fact, both particles move around a common center of masses. The ratio of the proton mass to the electron mass $m_p/m_e$ is 1835.55. Thus, the massive nucleus may by considered as almost completely stationary.

A coordinate of the electron is determined by the vector $\boldsymbol{r}(x, y, z)$. In the rectangular coordinate system one has $|\boldsymbol{r}| = r = \sqrt{x^2 + y^2 + z^2}$. The electron moves under the influence of the Coulomb potential given by

$$V(x, y, z) = -\frac{1}{4\pi\varepsilon_0}\frac{Ze^2}{\sqrt{x^2 + y^2 + z^2}}\,, \tag{3.1}$$

where $1/(4\pi\varepsilon_0)$ is the Coulomb law constant, $-e$ is the electron charge, $Z$ is the number of the element in the periodic table, $+Ze$ is the nucleus charge.

The time-independent Schrödinger equation for one-electron atom is expressed as

$$-\frac{\hbar^2}{2m}\nabla^2\psi(\boldsymbol{r}) + V(\boldsymbol{r})\psi(\boldsymbol{r}) = E\psi(\boldsymbol{r}) \tag{3.2}$$

or, as a functional equation,

$$H\psi = E\psi\,, \tag{3.3}$$

where $H$ is called the Hamiltonian. Hamiltonian is the operator associated with energy of the electron. Solutions of the Schrödinger equation (3.2) exist only for certain values of energy; these values are called eigenvalues of energy. To each eigenvalue of energy $E_i$ corresponds an eigenfunction $\psi_i$, so that

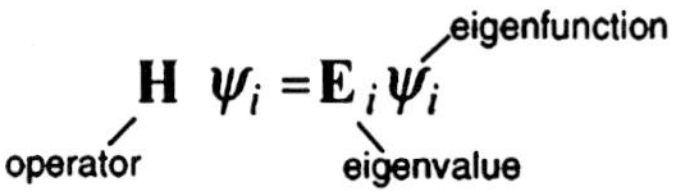

eigenfunction

$$\mathbf{H}\,\psi_i = \mathbf{E}_i\psi_i$$

operator eigenvalue

**Figure 3.1** Eigenvalue and eigenfunction in quantum physics.

As before, we require that an eigenfunction $\psi(x, y, z)$ and its derivatives $\psi'(x, y, z)$ must be finite, single valued and continuous. Such functions are called "well-behaved."

There are three unknown independent variables in (3.2), namely $x$, $y$, $z$. One should separate the variables in order to split the partial differential equation into a set of three ordinary differential equations, each involving only one coordinate. However, the separation of variables cannot be carried out when rectangular coordinates are employed because the Coulomb potential energy (3.1) could not be represented as a product of functions each depending on one variable only.

However, a solution can be found in spherical polar coordinates. This solution can be represented as a product of functions each dependent on one coordinate. These are the coordinates $r, \theta, \varphi$, where $r$ is the length of the segment connecting the electron with the nucleus (the origin), $\theta$ and $\varphi$ are the polar and azimuthal coordinates, respectively. Because of this simplification of the potential, it is possible to carry out the separation of variables in the time-independent Schrödinger equation. This now can be written instead of (3.2) as

$$-\frac{\hbar^2}{2m}\nabla^2\psi(r,\theta,\varphi) + V(r)\psi(r,\theta,\varphi) = E\psi(r,\theta,\varphi) . \tag{3.4}$$

The change of coordinates allows one to find solutions of the time-independent Schrödinger equation of the form

$$\psi(r,\theta,\varphi) = R(r)\Theta(\theta)\Phi(\varphi) . \tag{3.5}$$

Thus, the solutions $\psi = \psi(r,\theta,\varphi)$ are products of three functions, $R(r)$, $\Theta(\theta)$, and $\Phi(\varphi)$, each of which depends on only one of the coordinates.

The reader may refer to the book [2] for mathematical aspects of this approach. Here, we state the results.

The solutions of the ordinary differential equations in the assumed product form of (3.5) depend on three integer values that have certain physical meaning. These values are denoted by $n$, $l$ and $m_l$. They are called quantum numbers.

The quantum numbers characterize the following properties of the electron in atom, respectively:

- the energy of the electron;
- its angular momentum[1];
- the projection of the orbital angular momentum onto direction of the external magnetic field.

The total energy of the electron $E_n$ is found to be finite only if it has one of the values

$$E_n = -\left(\frac{1}{4\pi\varepsilon_0}\right)^2 \frac{m_e Z^2 e^4}{2\hbar^2}\frac{1}{n^2} , \tag{3.6}$$

where the principle quantum number $n = 1, 2, 3, 4, \ldots$ As $n$ is an integer the energy of the electron cannot be continuous. It can only take discrete values. The allowed energies of the electron relate as $1 : 1/4 : 1/9 : 1/16 \ldots$ The energy state of the one-electron atom at $n = 1$ is called the ground state. All other states are excited states. The total energy $E_n$ is negative because the Coulomb force between the nucleus and the electron is an attractive one.

Consequently, it takes a work $\Delta W_n$ to move the electron from atom to infinity. $\Delta E_n = -\Delta W_n$. A change in the energy is positive as the expended work is negative.

1) Recall that in classical physics, by definition, the angular momentum of a particle is the vector $\boldsymbol{L} = [\boldsymbol{r} \times m\boldsymbol{v}]$, where $\boldsymbol{r}$ is the position vector of the particle relative to the origin, $m$ is the mass of the particle, and $\boldsymbol{v}$ is the velocity of the particle.

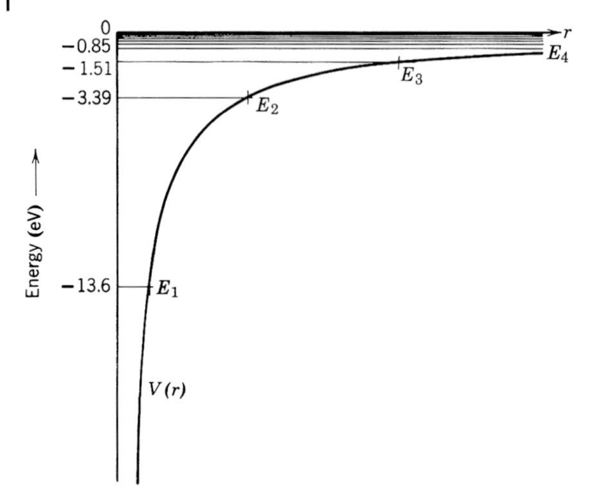


**Figure 3.2** The dependence of the Coulomb potential $V$ on distance from nucleus $r$ and the values of energy for an one-electron atom. $E_n$ is negative and increases with principal quantum number $n$ as $1/n^2$. The negative energy means that it takes the energy in order to remove the electron from the atom. $E_n \to 0$ if $n \to \infty$.

Figure 3.2 illustrates the dependence of energy of the electron in one-electron atom on its distance from the nucleus.

The energy required to remove an electron from an atom is called the ionization potential. The work needed to remove the electron, which is in the ground state, to infinity equals to $\Delta E_{n=1} = 13.605\,\text{eV} = 2.18 \times 10^{-18}\,\text{J}$. We have seen that the value of 13.605 eV = 1 Ry (Rydberg) is used as a unit of energy in quantum physics. 3.39 eV is sufficient to remove the electron from the second energy level.

Thus, the principal quantum number $n$ determines the energy of the electron.

The atom is a three-dimensional system. Consequently, the system has an angular momentum. The function $\Theta(\theta)$ turned out to be dependent on the quantum number $l$. $l$ is called the azimuthal quantum number. The azimuthal angular momentum of the atom depends on $l$, which may be $0, 1, 2, 3, \ldots, n-1$. The second quantum number $l$ determines the orbital angular momentum of the electron. The magnitude of the orbital momentum is given by

$$L = \sqrt{l(l+1)}\hbar \,. \tag{3.7}$$

In Table 3.1 the commonly used notation for the azimuthal quantum numbers $l$ is presented.

The third quantum number $m_l$ characterizes the behavior of the electron in an external magnetic field. $m_l$ is called the magnetic quantum number. $m_l$ takes values of $-l, -l+1, -l+2, \ldots, 0, 1, 2, +l-1, l$, that is $2l+1$ values total. The quantum number $m_l$ determines a projection of the orbital momentum onto the external magnetic field. This is $L_z$ component,

$$L_z = \hbar m_l \,. \tag{3.8}$$

**Table 3.1** The notations of the azimuthal quantum numbers and the corresponding electronic shells.

| Quantum number $l$ | 0 | 1 | 2 | 3 | 4 | 5 | 6 |
|---|---|---|---|---|---|---|---|
| Notation | s | p | d | f | g | h | i |

Eigenvalues of energy $E_i$ of the one-electron atom depend only on the principal quantum number $n$. The eigenfunctions $\psi_i$ depend on all three quantum numbers $n$, $l$, $m_l$. Gathering together the conditions that the quantum numbers satisfy, we get

$$\begin{aligned} n &= 1, 2, 3, \ldots ; \\ l &= 0, 1, 2, \ldots, n-1 ; \\ m_l &= -l, -l+1, \ldots, 0, 1, \ldots, l-1, l . \end{aligned}$$

There are several different possible values of $l$ and $m_l$ for a given value of $n$. One can see from (3.6) that the total energy of the electron does not depend on the quantum numbers $l$ and $m_l$. "As the eigenfunctions $\psi(r, \theta, \varphi)$ describe the behavior of the atom, we see that it has states with *completely different behavior* that nevertheless have the same total energy [2]."

As noted above there are several possible values of $l$ and $m_l$ for a given $n$. The same total energy of the electron corresponds to these different $l$, $m_l$. This phenomenon is called degeneracy of states. However, if an external magnetic field is applied to the one-electron atom, then its total energy is dependent on the orientation of the atom in the space because of interaction between the "electric current" in the atom and the applied field. In an external magnetic field the degeneracy with respect to $m_l$, disappears.

We illustrate the eigenfunction for $n = 1$, $l = 0$, $m_l = 0$:

$$\psi_{1,0,0} = \frac{1}{\sqrt{\pi}} \left(\frac{Z}{a_0}\right)^{\frac{3}{2}} \exp\left(-\frac{Zr}{a_0}\right) , \tag{3.9}$$

where $r$ is the distance from the nucleus center, $a_0$ is the Bohr radius.

The formula for $n = 2$, $l = 1$, $m_l = 0$ is

$$\psi_{2,1,0} = \frac{1}{4\sqrt{2\pi}} \left(\frac{Z}{a_0}\right)^{\frac{3}{2}} \frac{Zr}{a_0} \left[\exp\left(-\frac{Zr}{2a_0}\right)\right] \cos\theta . \tag{3.10}$$

There is an experimental evidence[2)] that the electron has also an intrinsic angular momentum $\boldsymbol{S}$ called the spin. It follows from experimental observations that the

2) As early as in 1925, Goudsmit and Uhlenbeck revealed that certain lines of the optical spectra of hydrogen and the alkali atoms are composed of a closely spaced pars of lines. They supposed that this fine structural splitting is related to an intrinsic angular momentum of an electron.

fourth quantum number $s$ may take only two values,

$$s = +\frac{1}{2}, \quad s = -\frac{1}{2}. \tag{3.11}$$

The projection of the spin momentum onto an arbitrary axis $z$ is

$$S_z = s\hbar. \tag{3.12}$$

Thus, two different states of the atom are possible for the given $l$ and $m_l$. They differ from each other in the values of $s$. Three quantum numbers determine the space variables of the electron, its energy and the $s$ quantum number determines a specific property of the electron, its spin.

Let us calculate the number of electrons having the same value of the principal quantum number $n$. If a $n$ value is given the repetition factor with respect to quantum numbers $l$, $m_l$, and $s$ is expressed as

$$\sum_{l=0}^{n-1} 2(2l+1) = 2\frac{1+[2(n-1)+1]}{2}n = 2n^2. \tag{3.13}$$

It is of interest to calculate the probability distribution of the electron in space. The corresponding probability density is a function of three polar coordinates and it can be expressed as

$$\psi(r,\theta,\varphi)\psi^*(r,\theta,\varphi) = R(r)R^*(r)\Theta(\theta)\Theta^*(\theta)\Phi(\varphi)\Phi^*(\varphi). \tag{3.14}$$

Solutions of (3.4) allow one to find the probability density of the electron in the space. The radial dependence of the probability density is treated in terms of $P(r)$ defined so that $P(r)dr$ is the probability of finding the electron at any location with the radial coordinate between $r$ and $r + dr$. A probability for the electron to locate within the volume $dV = 4\pi r^2 dr$ enclosed between spheres of radii $r$ and $r + dr$ is given by

$$P(r)dr = \psi\psi^* dV = R(r)R^*(r)4\pi r^2 dr, \tag{3.15}$$

where $\psi\psi^*$ is a probability per unit volume.

Figure 3.3 presents the probability density distribution for the one-electron atom. The distribution of the electron density is shown for the principal quantum numbers $n = 1, 2, 3$ and the $l = 0, 1, 2$. It is obvious that the radial probability density has significant values in restricted ranges of the radial coordinate. When the atom is in one of its quantum states the electron can probably be found within a certain so-called electronic shell. The characteristic radius of this shell is generally determined by the principle quantum number $n$. One can also see from Figure 3.3 that the quantum number $l$ affects the radial density distribution. Additional maxima appear for $n = 2, l = 0$; $n = 3, l = 0$ and 1. However, there are no such maxima when $l$ takes its largest possible value.

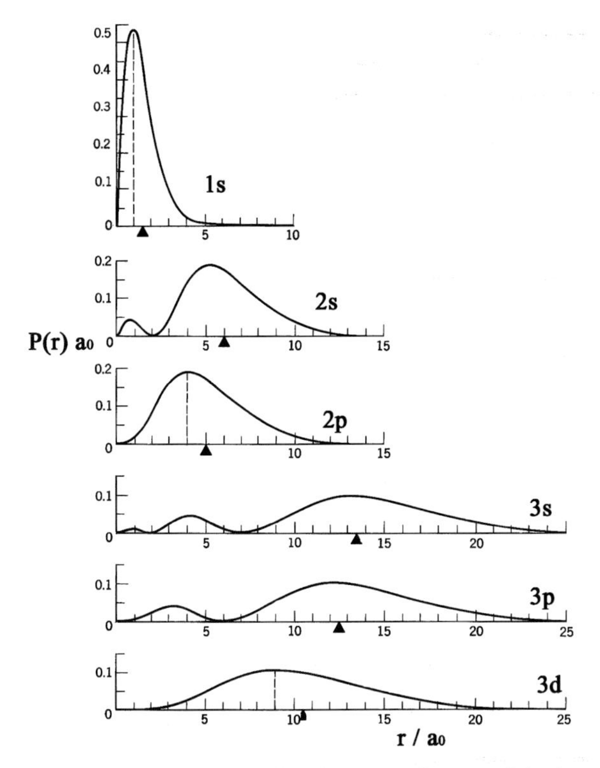


**Figure 3.3** The radial probability density as a function of the distance from the nucleus for the electron in an one-electron atom. $Z = 1$. $a_0$ is the Bohr radius. The triangles at each abscissa indicate the expectation value of the shell radius. The dashed lines indicate $n^2 a_0$ values.

Probability density functions are spherically symmetric for the *s* electrons. They vanish at one point for $n = 1$, at two points at $n = 2$, at three points at $n = 3$ and so on.

The one-electron atom in the ground state has quantum numbers $n = 1, l = 0, m_l = 0$. The probability density $\psi\psi^*$ depends on neither $\theta$ nor $\varphi$, and therefore the probability density is a spherically symmetrical function. For other states the probability density in the plane perpendicular to the *z* axis becomes more and more pronounced as *l* increases from 0 to 4 (Figure 3.4).

The electron charge looks as if it is smeared in space. One describes the electron distribution as an electron cloud of different densities. This representation is an intuitive and useful one but not exact. It is preferable to describe the electron distribution as probability density of the eigenfunction.

The probability density functions (3.14) have the angular dependence. The $\psi\psi^*$ value shifts from the vertical *z* axis in the plane perpendicular to *z* as the absolute value of $m_l$ number increases.

The probability density functions have a set of spherical and conical surfaces depending on certain values of *r* and $\theta$, at which they vanish. These nodal surfaces are characteristic for electrons, if their motion in the space is confined in an area

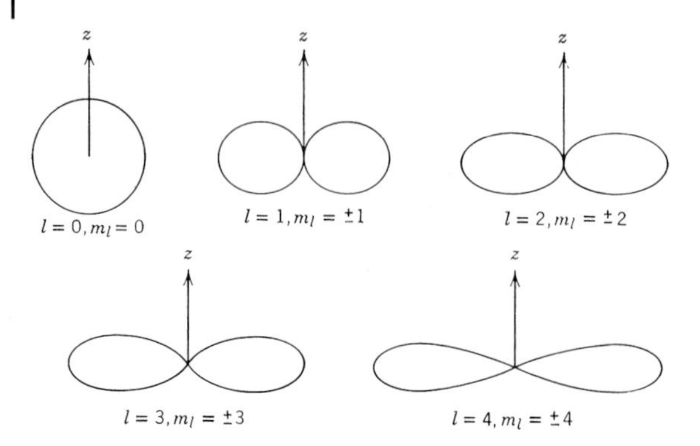


**Figure 3.4** Polar diagrams of the directional dependence of the probability densities for the one-electron atom. $l = 0, 1, 2, 3, 4$ and $m_l = \pm l$.

of the size of the atom. They are analogous to the nodal points, at which the probability density vanishes for an electron bound in an one-dimensional potential well. This can be seen in Figure 2.7. Figure 3.5 illustrates the probability densities for electrons with different values of quantum numbers $n$ and $l$.

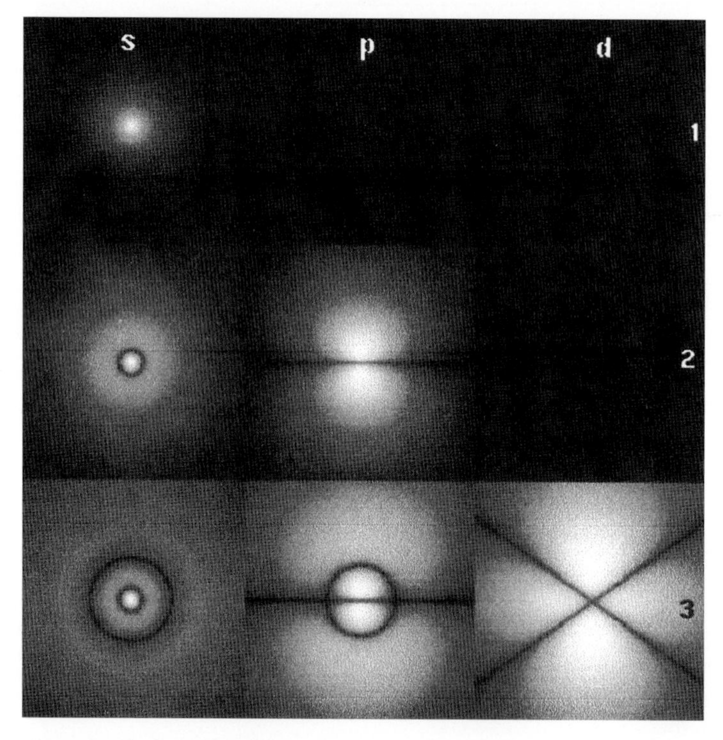


**Figure 3.5** The functions of the probability densities of the one-electron atoms for 1s; 2s, 2p; 3s, 3p, 3d electrons. The numbers on the right margin of the figure are the $n$ values. The $x0y$ plane coincides with the plane of the diagram.

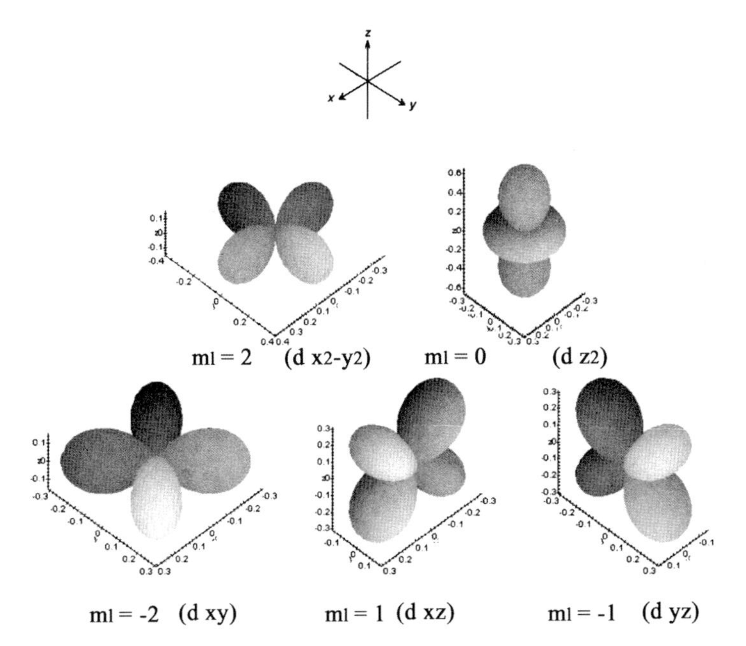


**Figure 3.6** The probability density functions of d electrons ($l = 2$) for the one-electron atom. The three-dimensional image, accepted notations of the electron orbitals are given.

The space probability density for the d electrons is shown in Figure 3.6.

One uses a term "orbital" for the space probability distribution of an electron. We would like to recall that an atomic orbital is the eigenfunction squared that describes the wave-like behavior of the electron in the atom. This function is used to calculate the probability of finding the electron in any specific region around the nucleus.

The angular character of the orbitals is determined by the appropriate spherical harmonic in equation

$$\psi_{nlm} = R_{nl}(r)\, Y_l^{m_l}(\theta, \phi) \,, \tag{3.16}$$

where $r$, $\theta$ and $\phi$ are spherical polar coordinates and $Y_l^{m}(\theta, \phi)$ are spherical harmonics. For $m_l = 0$ the first few spherical harmonics are given by

$$Y_0^0 = \frac{1}{\sqrt{4\pi}} \,; \tag{3.17}$$

$$Y_1^0 = \sqrt{\frac{3}{4\pi}} \cos\theta \,; \tag{3.18}$$

$$Y_2^0 = \sqrt{\frac{5}{16\pi}} (3\cos^2\theta - 1) \,. \tag{3.19}$$

Figure 3.7 shows these angular dependences. We see that the s state is spherically symmetric, taking the same positive value along all directions. The p state ($l =$

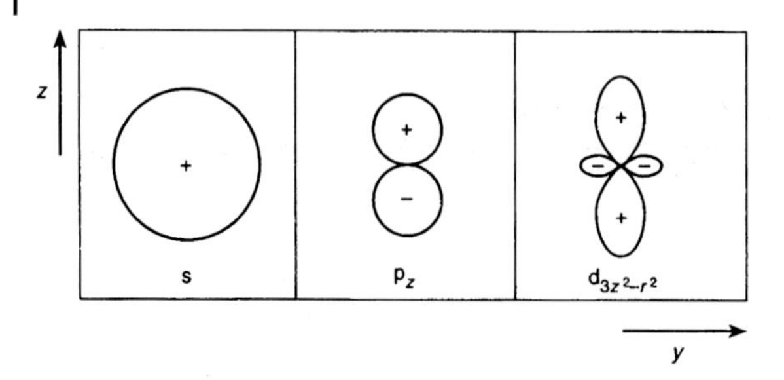


**Figure 3.7** The angular dependence of the s, $p_z$ and $d_{3z^2-r^2}$ probability density functions in the $y0z$ plane.

1, $m_l = 0$), on the other hand, has two lobes pointing out along the $z$ direction. These lobes are spherical. This state is referred to as a $p_z$ orbital.

The d state ($l = 2$, $m_l = 0$) has two positive lobes along $z$ axis and negative lobes in the equatorial $x0y$ plane. This state is referred to as $d_{3z^2-r^2}$ orbital, or simply $d_{z^2}$.

Figure 3.8 illustrates the probability densities $|Y_l^m(\theta), \phi|^2$ that give the angular dependence of the probability density $|\psi_{nlm}(\mathbf{r})|^2$. The s state is spherically symmetrical whereas the lobes of the $p_z$ state are distorted from sphericity along the $z$ axis, reflecting the change from $\cos\theta$ to $\cos^2\theta$, (3.18), (3.19). Since we often deal with atoms in a cubic environment, it is customary to form $p_x$ and $p_y$ orbitals by taking the linear combination of two remaining $l = 1$ states corresponding to $m_l = \pm 1$. They are illustrated in Figure 3.8b. A full p shell has spherical symmetry since $x^2 + y^2 + z^2 = r^2$. The probability densities of the five d orbitals are shown in Figure 3.8c.

## 3.3 Multi-Electron Atoms

A multi-electron atom consists of a nucleus of a charge of $+Ze$ and $Z$ electrons each of charge $-e$.

In the helium atom, the eigenfunctions of two electrons overlap highly in any quantum states, and so the electrons cannot be distinguished. Indistinguishability of electrons must be taken into account when describing multi-electron atoms.

Considering multi-electron atoms we should bear in mind the Pauli exclusion principle. For electrons in a single atom, it states that no two electrons can have all of the four corresponding quantum numbers equal. That is, for example, if $n$, $l$ and $m_l$ are the same, $s = \pm 1/2$ must be different. So two electrons have opposite spins. In any atomic system, for two identical electrons the total eigenfunction $\psi$ is antisymmetric. The Pauli principle requires the sign of eigenfunction to change when any two electrons swap.

Each of $Z$ moving electrons is affected by the Coulomb attraction of nucleus and by the Coulomb repulsion of the other $Z - 1$ electrons. Due to the magnitude of

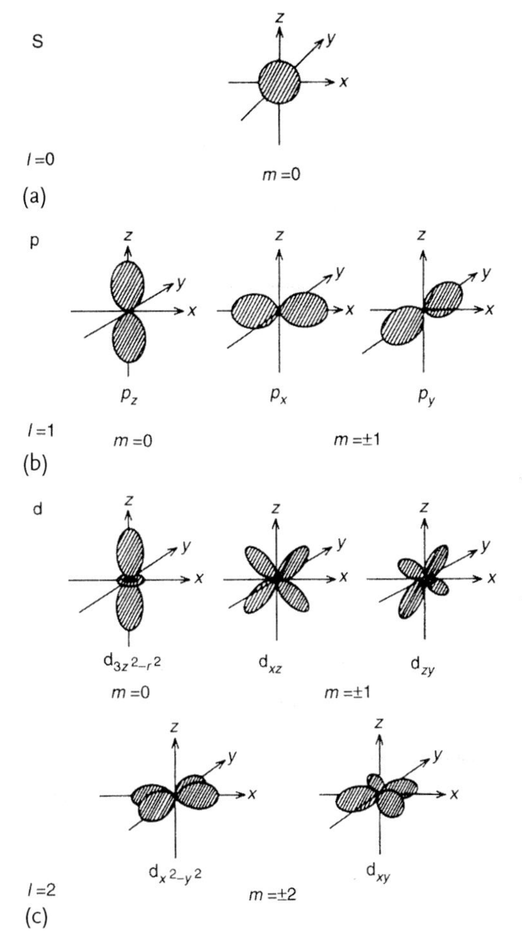


**Figure 3.8** The probability densities corresponding to s orbital (a), p orbitals (b), and d orbitals (c). (After [3] by permission of Oxford University Press.).

the nuclear charge, this is the strongest single interaction felt by each electron. There are also interactions between each electron and all the other electrons in the atom. Figure 3.9 illustrates the fact that the closer an electron to the nucleus the stronger the attraction of the nucleus exerts on the electron. More distant electrons are shielded by electrons located closer to the nucleus. The nearest electrons to the nucleus feel the full Coulomb attraction of its charge $+Ze$, while an electron being very far from the nucleus feels a smaller charge because the nuclear positive charge is shielded by the charge $-(Z-1)e$ of the other electrons surrounding the nucleus.

Energy of an electron in the multi-electron atom is given by

$$E_{nl} = -\left(\frac{1}{4\pi\varepsilon_0}\right)^2 \frac{m_e(Z-\sigma_{n,l})^2 e^4}{2\hbar^2}\frac{1}{n^2}, \tag{3.20}$$

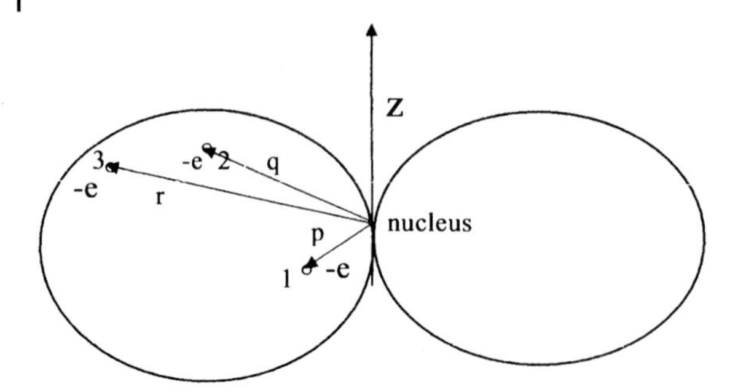


**Figure 3.9** A three-electron atom. Distance of the electrons from the nucleus $p < q < r$. The nucleus exerts a strong attractive force on electron 1. The effect of the nucleus on electron 3 is weakened by the electric fields of electrons 1 and 2. Consequently, electron 3 is shielded by fields of electrons 1 and 2.

where $\sigma_{n,l}$ is a constant which takes shielding into account. $\sigma_{n,l}$ increases with increase of $n$. At given $n$ the more quantum number $l$ the more constant shielding. Energy of the electron in a multi-electron atom turns out to be dependent on $n$ and $l$,

$$E_{ns} < E_{np} < E_{nd} < E_{nf} . \tag{3.21}$$

The electrons with smaller values of $l$ are bounded with the nucleus stronger that the electrons with greater values of $l$.

For an atom containing two electrons, 1 and 2, one shortens the notations by writing the eigenfunctions as $\psi_\alpha(1)$, $\psi_\beta(2)$. Subscripts $\alpha$, $\beta$ describe the state of electrons. These states differ in four quantum numbers.

The antisymmetry of the electron eigenfunction is the fundamental manifestation of the Pauli principle. To illustrate the nature of the Pauli exclusion principle let us suppose that electrons 1 and 2 are in states $\alpha$ and $\beta$, respectively. The eigenfunction for the two electron system A seems to be

$$\psi_A = \psi_\alpha(1)\psi_\beta(2) . \tag{3.22}$$

However, this wave function is unacceptable because the electrons are identical and indistinguishable. The function (3.22) does not change its sign if we swap $\alpha$ and $\beta$ or 1 and 2.

To take into consideration this indistinguishability one ought to take a linear combination of two eigenfunctions: one for atom 1 in state $\alpha$ and another eigenfunction for atom 1 in state $\beta$. For the two-electron atom the eigenfunction $\psi_A$ can be expressed as

$$\psi_A = \frac{1}{\sqrt{2}}\left[\psi_\alpha(1)\psi_\beta(2) - \psi_\beta(1)\psi_\alpha(2)\right] . \tag{3.23}$$

It is readily seen from (3.23) that the antisymmetric eigenfunction $\psi_A$ changes its sign when one swaps 1 and 2 or $\alpha$ and $\beta$. The minus sign in the above relationship forces the eigenfunction to vanish identically if both states are $\alpha$ or $\beta$, implying that it is impossible for both electrons to occupy the same state.

The formula (3.23) may be written as a so-called Slater determinant

$$\psi_A = \frac{1}{\sqrt{2!}} \begin{vmatrix} \psi_\alpha(1) & \psi_\alpha(2) \\ \psi_\beta(1) & \psi_\beta(2) \end{vmatrix} . \tag{3.24}$$

The exchange of labels of the eigenfunctions in (3.23) or (3.24) leads to a coupling between space variables and spins. It is of importance that electrons also move under the influence of an additional force whose sign depends on the relative orientation of their spins. This force is called the exchange force and the interaction energy is called exchange energy. It is a purely quantum effect and it has no classical analogy.

In determinant form, the eigenfunction for an atom with three electrons is

$$\psi_A = \frac{1}{\sqrt{3!}} \begin{vmatrix} \psi_\alpha(1) & \psi_\alpha(2) & \psi_\alpha(3) \\ \psi_\beta(1) & \psi_\beta(2) & \psi_\beta(3) \\ \psi_\gamma(1) & \psi_\gamma(2) & \psi_\gamma(3) \end{vmatrix} . \tag{3.25}$$

The fact that the sign of a determinant changes when any two columns are interchanged ensures that the condition of the Pauli principle holds. By analogy, for $N$ electrons the antisymmetrical expression can be written compactly as the $N \times N$ Slater determinant

$$\psi_A = \frac{1}{\sqrt{N!}} \begin{vmatrix} \psi_\alpha(1) & \psi_\alpha(2) & \dots & \psi_\alpha(N) \\ \psi_\beta(1) & \psi_\beta(2) & \dots & \psi_\beta(N) \\ \vdots & \vdots & \vdots & \vdots \\ \psi_\omega(1) & \psi_\omega(2) & \dots & \psi_\omega(N) \end{vmatrix} . \tag{3.26}$$

## 3.4 The Hartree Theory

The Schrödinger equation can be solved exactly for the hydrogen atom. Its solution for a multi-electron atom is a many-body problem, which generally cannot be solved exactly or analytically because of their complexity. The eigenfunction is no longer dependent on the coordinates of a single electron only but it depends now on all $N$ present electrons. Owing to a large difference between masses of one electron and the nucleus it is possible to employ a so-called adiabatic approximation. In this approximation the electrons are considered as moving in a static frame of nuclei.

The Schrödinger equation for a multi-electron atom can be written in the form

$$-\frac{\hbar^2}{2m}\sum_{i=1}^{N} \nabla_i^2 \psi - \frac{1}{4\pi\varepsilon_0}\sum_{i=1}^{N}\sum_{R} \frac{Ze^2}{|\boldsymbol{r}_i - \boldsymbol{R}|}\psi + \frac{1}{2}\frac{1}{4\pi\varepsilon_0}\sum_{i\neq j} \frac{e^2}{|\boldsymbol{r}_i - \boldsymbol{r}_j|}\psi = E\psi . \tag{3.27}$$

The first term on the left-hand side of (3.27) is the sum of kinetic energies of individual electrons. $-1/(4\pi\varepsilon_0)\sum_R Ze^2/|\boldsymbol{r}_i - \boldsymbol{R}|$ is the potential energy of the interaction of $i$th electron with all nuclei $\boldsymbol{R}$. The second term $-(1)/(4\pi\varepsilon_0)\sum_{i=1}^{N}\sum_R Ze^2/|\boldsymbol{r}_i - \boldsymbol{R}|$ represents potential energy of interaction of all electrons with all nuclei. The last term describes the Coulomb electron–electron interactions. Calculation of this term presents great difficulties. We must consider the Coulomb interactions between each electron and all other electrons in the atom. The determination of the ground-state eigenfunction and the ground-state energy of a quantum many-body system appears to be a formidable problem.

To describe completely the quantum mechanical behavior of electrons in solids it is necessary to calculate the many-electron eigenfunction for the system. In principle this may be obtained from the time-independent Schrödinger equation, but in practice the potential experienced by each electron is dictated by the behavior of all the other electrons in the solid. Of course, the influence of nearby electrons will be much stronger than that of far-away electrons since the interaction is electrostatic in nature, but the fact remains that the motion of any one electron is strongly coupled to the motion of all other electrons in the system.

As early as 1927, Duglas Hartree with collaborators studied a multi-electron system and proposed an approach that keeps its importance to this day. It involves solving the time-independent Schrödinger equation for a system of $Z$ electrons moving independently from each other in the atom. Each electron feels the average total electrostatic field of all other electrons in addition to the potential from the nucleus lattice. This average field has to be determined self-consistently in that the input charge-density, which enters the electrostatic potential on the left-hand side of the Schrödinger equation, depends on the eigenfunction of individual electrons that is predicted as output. Hartree suggested to reduce the multi-electron problem to the one-electron problem, in which the effective single-particle field for the electrons can be replaced by the ordinary atomic potential plus an averaged field. This field is called the Hartree self-consistent field and represents the distribution of all other electrons. Hartree introduced a procedure, which he called the self-consistent field method (SCF), to calculate approximations of eigenfunctions and energies for atoms and nuclei.

A typical time-independent Schrödinger equation for one electron is

$$\left[-\frac{\hbar^2}{2m}\nabla^2 + V(r)\right]\psi(r,\theta,\varphi) = E\psi(r,\theta,\varphi)\,, \tag{3.28}$$

where $r, \theta, \varphi$ are the spherical polar coordinates of the typical electron, $\nabla^2$ is the Laplacian operator in these coordinates, $E$ is the total energy of the electron, $V(r)$ is the net potential, and $\psi(r,\theta,\varphi)$ is the eigenfunction of the electron. According to Hartree, the total eigenfunction of the atom is composed of products of $Z$ these eigenfunctions that describe the independently moving electrons.

The Hartree theory involves solving (3.28) for a system of $Z$ electrons moving independently in the atom. The total potential of the atom can be written as a sum of a set of $Z$ identical potentials $V(r_i)$, each depending on the radial coordinate $r$ of one electron only. Consequently, the equation can be separated into a set of $Z$ time-independent Schrödinger equations, all of which have the same form, and each of which describes one electron moving in its net potential. The total energy of the atom is then the sum of $Z$ of energies of individual electrons. The corresponding eigenfunction for the atom is the product of all $Z$ eigenfunctions describing the independently moving electrons. The exact form of the net potential $V(r)$ is initially not known. It can be approximated iteratively. At first step one takes the potential as

$$V(r) = -\frac{Ze^2}{4\pi\varepsilon_0 r}\,, \quad r \to 0 \tag{3.29}$$

and

$$V(r) = -\frac{e^2}{4\pi\varepsilon_0 r}\,, \quad r \to \infty \tag{3.30}$$

for some reasonable interpolation values of $r$. This step is based on the idea that an electron far from the nucleus is shielded by other $Z - 1$ electrons (Figure 3.9).

The Schrödinger equation (3.28) is solved for the net potential obtained in the previous step.

One denotes the eigenfunctions found in this step as $\psi_\alpha(r,\theta,\varphi)$, $\psi_\beta(r,\theta,\varphi)$, $\psi_\gamma(r,\theta,\varphi)$, ... and the total energies of the electrons as $E_\alpha$, $E_\beta$, $E_\gamma$, ... Each of the symbols $\alpha, \beta, \gamma, \ldots$ stands for a complete set of the quantum numbers $n$, $l$, $m_l$, $s$.

Further, one adds one electron at the next state where states are processed in the ascending energy order. The electron charge distributions of the atom are then evaluated from the eigenfunctions specified in the previous step.

The Gauss law of electrostatics is used to calculate the electric field produced by the total charge distribution obtained in the previous step. The new function $V(r)$ generally differs from the estimation made in the first step. The procedure is repeated until two consecutive values of $V(r)$ become essentially the same. Then the obtained self-consistent solution describes the electrons in the ground state of the multi-electron atom.

## 3.5 Results of the Hartree Theory

Here we consider the results of the Hartree theory of multi-electron atoms given in [2].

The Hartree eigenfunctions for the electron in the multi-electron atom can be written

$$\psi_{nlm_ls} = R_{nl}(r)\Theta_{lm_l}(\theta)\Phi_{ml}(\varphi)(s)\,, \tag{3.31}$$

where the eigenfunctions are labeled by the same set of quantum numbers $n$, $l$, $m_l$, $s$ as are used for the one-electron eigenfunctions. The spin eigenfunction $(s)$

is exactly the same as for one-electron atoms. Furthermore, functions describing the angular dependence $\Theta_{lm_l}(\theta)$ and $\Phi_{ml}(\varphi)$ are also exactly the same. The reason is that the Schrödinger equation (3.28) for an electron in a spherically symmetrical net potential is of exactly the same form as the time-independent Schrödinger equation for an electron in the spherically symmetrical Coulomb potential (3.4).

However, the radial dependence of the eigenfunctions for an electron in a multi-electron atom is not the same as for an electron in an one-electron atom. The reason is that the net potential $V(r)$, which enters the differential equation that determines the functions $R_{nl}(r)$, does not have the same $r$ dependence as the Coulomb potential. The innermost electron shells of an atom shield the outermost shells from parts of the nuclear charge $+Ze$ and reduce it.

Figure 3.10 presents the radial behavior of the multi-electron atom eigenfunctions for an argon atom, $Z = 18$. The innermost electron shell 1s has a much smaller radius than the hydrogen shell because the ratio of their nuclei charges is 18 : 1. For argon, the radius of the innermost shell ($n = 1$) is 0.07 $a_0$. The Hartree theory predicts that the radius of the $n = 1$ shell for multi-electron atoms is smaller than that of the $n = 1$ shell for the one-electron atoms by the factor $1/(Z - 2)$. In multi-electron atoms, the inner shells of $n = 1$, $n = 2$ have relatively small radii because for these shells there is little shielding, and the electrons feel the full Coulomb attraction.

The characteristic radius of the outermost shell of argon (the principal quantum number $n = 3$) has the ratio $r/a_0$ which is equal to 1.5.

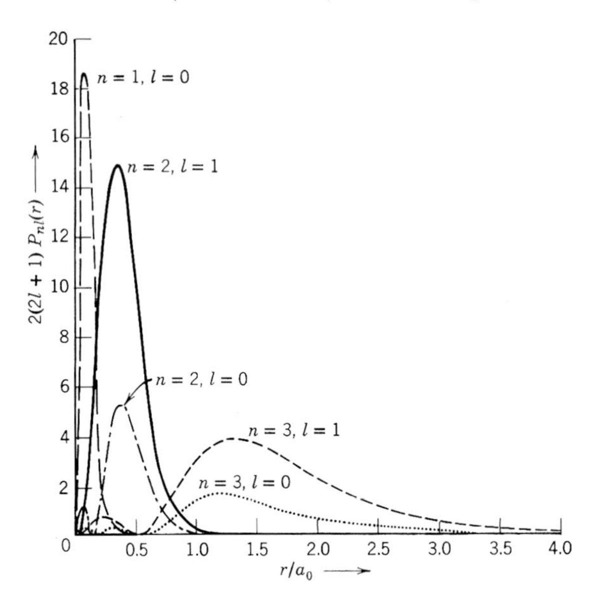


**Figure 3.10** The total radial probability densities for filled quantum states of the argon atom in accordance with the Hartree theory. The radial coordinate is plotted in units of the Bohr radius $a_0$. In the ordinate axis is plotted the radial probability density for each state times the number of electrons, which it contains.

A Coulomb potential for the Hartree theory can be expressed as

$$V_n(r) = -\frac{1}{4\pi\varepsilon_0} \cdot \frac{Z_n e^2}{r} \,, \tag{3.32}$$

where $Z_n$ is the effective charge for the shell. $Z_n < Z$. Estimated values of the effective charges for the argon atom ($Z = 18$) are 16, 6, and 3 for $Z_{n=1}$, $Z_{n=2}$, $Z_{n=3}$, respectively. Energy of inner electrons ($n = 1$) has been estimated to be equal to $-3500\,\text{eV}$. Energy of outermost electrons ($n = 3$) equals to $-14\,\text{eV}$. We see that the difference in the energy of electrons is two orders of magnitude due to the shielding. The magnitude of the total energy of an electron in the $n = 1$ shell is less than that of an electron in $n = 1$ shell of hydrogen by approximately a factor of $(Z - 2)^2$:

$$\frac{E_{Z,n=1}}{E_H} \simeq (Z - 2)^2 \,. \tag{3.33}$$

Indeed, for argon we obtain $-3500/-14 = 250$; $(18-2)^2 = 256$. Note that (3.33) is not applicable for atoms with high $Z$ because of relativistic effects.

According to the Hartree results for all the electrons with common values of $n$ the probability densities are relatively large only in essentially the same range of distances $r$. One says all these electrons are in the same shell. The charge $Z(r)$ is large within the shell, thus the thickness of each shell is well defined.

This is seen in Figure 3.11 that illustrates radial eigenfunctions for the manganese atom. The distance from the nucleus at which the probability to find 1s, 2s, 2p electrons is less than the Bohr radius. On the contrary, a 4s electron is located far off the nucleus because of shielding by inner electron shells.

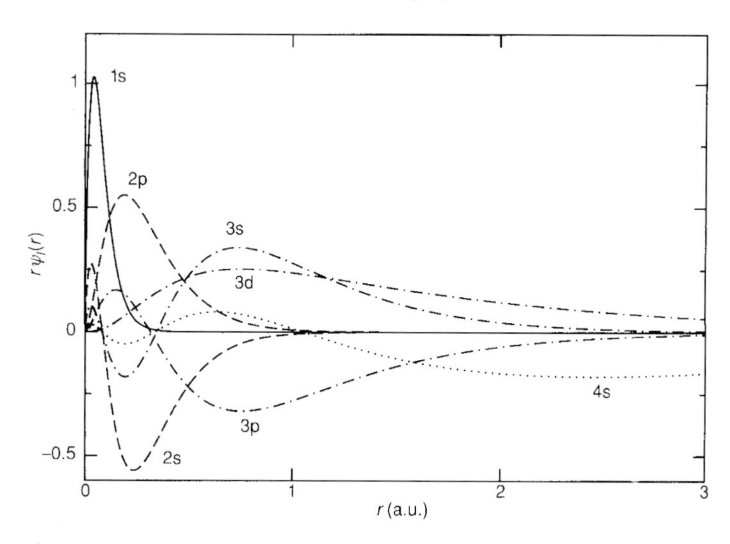


**Figure 3.11** Radial eigenfunctions $r\psi_l(r)$ for the manganese atom ($Z = 25$). The extremum in the 3d is close to those of the 3s and 3p shells even though these are much lower in energy. Note that the 4s states are much more delocalized than the 3d states although they have similar energies. (After [1] by permission of Cambridge University Press.)

Electrons in the outer shells of large $n$ are almost completely shielded from the nucleus. As a consequence, the total energy of an electron in the outermost populated shell of any atom is comparable to that of an electron in the ground state of one-electron atom. The basic reason for this is the shielding of the outer shell electron from the full nuclear charge by the charges of the inner shell electrons.

The radius of the outermost populated shell increases only very slowly by increasing the atomic number. It is given by

$$r_{\text{out}} \simeq n a_0 \tag{3.34}$$

The Hartree results show that the radius of the outermost shell is only about three times larger for elements of the highest atomic number than that of hydrogen.

The total energy of an electron in a multi-electron atom increases very rapidly with increasing $n$ for small $n$, but far less rapidly for large $n$.

The total energy of an electron in the multi-electron atom is negative as the electron is bounded to the nucleus. The total energy is the sum of kinetic energy and potential energy, $E = K + V$ where $E$ is negative. The less $E$ the stronger the bonding and the more work is needed to remove the electron out of the atom. $E$ depends on the principal quantum number $n$ according to (3.6). However, the electron energy is also affected by the orbital quantum number $l$. This is because the angular momentum retains a constant value. An electron with the orbital quantum number $l$ has an orbital angular momentum of a constant magnitude $L = \sqrt{l(l+1)}\hbar$. But at the same time $L = r p_\perp$ where $p_\perp$ is the magnitude of the component of linear momentum perpendicular to its radii coordinate vector whose length is $r$. Kinetic energy of the electron is $K = p_\perp^2/2m$. If the electron moves into a region where $r$ becomes small, then $p_\perp$ must become large (Figure 3.12). Since kinetic energy $K$ of the electron contains a term proportional to $p_\perp^2$, it increases when $r$ decreases in proportion to $1/r^2$ for small $r$.

The potential energy of the electron $V \sim -1/r$. As $r \to 0$ its kinetic energy increases more rapidly than its potential energy decreases. Thus, the electron avoids this region because there it cannot maintain a constant value of its total energy $E = K + V$, as is required by the law of energy conservation. However, the tenden-

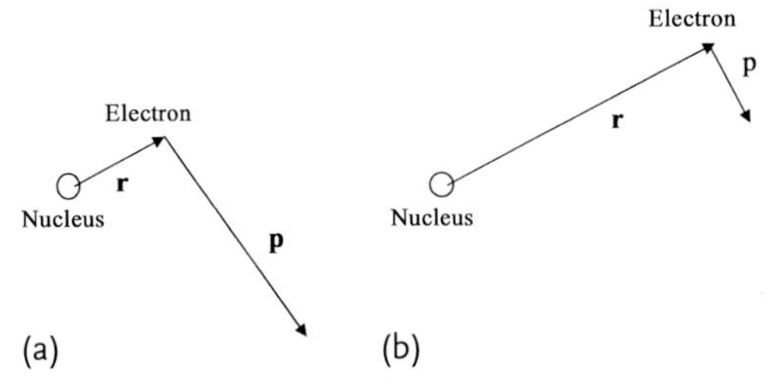


**Figure 3.12** The electron at a small distance from nucleus (a) and at a large distance (b). $\boldsymbol{p}$ is the linear momentum component perpendicular to the radial vector $\boldsymbol{r}$. The magnitude of the angle momentum $[\boldsymbol{r} \times \boldsymbol{p}]$ for the electron is a constant value. Therefore the kinetic energy of the electron is large if it is near the nucleus.

cy to avoid the region of small $r$ is not present for s-electrons with $l = 0$ since then $L = 0$. So there is much more chance of finding an $l = 0$ electron at a small value of $r$ than of finding an $l = 1$ electron in the same region. The tendency to avoid small $r$ is more distinct for $l = 2$, $l = 3$ and so on.

## 3.6 The Hartree–Fock Approximation

In 1930, Fock extended Hartree's approach to the problem by including into the consideration, the exchange energy of interacting electrons. This exchange interaction results in the decreasing of the total binding energy by keeping parallel spin electrons apart according to the Pauli exclusion principle. The effect of exchange is for electrons having the same spin to avoid each other with a result that as if the area around an electron is depleted of the negative charge. Each electron of a given spin is consequently surrounded by a positive "exchange hole," a small volume around the electron which is avoided by electrons with the same spin and the same positive "exchange hole." Fock developed Hartree's results using the variational procedure in which the trial eigenfunction was a product of one-electron eigenfunctions. This led Fock to the observation that the Hartree trial function did not incorporate the Pauli principle which would require the multi-electron eigenfunction to be antisymmetric with respect to permutations of all electron coordinates. From the Hartree–Fock theory approximation arises the concept of exchange interaction.

The Hartree–Fock method assumes that the exact $N$-electron eigenfunction of the system can be approximated by a single Slater determinant of $N$ orbitals (3.26). Applying the variational principle, one can derive a set of $N$-coupled equations for the $N$ orbitals. Solutions of these equations enables to determine the Hartree–Fock eigenfunction and energy of the system.

For molecules, Hartree–Fock approximation is the central starting point for most *ab initio* quantum chemistry methods. It was then shown by Fock that a Slater determinant, a determinant of one-particle orbitals first used by Heisenberg and Dirac in 1926, has the same antisymmetric property as the exact solution and hence is a suitable ansatz[3] for applying the variational principle.

The original Hartree method can then be viewed as an approximation to the Hartree–Fock method by neglecting the exchange energy. The Fock original method relied heavily on the group theory and was too abstract for contemporary physicists to understand and implement. In 1935 Hartree reformulated the method to make it more suitable for calculations. The Hartree–Fock method, in despite of its more accurate physical picture, was little used until the advent of computers in the 1950s because it required much more computational power than the original Hartree method.

3) Approach.

Initially, both the Hartree method and the Hartree–Fock method were applied exclusively to atoms, where the spherical symmetry of the system allowed one to simplify the problem considerably. These approximate methods were (and still are) often used together with the central field approximation, to enforce the condition that electrons in the same shell have the same radial part, and to restrict the variational solution to be a spin eigenfunction.

On the other hand, the linear combination of atomic orbitals – molecular orbital (LCAO-MO) theory, is actually the same as Hartree–Fock theory. The basic idea of this theory is that a molecular orbital is made of a linear combination of atom-centered basis functions describing the atomic orbitals. The Hartree–Fock procedure simply determines the linear expansion coefficients of the linear combination. The variables in the Hartree–Fock equations are recursively defined, that is, they depend on themselves, so the equations are solved by an iterative procedure. In typical cases, the Hartree–Fock solutions can be obtained in roughly 10 iterations. For tricky cases, convergence may be improved by changing the form of the initial guess. Since the equations are solved self-consistently, Hartree–Fock is an example of a self-consistent field (SCF) method.

The molecular orbitals are not something real, they are just models of moving electrons. The notion of molecular orbitals is an essential part of the Hartree–Fock theory and this theory is an approximation of the solution to the electronic Schrödinger equation. The approximation means that one assumes that each electron feels only the average Coulomb repulsion of all the other electrons. This approximation makes the Hartree–Fock theory much simpler to solve numerically than the original $N$-body problem. Unfortunately, in many cases iterative procedures based on this approximation diverge rather seriously from the reality and thus give incorrect results.

To solve the Schrödinger equation for solids would require one to solve a system of an enormous number of differential equations. Such volume of calculations is far beyond the capabilities of present-day computers, and it is likely to remain so for any foreseeable future. The problem can be solved by explicitly taking into consideration a correlation between electrons as the density functional theory (DFT) does. We shall consider the density functional theory and its application to solids in Chapter 8.

## 3.7
## Multi-Electron Atoms in the Mendeleev Periodic Table

The Table 3.2 shows the capacity of the electron shells of elements. Note that the values in the third column are not ordered by $n$, there are some swaps. For example, a 4s-electron has less energy than a 3d-electron and a 5s-electron has less energy than a 4d-electron. One can say the same about 6s and 4f, 6s and 5d, 7s and 6d. This fact is due to the effect of orbital quantum number $l = 0$ on the electron energy, which we have considered in previous section.

**Table 3.2** The number of electrons in the shells. The energy of the electron is negative and increases from top to bottom.

| $n$ | $l$ | Notation of shell | $2(2l+1)$ |
|---|---|---|---|
| 1 | 0 | 1s | 2 |
| 2 | 0 | 2s | 2 |
| 2 | 1 | 2p | 6 |
| 3 | 0 | 3s | 2 |
| 3 | 1 | 3p | 6 |
| 4 | 0 | 4s | 2 |
| 3 | 2 | 3d | 10 |
| 4 | 1 | 4p | 6 |
| 5 | 0 | 5s | 2 |
| 4 | 2 | 4d | 10 |
| 5 | 1 | 5p | 6 |
| 6 | 0 | 6s | 2 |
| 4 | 3 | 4f | 14 |
| 5 | 2 | 5d | 10 |
| 6 | 1 | 6p | 6 |
| 7 | 0 | 7s | 2 |
| 5 | 3 | 5f | 14 |
| 6 | 2 | 6d | 10 |

Figure 3.13 illustrates the electron shells of chemical elements. Electron energy increases from inner shells to outer shells as can be seen in Table 3.2.

In the hydrogen atom, the single electron occupies the 1s shell. For the helium atom both electrons are in the 1s shell. The spins of electrons are directed oppositely. The 2s shell is filled in lithium and beryllium.

These configurations are written as

$$
\begin{aligned}
&{}^{1}\mathrm{H}\colon 1\mathrm{s}^1 , \\
&{}^{2}\mathrm{He}\colon 1\mathrm{s}^2 , \\
&{}^{3}\mathrm{Li}\colon 1\mathrm{s}^2 2\mathrm{s}^1 , \\
&{}^{4}\mathrm{Be}\colon 1\mathrm{s}^2 2\mathrm{s}^2 .
\end{aligned}
$$

The upper index on the left of the element symbol is its number in the Mendeleev periodic system as well as the number $Z$ of protons in the nucleus.

The electrons fill the 2p-shell in sequence in the next six elements,

$$
\begin{aligned}
&{}^{5}\mathrm{B}\colon 1\mathrm{s}^2 2\mathrm{s}^2 2\mathrm{p}^1 , \quad &{}^{6}\mathrm{C}\colon 1\mathrm{s}^2 2\mathrm{s}^2 2\mathrm{p}^2 , \quad &{}^{7}\mathrm{N}\colon 1\mathrm{s}^2 2\mathrm{s}^2 2\mathrm{p}^3 , \\
&{}^{8}\mathrm{O}\colon 1\mathrm{s}^2 2\mathrm{s}^2 2\mathrm{p}^4 , \quad &{}^{9}\mathrm{F}\colon 1\mathrm{s}^2 2\mathrm{s}^2 2\mathrm{p}^5 , \quad &{}^{10}\mathrm{Ne}\colon 1\mathrm{s}^2 2\mathrm{s}^2 2\mathrm{p}^6 .
\end{aligned}
$$

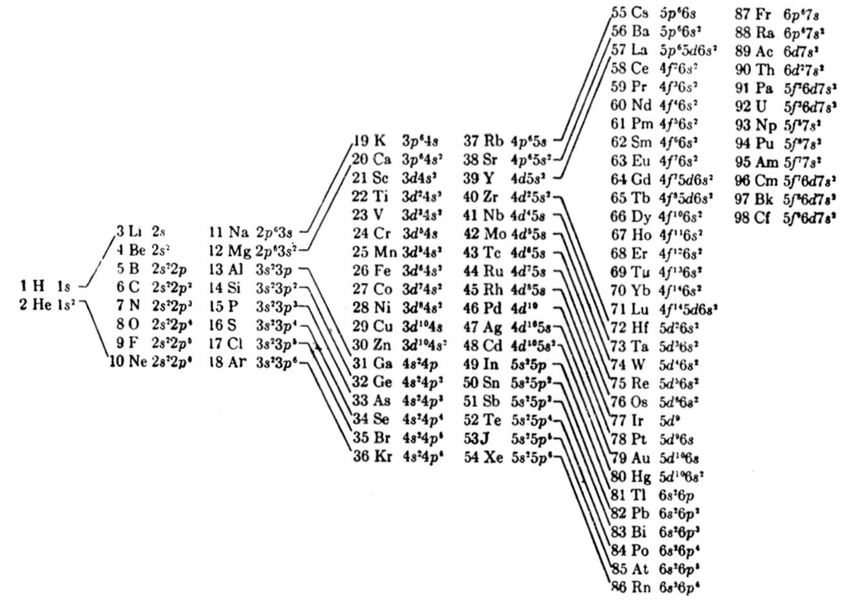


**Figure 3.13** The electron shells of the free atoms of chemical elements.

Neon is a noble gas. The electrons continue to fill 3s, 3p shells for elements from $^{11}$Na to $^{18}$Ar. These configurations are

$$^{11}\text{Na: } 1s^22s^22p^63s^1 , \quad ^{12}\text{Mg: } 1s^22s^22p^63s^2 , \quad ^{13}\text{Al: } 1s^22s^22p^63s^23p^1 ,$$
$$^{14}\text{Si: } 1s^22s^22p^63s^23p^2 , \quad ^{15}\text{P: } 1s^22s^22p^63s^23p^3 , \quad ^{16}\text{S: } 1s^22s^22p^63s^23p^4 ,$$
$$^{17}\text{Cl: } 1s^22s^22p^63s^23p^5 , \quad ^{18}\text{Ar: } 1s^22s^22p^63s^23p^6 ,$$

and further

$$^{19}\text{K: } 1s^22s^22p^63s^23p^64s^1 , \quad ^{20}\text{Ca: } 1s^22s^22p^63s^23p^64s^2 ,$$
$$^{21}\text{Sc: } 1s^22s^22p^63s^23p^63d^14s^2 , \quad ^{22}\text{Ti: } 1s^22s^22p^63s^23p^63d^24s^2$$
$$^{23}\text{V: } 1s^22s^22p^63s^23p^63d^34s^2 , \quad ^{24}\text{Cr: } 1s^22s^22p^63s^23p^63d^54s^1 ,$$
$$^{25}\text{Mn: } 1s^22s^22p^63s^23p^63d^54s^2 , \quad ^{26}\text{Fe: } 1s^22s^22p^63s^23p^63d^64s^1 ,$$
$$^{27}\text{Co: } 1s^22s^22p^63s^23p^63d^74s^2 , \quad ^{28}\text{Ni: } 1s^22s^22p^63s^23p^63d^84s^2 , \ldots$$

Let us now focus our attention to the outer shells of chemical elements starting from Sc. The 4s shell is filled by electrons whereas 3d shell is not yet filled. It is seen the same for V, Cr, Mn, Fe, Co, Ni. These metals are called the transition metals.

For atoms with larger $Z$ this effect takes place for 4d and 5d shells, too. A transition metal is defined as an element whose atom has an incomplete d shell. The elements in the periodic table from yttrium to silver (39–47), and from lanthanum to gold (57–79, including the lanthanide series) are frequently referred to as three main transition series. (Those in the actinide series and beyond, 89–111, also qualify.) All these elements are metals, many of them are of economic or industrial

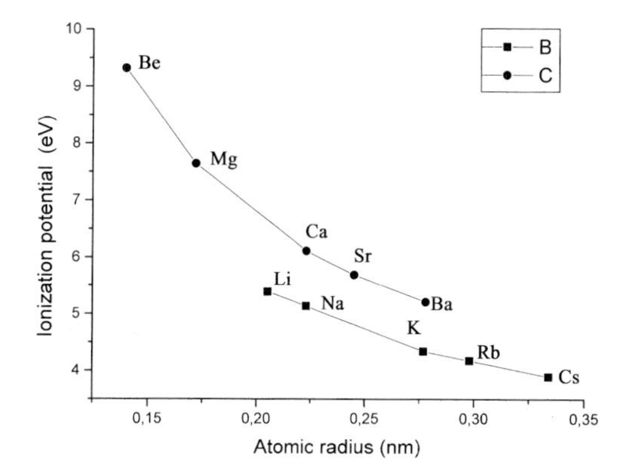


**Figure 3.14** The correlation between the first ionization potential and the atomic radii for free atoms of alkali (B) and alkali-earth (C) metals.

importance, such as titanium, vanadium, iron, cobalt, nickel, zirconium, niobium, molybdenum, tantalum, tungsten.

One can conclude that the energy difference between the 3d and 4s, the 4d and 5s, the 5d and 6s shells is very small.

There is a correlation between the atomic radius and the first ionization potential for alkali and alkali-earth metals (Figure 3.14). The larger the atomic radius the less the ionization potential.

The first ionization potential is affected by shielding of outer electrons. From lithium ($Z = 3$) to cesium ($Z = 35$) the ionization potential of the external $s$ electron decreases from 5.4 to 4.0 eV. From beryllium ($Z = 4$) to barium ($Z = 56$) the ionization potential decreases from 9.4 to 5.4 eV.

It is worth mentioning that the atomic radius in quantum physics is a physical abstraction and does not correspond to any physical observable value. Also the electronic atomic shells have no well-distinguishable boundaries and are not necessarily spherical.

## 3.8 Diatomic Molecules

Two atoms covalently bonded form a diatomic molecule. A covalent bond can be interpreted as a shearing of electrons by both atoms. The diatomic molecule is a much simpler system than a solid and is well described within the molecular orbital (MO) framework, which is based on the concept of overlapping of atomic eigenfunctions [3].

The atomic orbital of a free atom A is characterized by the electronic eigenfunction $\psi_A$. Correspondingly, the B orbital is characterized by the eigenfunction $\psi_B$. An eigenfunction $\psi_{AB}$ will determine the state of the electron in the molecule. One

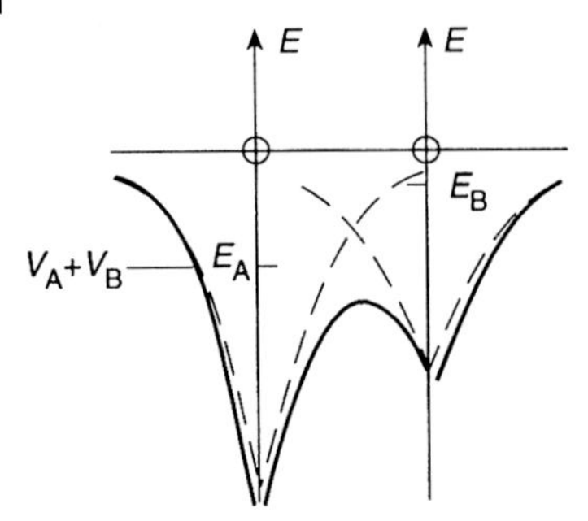


**Figure 3.15** The potential of a diatomic molecule AB as a function of the distance.

should solve the one-electron Schrödinger equation for the AB molecule, namely

$$-\frac{\hbar^2}{2m}\nabla^2\psi_{AB}(\boldsymbol{r}) + V_{AB}(r)\psi_{AB}(\boldsymbol{r}) = E\psi_{AB}(\boldsymbol{r}) , \tag{3.35}$$

where $V_{AB}(r)$ is the molecular potential. Both the Hartree and exchange-correlation potentials (see Section 5.6) enter as input to the left-hand side of (3.35). The potential $V_{AB}(r)$ depends on the output electronic eigenfunction $\psi_{AB}$ and should be determined self-consistently. One approximates this self-consistent molecular potential for covalent bonds by the sum of the individual free atomic potentials, so that

$$V_{AB} = V_A + V_B \tag{3.36}$$

as shown in Figure 3.15. This is a good approximation for the covalently bonded systems. However, it is a poorer approximation for systems towards the ionic end of the bonding spectrum where explicit shifts in the energy levels due to the flow of electrons from one atom to another occurs. This shift of the charge must be incorporated into the model.

In general, an exact solution to (3.35) cannot be calculated because the potential $V_{AB}(r)$ is unknown. One looks instead for an approximate solution that is given by some linear combination of atomic orbitals (LCAO). Considering Figure 3.15 we write

$$\psi_{AB} = c_A\psi_A + c_B\psi_B , \tag{3.37}$$

where $c_A$ and $c_B$ are two constants to be determined.

It follows from the Schrödinger equation (3.35) that

$$\left(-\frac{\hbar^2}{2m}\nabla^2 + V_{AB}\right)(c_A\psi_A + c_B\psi_B) = E(c_A\psi_A + c_B\psi_B) . \tag{3.38}$$

The molecular Schrödinger equation can be solved exactly for the case of $H_2^+$, for the hydrogen ion. It is a case of the true covalent bond. Here we have two nuclei repulsing each other, and both exerting a Coulomb attraction on the single electron.

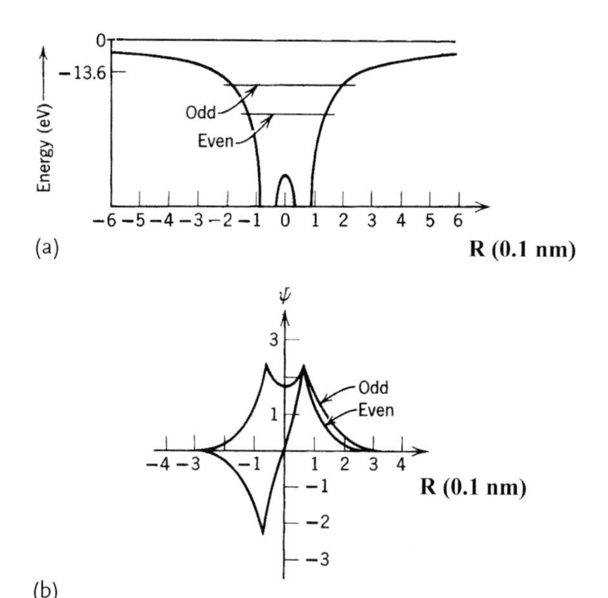


**Figure 3.16** The potential function for an electron in $H_2^+$ molecule (a); the even and odd eigenfunctions corresponding to the two energy levels (b). The internuclear separation $R_0$ equals to 0.11 nm. Near each nucleus, both eigenfunctions have magnitudes that are decreasing exponentially of the distance from the nucleus. (After [2]).

Since the electron motion is very rapid compared with the nuclear motions, the procedure is to assume that the nuclei are at rest a distance $R$ apart from each other with the single electron moving in their Coulomb fields. Next, one can treat $R$ as a variable and consider both the electron energy and the internuclear Coulomb repulsion energy, as a function of the internuclear separation. The total energy of the system is the sum of these two energies, and the system will be bound if the total energy $E$ exhibits a minimum at some value of internuclear separation.

Figure 3.16a indicates the potential energy in which the electron moves. The potential energy is symmetrical with respect to a plane perpendicular to the line connecting the two nuclei and passing through its middle.

Because the motion of the electron in a bound state has the same symmetry, the probability densities of the electron has equal values at two points on either site of the plane and equidistant from it, $(\psi\psi^*)_x = (\psi\psi^*)_{-x}$. This requires each of eigenfunctions of the electron $\psi$ to have either the same value at the two points $x$ and $-x$, or to have values summing up to zero at these points (that is, one is an additive inverse of another). Thus, the eigenfunctions that are solutions of (3.38) must be either even or odd with respect to reflection in the plane.

The trend of the eigenfunctions is shown in Figure 3.16b. The odd eigenfunction vanishes at the center of a line passing through two nuclei since it satisfies the equation $\psi(-x) = -\psi(x)$, which would otherwise be internally inconsistent at the center where $x = 0$. But the even eigenfunction $\psi(-x) = \psi(x)$ is not so constrained, and has a non-zero value at $x = 0$.

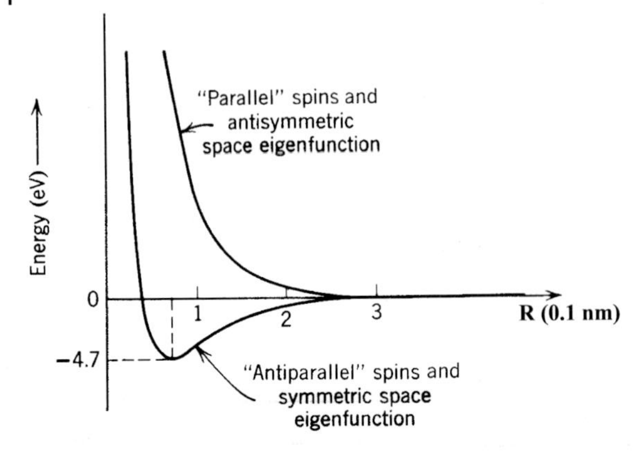


**Figure 3.17** The dependence of the total energy of the $H_2$ molecule on internuclear separation. The nuclei are bound only in the state where the electron spins are antiparallel.

An electron with probability density $\psi\psi^*$ for the odd eigenfunction must avoid the center of the molecule. Since the integral of $\psi\psi^*$ over the whole space is one, if that quantity is relatively small in the region between nuclei, it must be large in the regions outside the nuclei. The odd eigenfunction can be more tightly concentrated in the regions near the nuclei, while still being zero at the center, but only if its curvature is higher. Since higher curvature requires higher kinetic energy, this would not decrease the total energy of the electron. An electron whose behavior is described by the probability density having an even eigenfunction has a relatively high probability of being found in a region between the nuclei. Thus such an electron is relatively tightly bound. The lowest of these quantum states is the state in which the eigenfunction is even (Figure 3.16a). The energies of the electron (the eigenvalues) equal to −16.3 and −13.6 eV in states that are described by even and odd eigenfunctions, respectively.

For the hydrogen molecule $H_2$ that contains two electrons, the internuclear distance is about 0.07 nm. The energy binding the electrons with antiparallel spins equals to −4.7 eV. This is the correlation energy. In the lowest energy state of $H_2$ both electrons are in a state with the same eigenfunction, and that eigenfunction is even with respect to reflection in the plane halfway between the two nuclei. So for both electrons the probability density shows the some concentration in the region between the two nuclei. According to the Pauli exclusion principle, the two electrons must have spins with opposite $z$ components. Thus, the shared pair of electrons with "antiparallel" spins ($s = +1/2$ and $-1/2$) form a covalent bond. Figure 3.17 shows total energy of the molecule as a function of internuclear separation.

There is an atomic energy-level mismatch for diatomic molecules, $V_A \neq V_B$, $E_A \neq E_B$, Figure 3.15. According to solutions to (3.38) a given atomic energy level splits into bonding and antibonding states (which even and odd eigenfunctions correspond to) separated by $2|h|$ where $h$ is the bonding integral that couples $\psi_A$ and $\psi_B$ together through the average molecular potential $\overline{V}$.

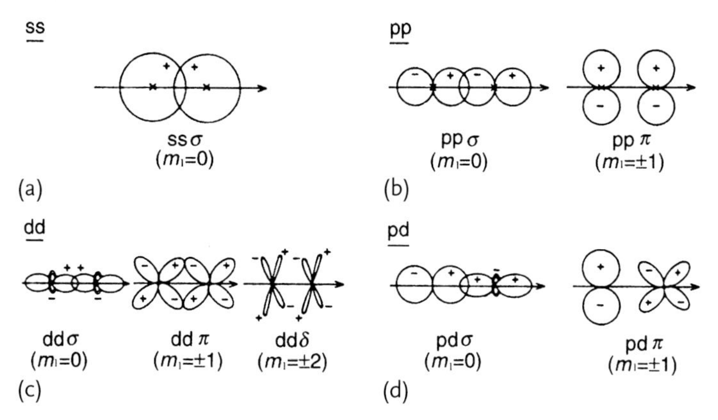


**Figure 3.18** The formation of $\sigma$, $\pi$ and $\delta$ bonds. (a) ss$\sigma$ bond is formed by two s-electrons; (b) pp$\sigma$ and pp$\pi$ bonds that are formed by p-electrons; (c) dd$\sigma$, dd$\pi$ and dd$\delta$ bonds are formed by d-electrons; (d) p- and d-electrons form pd$\sigma$ and pd$\pi$ bonds. In this Figure the signs + and − denote conditionally the distribution of $\psi^2$ function ("electron cloud") relatively to coordinate axes, see Figure 3.8. (After [3] by permission of Oxford University Press.)

Two values appear in solution of (3.35) namely $h = \int \psi_A(V_A + V_B)/2\psi_B d\mathbf{r}$ and $S = \int \psi_A \psi_B d\mathbf{r}$ [3]. These values are the bonding integral and the overlap integral, respectively. The bonding integral is negative, since the two positive s orbitals are overlapping in the negative molecular potential.

The covalent bond has two key ingredients: an attractive bonding energy pulling atoms together and a repulsive overlap potential keeping the atoms apart. *s*-valent diatomic molecules are characterized by bonding and antibonding states that differ in energy by the amount $W_{AB}$, such that

$$W_{AB} = \sqrt{4h^2 + (E_B - E_A)^2}\ . \tag{3.39}$$

A diatomic molecule has a cylindrical symmetry with respect to the internuclear axis, so that the angular momentum is conserved in this direction. This implies that the state of the molecule is characterized by the quantum number $m_l$ where $m_l\hbar$ gives the component of the angular momentum along the molecular axis. However, unlike the free atom, in which the $(2l + 1)$ different $m_l$ values are degenerate, the degeneracy is removed in the molecule. By analogy with the s, p, d, ... states of a free atom representing the orbital quantum numbers $l = 0, 1, 2, \ldots$ it is customary to refer to $\sigma, \pi, \delta, \ldots$ states of a molecule as those corresponding to $m_l = 0, \pm1, \pm2, \ldots$, respectively.

Figure 3.18 illustrates the different characteristics of the $\sigma$, $\pi$, and $\delta$ bondings.

If $\psi_A$ and $\psi_B$ are spherically symmetric s orbitals, then a ss$\sigma$ bonding is formed as shown schematically in Figure 3.18a. The areas of the spherical probability density overlap partially. If $\psi_A$ and $\psi_B$ are p orbitals (see Figures 3.4 and 3.5), then the bondings pp$\sigma$ and pp$\pi$ occur. If $\psi_A$ and $\psi_B$ are d orbitals whose electronic probability densities are sketched in Figure 3.5 then electronic distributions for molecules look like those in Figure 3.18c. A pd bonding is formed from pd$\sigma$ and pd$\pi$ states as illustrated in Figure 3.18d.

A $\sigma$ bonding is relatively strong, since the angular lobes point along the molecular axis and can give rise to a large overlap in the bonding region. On the other hand, the pp$\sigma$ and dd$\delta$ bondings are much weaker since angular lobes extend in the plane perpendicular to the molecular axes.

For carbon and silicon, the bonding integrals at their equilibrium separation equal to

$$\mathrm{pp}\sigma \approx \mathrm{sp}\sigma \approx |\mathrm{ss}\sigma| \tag{3.40}$$

and

$$|\mathrm{pp}\pi| \approx \frac{1}{2}|\mathrm{ss}\sigma| \,. \tag{3.41}$$

# 4
# The Crystal Lattice

An ideal crystal is an ordered state of matter in which the positions of atoms (or nuclei or ions or molecules) are repeated periodically in space. The periodicity of the atomic positions can also be described by means of a unit cell. This is a geometric figure constructed so that when a large number of them are placed with the same periodicity as the atoms, they fill the space with no overlap and without any space between. The positions and types of atoms in the primitive unit cell are called the basis. The set of a translation, which generate the entire periodic crystal by repeating the basis, is a lattice of points in space called the Bravais lattice. There are 14 different types of the Bravais lattice. They are presented in Figure 4.1.

The crystal lattice consists of points given by positions of the vectors $\boldsymbol{r}$

$$\boldsymbol{r} = u\boldsymbol{a}_1 + v\boldsymbol{a}_2 + w\boldsymbol{a}_3 \,, \tag{4.1}$$

where $u, v, w$ are integers, $\boldsymbol{a}_1$, $\boldsymbol{a}_2$, $\boldsymbol{a}_3$ are the translation vectors of the unit cell.

One says that vectors $\boldsymbol{a}_i$ generate the primitive crystal lattice. Magnitudes of vectors $\boldsymbol{a}_i$ and angles between them determine a symmetry class of the crystal. It is convenient to consider the crystal structure in Cartesian coordinates. Vectors $\boldsymbol{a}_i$ are chosen perpendicular to each other, if possible.

In most cases a crystal structure consists of identical copies of the same physical unit called the basis, not the primitive cell. For example, the body-centered cubic structure consists of two primitive cubic cells displaced along the body diagonal of the cubic cell by one half of the length of the diagonal.

## 4.1
## Close-Packed Structures

A close-packed structure is defined as a structure in which hard spheres that represent atoms can be placed with the maximum filling of space. In the plane, each atom has six neighbors in a hexagonal arrangement. The adjustment layer can be stacked in one of two ways (Figure 4.2): if the given plane is labeled A, then two possible positions for the next plane can be labeled as B and C. The face-centered cubic structure (fcc) is the cubic close-packed structure which can be viewed as

*Interatomic Bonding in Solids: Fundamentals,Simulation,andApplications*, First Edition. Valim Levitin.

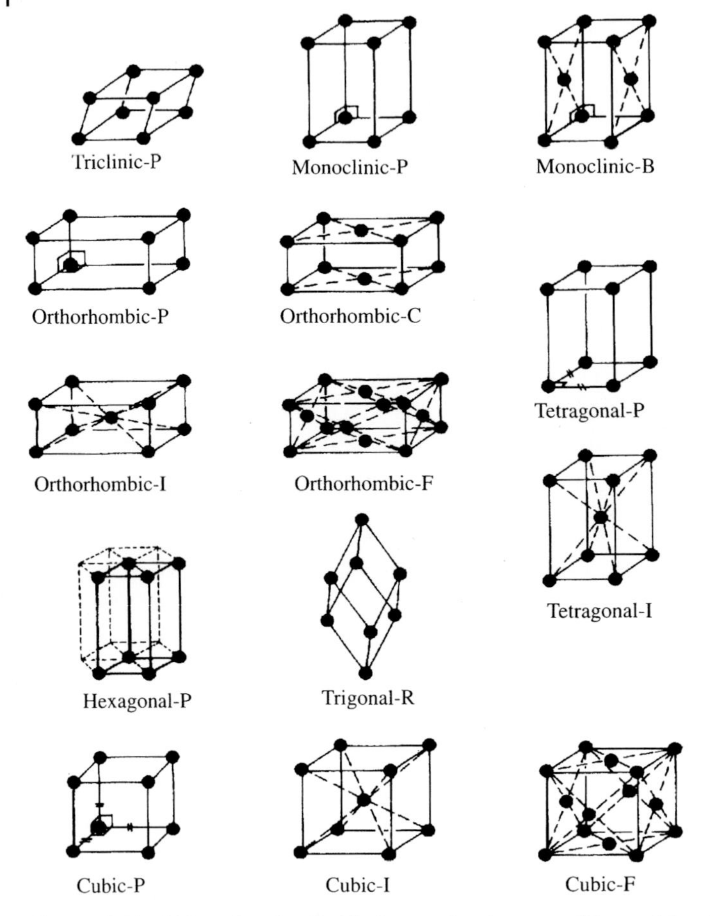


**Figure 4.1** The 14 Bravais unit cells. The letters after the name of the crystal system denote: P is the primitive unit cell; B and C, the cells are centered on two faces; I is the body-centered cell; F is the face-centered cell.

a sequence of close-packed planes in a sequence ...ABCABCABC... The hexagonal close-packed structure (hcp) consists of close-packed planes in a sequence ...ABABAB...

## 4.2 Some Examples of Crystal Structures

The structure of diamond is presented in Figure 4.3. The crystal lattice consists of two interpenetrating face-centered cubic lattices, displaced along the body diagonal of the cubic cell by one quarter of the length of the diagonal. It is obvious that the

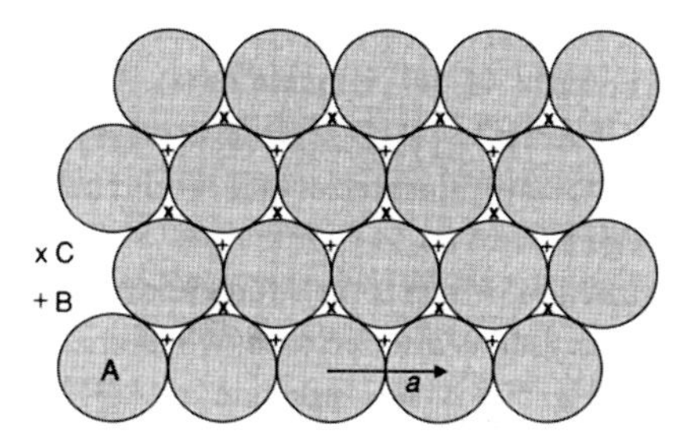


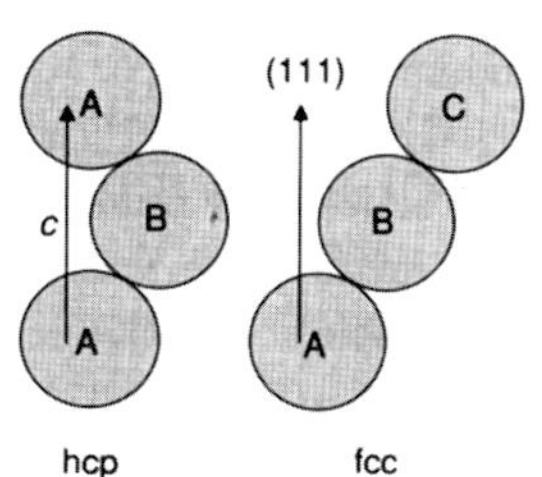


**Figure 4.2** Stacking of close-packed planes to create close-packed three-dimensional lattices. The two-dimensional close-packed is the layer of spheres labeled A, with a lattice constant $a$. Each atom has six neighbors in the plane A. Right: three-dimensional stacking can have the next layers in either B or C positions. The ...ABCABC... stacking forms the fcc lattice and ...ABAB... forms hcp lattice. In these structures each atom has 12 nearest neighbors, that is, the coordination number equals to 12.

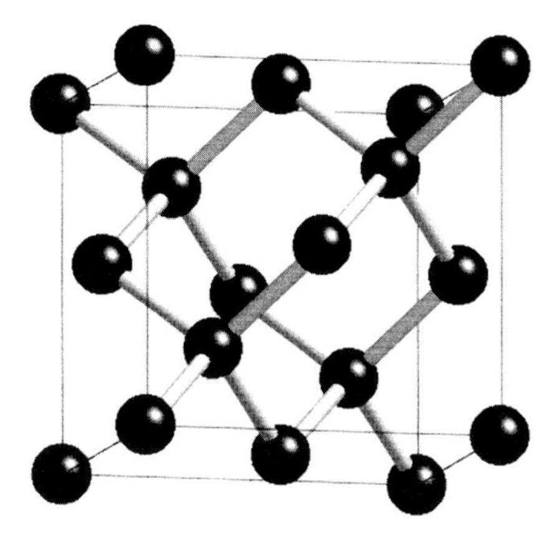

**Figure 4.3** The crystal structure of diamond. The semiconductors silicon and germanium have the same crystal lattice.

unit cell of diamond contains eight atoms. The coordinates of atoms are

$$[000], \quad [1/2\,1/2\,0], \quad [1/2\,0\,1/2], \quad [0\,1/2\,1/2],$$
$$[1/4\,1/4\,3/4], \quad [1/4\,3/4\,1/4], \quad [3/4\,1/4\,1/4], \quad [3/4\,3/4\,3/4].$$

In the lattice of diamond, there are four equal covalent bonds located so that they form an angle of $109^\circ\ 28'$ to each other (Figure 4.4); each one is formed by a pair of electrons with oppositely directed spins. In diamond lattice (formed by the carbon atoms) the four nearest-neighbor bonds form the vertices of a regular tetrahedron.

In the zinc-blende ZnS structure, the sites inside the cell are occupied by sulfur atoms, and other sites are occupied by zinc atoms (Figure 4.5). The atoms have the

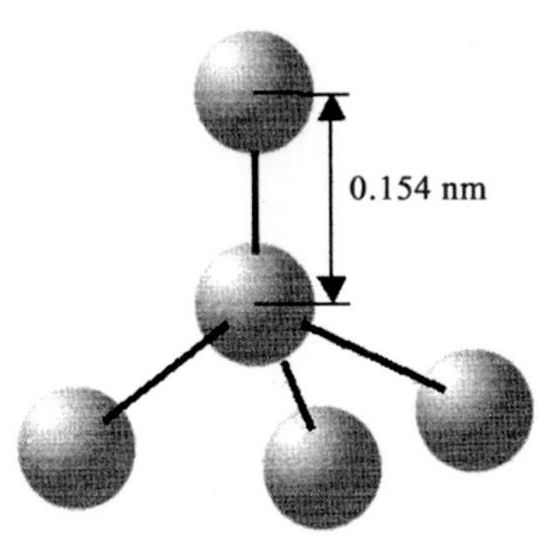


**Figure 4.4** The neighboring atoms in diamond unit cell.

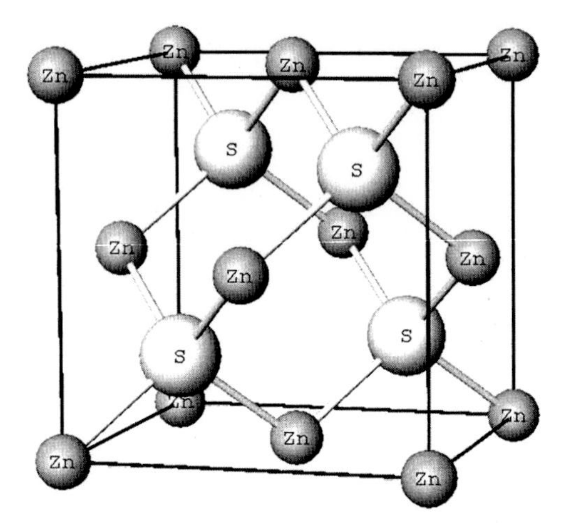


**Figure 4.5** The crystal structure of zinc-blende (sphalerite) ZnS. The compounds AgI, BC, BN, CuCl, CdS, MnSe, SiC, ZnSe, and XY have the same structure (X = Al, Ga, In; Y = As, P, Sb).

coordinates as follows:

$$\begin{aligned} \text{Zn}: &\quad [000], \quad [0\,1/2\,1/2], \quad [1/2\,0\,1/2], \quad [1/2\,1/2\,0]; \\ \text{S}: &\quad [1/4\,1/4\,3/4], \quad [1/4\,3/4\,1/4], \quad [3/4\,1/4\,1/4], \quad [3/4\,3/4\,3/4]. \end{aligned}$$

Figure 4.6 presents the structure of calcium fluoride $CaF_2$. The unit cell contains four calcium atoms and eight fluoride atoms. Calcium atoms are located in the neighborhood of fluoride atoms that are situated in corners of a cube, and every fluoride atom is located in the center of a tetrahedron in vertices of which calcium atoms are situated.

The unit cell of AuCd contains one molecule (Figure 4.7). The coordinates of atoms are

$$\begin{aligned} \text{Au}: &\quad [1/2\,1/2\,1/2]; \\ \text{Cd}: &\quad [000]. \end{aligned}$$

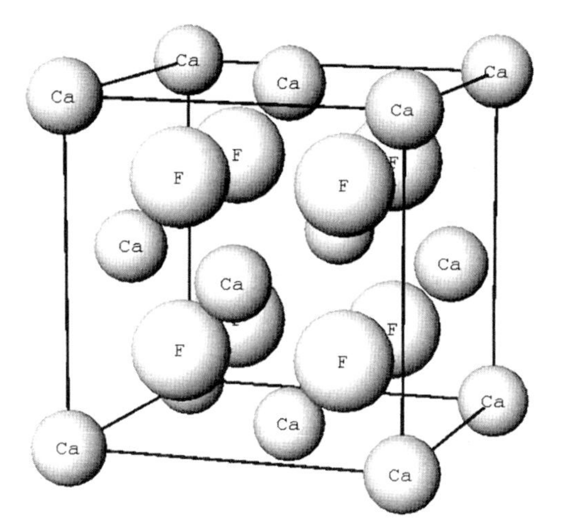


**Figure 4.6** The crystal structure of $CaF_2$. The compounds $AuAl_2$, $BaF_2$, $Ir_2P$, $Li_2Te$, $PbMg_2$, $PtGa_2$, $SiMg_2$, $UO_2$ have the same structure.

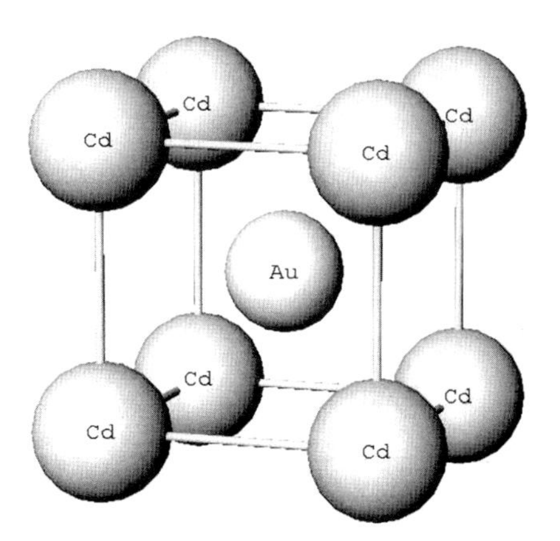


**Figure 4.7** The crystal structure of AuCd. The compounds AgMg, AlNi, BeCu, CsCl, CuZn, TiN have the same structure.

## 4.3
## The Wigner–Seitz Cell

Let us suppose that we are mainly interested in electrons that inhabit the crystal. As we know, electrons glue particles of a solid. As far as electrons are concerned, it is convenient to describe the lattice by the primitive lattice translation vectors. A primitive unit cell, which can fill up all space, is important in this case. Such a unit cell in the real space is called the Wigner–Seitz cell.

The choice of a unit primitive cell, that is, a cell of the smallest possible volume for a given lattice structure is somewhat arbitrary. Instead of using a conventional

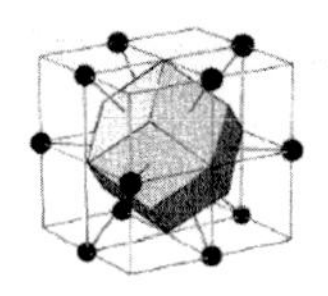


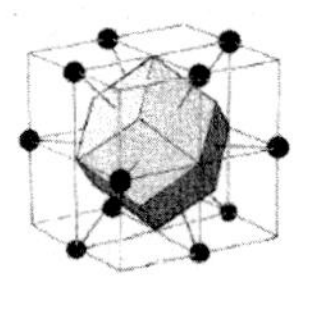

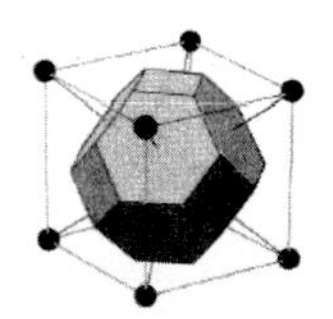

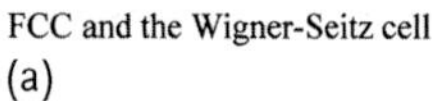


**Figure 4.8** The Wigner–Seitz cells (a) and the first Brillouin zones (b) for the face-centered cubic (fcc) and body-centered cubic (bcc) crystal lattices.

cubic unit cell, one may use a Wigner–Seitz cell. The Wigner–Seitz cell about a lattice site is a region of the space that is closer to a given site than to any other lattice site. The cell is obtained by starting at any lattice site (which then becomes the origin of the cell) and constructing vectors to all neighboring lattice points. Planes are then constructed perpendicular to and passing through the midpoints of these vectors. The polyhedron bounded by these planes is the Wigner–Seitz cell. The Wigner–Seitz cell, when translated through all lattice vectors, will just fill the whole space without overlapping. It is the primitive cell with the smallest possible volume about the origin bounded by these planes.

Figure 4.8a, illustrates the construction of the Wigner–Seitz cells for the face-centered and the body-centered cubic crystal lattices. An atom is located at the center of each polyhedron.

## 4.4 Reciprocal Lattice

Conception of the reciprocal lattice is of great importance in studies of crystal lattices. In solid state physics, it is convenient in most applications to define another space that is unambiguously related to the real space. This notion, though being a mathematical abstraction, is useful in many applications. In short, the reciprocal lattice is a lattice defined in a particular way based on the real space.

A number of physical phenomena lead us to the concept of reciprocal lattice:

- diffraction of electromagnetic waves passing through crystals;
- movement of electrons in a field of a periodic potential in the crystal lattice;
- a close connection between a set of points constituting a real crystal lattice and a set of plane waves spreading in this lattice.

There are several possible analytical treatments for introduction of the basic concept of the reciprocal lattice and the reciprocal space.

## Formal description

The formal relation between a real and a reciprocal lattices is as follows. The reciprocal lattice is a set of imaginary points constructed in such a way that the direction of vectors in this lattice from one point to another coincides with the direction of normals to planes in the real crystal lattice. The magnitude of the reciprocal lattice vector **g** is equal to the reciprocal of the interplanar spacing $1/d$ in the real lattice multiplied by $2\pi$.

The crystal planes hkl in the real crystal lattice define the coordinates of points of the reciprocal lattice space, also called $\boldsymbol{k}$-space. A plane in the real-space maps to a point in the reciprocal space and on the contrary, so there is one-to-one correspondence between planes in the real space and points in the reciprocal space.

A unit cell in a Bravais lattice is described by a set of vectors $\boldsymbol{a}_1, \boldsymbol{a}_2, \boldsymbol{a}_3$. The position of any atom in a real (direct) crystal lattice can be determined by the lattice translation vectors $\boldsymbol{r}$,

$$\boldsymbol{r} = u\boldsymbol{a}_1 + v\boldsymbol{a}_2 + w\boldsymbol{a}_3 \,, \tag{4.2}$$

where $u, v, w$ are integers.

A vector of the reciprocal lattice that determines position of a point is given by

$$\boldsymbol{g} = h\boldsymbol{b}_1 + k\boldsymbol{b}_2 + l\boldsymbol{b}_3 \,, \tag{4.3}$$

where $h, k, l$ are integers and $\boldsymbol{b}_1, \boldsymbol{b}_2, \boldsymbol{b}_3$ are the translation vectors of the reciprocal lattice. These values are expressed in terms of the direct lattice as

$$\boldsymbol{b}_1 = 2\pi \frac{[\boldsymbol{a}_2 \times \boldsymbol{a}_3]}{(\boldsymbol{a}_1 \cdot [\boldsymbol{a}_2 \times \boldsymbol{a}_3])} \,; \tag{4.4}$$

$$\boldsymbol{b}_2 = 2\pi \frac{[\boldsymbol{a}_3 \times \boldsymbol{a}_1]}{(\boldsymbol{a}_1 \cdot [\boldsymbol{a}_2 \times \boldsymbol{a}_3])} \,; \tag{4.5}$$

$$\boldsymbol{b}_3 = 2\pi \frac{[\boldsymbol{a}_1 \times \boldsymbol{a}_2]}{(\boldsymbol{a}_1 \cdot [\boldsymbol{a}_2 \times \boldsymbol{a}_3])} \,. \tag{4.6}$$

The common denominator in each equation (4.4)–(4.6) is the volume $V$ of the primitive cell of the real lattice. The $\boldsymbol{a}_i$ and $\boldsymbol{b}_j$ vectors obey the relations

$$(\boldsymbol{b}_1 \cdot \boldsymbol{a}_1) = (\boldsymbol{b}_2 \cdot \boldsymbol{a}_2) = (\boldsymbol{b}_3 \cdot \boldsymbol{a}_3) = 2\pi \,; \tag{4.7}$$

$$(\boldsymbol{b}_1 \cdot \boldsymbol{a}_2) = (\boldsymbol{b}_1 \cdot \boldsymbol{a}_3) = (\boldsymbol{b}_2 \cdot \boldsymbol{a}_3) = (\boldsymbol{b}_2 \cdot \boldsymbol{a}_1) = (\boldsymbol{b}_3 \cdot \boldsymbol{a}_1) = (\boldsymbol{b}_3 \cdot \boldsymbol{a}_2) = 0 \,. \tag{4.8}$$

From (4.8) it is evident that $\boldsymbol{b}_1 \perp \boldsymbol{a}_2$, $\boldsymbol{b}_1 \perp \boldsymbol{a}_3$, $\boldsymbol{b}_2 \perp \boldsymbol{a}_1$, $\boldsymbol{b}_2 \perp \boldsymbol{a}_3$, $\boldsymbol{b}_3 \perp \boldsymbol{a}_1$, $\boldsymbol{b}_3 \perp \boldsymbol{a}_2$.

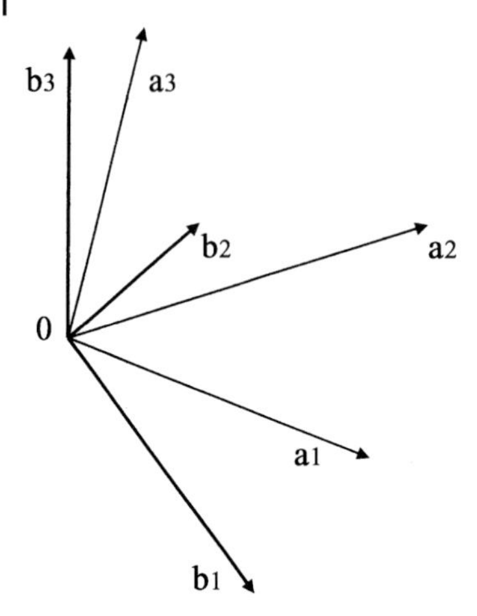


**Figure 4.9** The construction of the reciprocal lattice. $\boldsymbol{a}_1, \boldsymbol{a}_2, \boldsymbol{a}_3$ are the vectors of a real crystal lattice. $\boldsymbol{b}_1, \boldsymbol{b}_2, \boldsymbol{b}_3$ are the vectors of the reciprocal lattice. Units of the vector magnitudes are m and rad m$^{-1}$, respectively.

Figure 4.9 illustrates the construction of reciprocal lattice. The relationship between the reciprocal lattice vector and the interplanar distance in real-space is expressed as

$$|g_{hkl}| = \frac{1}{d_{hkl}} 2\pi \ . \tag{4.9}$$

Thus, one can formally generate the reciprocal lattice from the real lattice and vice versa.

### Description based on plane waves

Let us suppose that a real crystal lattice consists of a set of points $\boldsymbol{R}$ that are determined by vectors $\boldsymbol{r}$. Any property $\rho$ of the crystal lattice is periodic, $\rho(\boldsymbol{r}) = \rho(\boldsymbol{r} + \boldsymbol{R})$, where $\boldsymbol{R}$ is the lattice translation vector.

On the other hand, consider a set of plane waves $\exp i(\boldsymbol{k} \cdot \boldsymbol{r})$ which propagates in this crystal lattice. In general (that is for "randomly" chosen $\boldsymbol{k}$ wave vectors), a plane wave will not have the periodicity of the given lattice. But it will for a certain special choice of the wave vector $\boldsymbol{k}$, especially if we choose the wave vector equal to a new vector $\mathbf{g}$ and the wave has a periodicity of the given crystal lattice.

The equation defining this wave can be written as

$$\exp i(\boldsymbol{g} \cdot \boldsymbol{r}) = \cos(\boldsymbol{g} \cdot \boldsymbol{r}) + i \sin(\boldsymbol{g} \cdot \boldsymbol{r}) \ . \tag{4.10}$$

As this plane wave has the same periodicity as the real crystal lattice, then it satisfies the equation

$$\exp i(\boldsymbol{g} \cdot \boldsymbol{r}) = \exp i(\boldsymbol{g} \cdot (\boldsymbol{r} + \boldsymbol{R})) = \exp i(\boldsymbol{g} \cdot \boldsymbol{r}) \exp i(\boldsymbol{g} \cdot \boldsymbol{R}) \ . \tag{4.11}$$

And consequently

$$\exp i(g \cdot \boldsymbol{R}) = 1; \quad \cos(g \cdot \boldsymbol{R}) + i \sin(g \cdot \boldsymbol{R}) = 1 \tag{4.12}$$

for all $\boldsymbol{R}$ in the real lattice. From (4.12) we arrive at

$$(g \cdot \boldsymbol{R}) = 2\pi M , \tag{4.13}$$

where $M$ is an integer.

Thus, the set of plane waves $\exp i(g \cdot \boldsymbol{r})$ with the periodicity of the direct lattice generate the reciprocal lattice.

It is seen that (4.2)–(4.8) satisfy (4.13). It follows from these equations that

$$(g \cdot \boldsymbol{R}) = 2\pi(uh + vk + wl) . \tag{4.14}$$

### Diffraction at reciprocal lattice

Using of the reciprocal lattice let us better understand and interpret the results of diffraction experiments and thus obtain useful information about the internal structure of crystalline matter. It is employed by the so-called Evald construction, which allows one to translate the wavelength and the direction of the incident radiation into the reciprocal lattice and then determine the diffraction pattern in a relatively straightforward way.

The Evald sphere in the reciprocal space is presented in Figure 4.10.

It is seen from Figure 4.10 that $|g| = |\boldsymbol{k}_0| \cdot 2 \sin\theta$. As $|g| = 2\pi/d$ and $|\boldsymbol{k}_0| = 2\pi/\lambda$ we arrive at

$$\lambda = 2d \sin\theta , \tag{4.15}$$

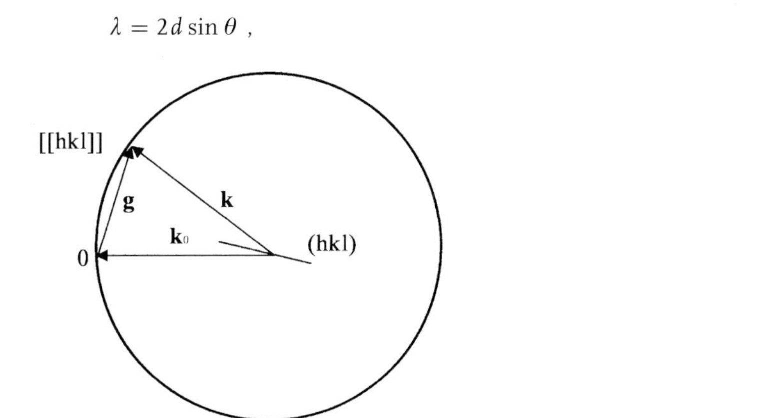


**Figure 4.10** The Evald sphere of a radius of $2\pi/\lambda$, where $\lambda$ is a wavelength of the irradiation. 0 is the origin of the reciprocal lattice, $[hkl]$ is a site in the reciprocal lattice. $\boldsymbol{k}_0$ is the wave vector of the incident beam, $\boldsymbol{k}$ is the wave vector of the diffracted beam. $(hkl)$ is a reflecting plane in the real crystal lattice. $g = \boldsymbol{k} - \boldsymbol{k}_0$ is the reciprocal lattice vector. The diffraction occurs in the direction where the Evald sphere intersects the reciprocal lattice site $[hkl]$. The angle $2\theta$ between vectors $\boldsymbol{k}_0$ and $\boldsymbol{k}$ is the double Bragg angle.

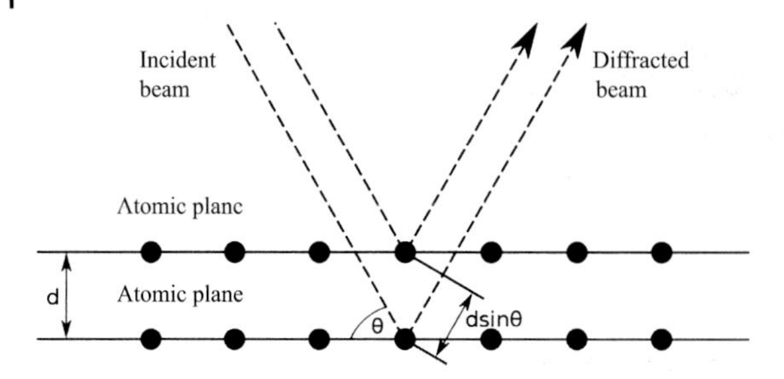


**Figure 4.11** The scheme of the Bragg diffraction experiment. The X-rays of the $\lambda$ wavelength are diffracted due to scattering by the electron shells of atoms.

where $\lambda$ is the wavelength of the electromagnetic radiation, $d$ is the interplanar spacing in the crystal, and $\theta$ is the Bragg angle (the angle of scattering). Equation (4.15) is the Bragg formula. The Bragg law is known to be a powerful tool for studying crystals by the X-ray diffraction technique. Figure 4.11 shows a conventional scheme of the X-ray investigation of a solid. X-ray structural analysis allows one to determine the structure of a crystal, parameters of the crystal lattice, and a high-accuracy measurement of the interatomic distances.

The diffraction will occur provided that the change in the wave vector equals to a vector of the reciprocal lattice. This is called the Laue condition,

$$g = \boldsymbol{k} - \boldsymbol{k}_0 \,. \tag{4.16}$$

We see that the conception of the reciprocal lattice is closely related to a conception of the plane wave $\exp i(\boldsymbol{k} \cdot \boldsymbol{r})$.

**Reciprocal space**

Reciprocal space (also called "$k$-space") can be defined as the Fourier transform of real-space. The Fourier transform can be expressed as

$$F(\mathrm{g}) = \int_V f(\boldsymbol{r}) \exp(-2\pi i \mathbf{g} \cdot \boldsymbol{r}) d^3 r \,. \tag{4.17}$$

The inverse transformation takes us from reciprocal space to real space.

A reciprocal lattice is a periodic set of points in this space, and it contains the points that compose the Fourier transform of a periodic spatial lattice. The Brillouin zone is a volume within this space that contains all the unique $\boldsymbol{k}$-vectors that represent the periodicity of classical or quantum waves allowed in a periodic structure.

The reciprocal lattice is a lattice in the Fourier space associated with the crystal. Wave vectors $\boldsymbol{k}$ exist in Fourier space, so that every position in Fourier space may be interpreted as a description of a wave, but there is a special significance to the points defined by the set of the $\mathbf{g}$ vectors associated with a crystal structure. However, points defined by the set of $\mathbf{g}$ vectors associated with a crystal structure have special significance. The set of reciprocal lattice vectors determines the possible X-ray reflections (Bragg reflections). This makes it possible to study the reciprocal lattice structure using X-ray diffraction.

Thus, one can write the transition schematically as

$$\boldsymbol{R}\text{-vectors} \Longleftrightarrow \boldsymbol{g}\text{-vectors},$$

$$\text{direct lattice, } \boldsymbol{r}\text{-space} \Longleftrightarrow \text{reciprocal lattice, } \boldsymbol{k}\text{-space} .$$

## 4.5 The Brillouin Zone

The state of an electron in a periodic crystal lattice can be described by a Bloch plane wave

$$\psi_{nk} = u_{n,k} \exp i(\boldsymbol{k} \cdot \boldsymbol{r}) , \tag{4.18}$$

where the amplitude $u_{n,k}(\boldsymbol{r}) = u_{n,k}(\boldsymbol{r} + \boldsymbol{R})$ for every site $\boldsymbol{R}$ in the Bravais lattice.

The electronic motion is extraordinarily related to the reciprocal lattice. Instead of dealing with just one electron dispersion relationship $E(\boldsymbol{k})$ there is an infinite number of equivalent dispersion relationships (see Figure 4.12) such that energy $E(\boldsymbol{k}) = E(\boldsymbol{k} + \boldsymbol{g})$ for all $\boldsymbol{g}$. However, because of the $\boldsymbol{k}$-space periodicity it is enough to only study the primitive unit cell of the reciprocal lattice, known as the first Brillouin zone, all other cells have the same characteristics.

The wave vector $\boldsymbol{k}$ appearing in (4.18) can always be confined to the first Brillouin zone in reciprocal lattice. The first Brillouin zone has a $k$-space volume

$$V_k = (\boldsymbol{b}_1 \cdot [\boldsymbol{b}_2 \times \boldsymbol{b}_3]) . \tag{4.19}$$

The first Brillouin zone is enclosed by planes normal to the lines that connect the neighboring points and pass through the midpoints of these lines (Figure 4.8).

The term "first Brillouin zone" is applied only to the $\boldsymbol{k}$-space cell. Because the reciprocal of the body-centered cubic lattice is a face-centered cubic lattice, the first Brillouin zone of the bcc is just the fcc Wigner–Seitz cell (Figure 4.8). Conversely, the first Brillouin zone of the fcc lattice is just the bcc Wigner–Seitz cell.

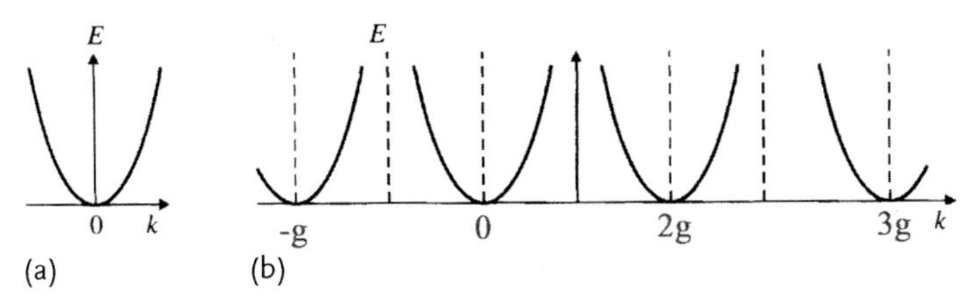


**Figure 4.12** The effect of introducing the reciprocal lattice; instead of dealing with just one electron dispersion relationship (a) we have an infinite number of copies of it (b) (after [4]).

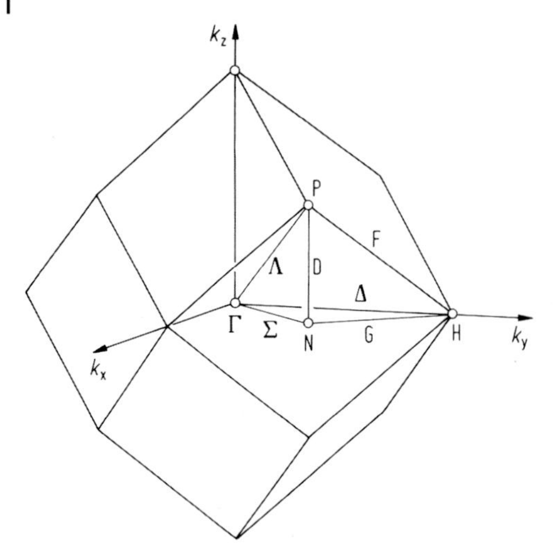


**Figure 4.13** First Brillouin zone of the body-centered cubic crystal lattice. $k_x$, $k_y$, $k_z$ are the axes of a Cartesian coordinate system in $\boldsymbol{k}$-space. The symmetry points and symmetry lines are indicated. See Table 4.1 for details.

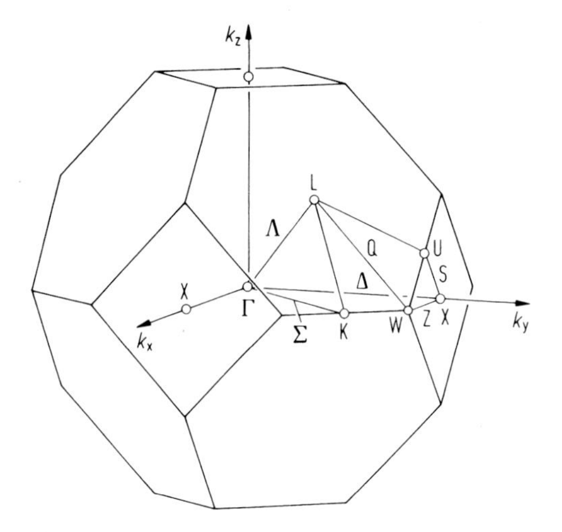


**Figure 4.14** First Brillouin zone of the face-centered cubic crystal lattice. $k_x$, $k_y$, $k_z$ are the axes of the Cartesian coordinate system in $\boldsymbol{k}$-space. The symmetry points and symmetry lines are indicated. See Table 4.1 for details.

Figures 4.13 and 4.14 present the Brillouin zones for cubic crystal lattices. One can see the Brillouin zone for hexagonal close-packed crystal lattice in Figure 4.15.

In Table 4.1 we present the symbols denoting the points and crystal directions in the first Brillouin zone for a number of the most important crystal lattices. Such points and directions are of a high symmetry and play an important role in the computation of the vibration spectrum and the electronic band-structure of the

**Table 4.1** Points and directions of high symmetry in the first Brillouin zones. fcc is the face-centered cubic crystal lattice; bcc is the body-centered cubic crystal lattice; hcp is the hexagonal close-packed crystal lattice.

| Lattice | Point | Description | Symbols of direction | Crystal direction |
|---|---|---|---|---|
| All | $\Gamma$ | Center of the Brillouin zone | – | – |
| fcc | $K$ | Middle of an edge joining two hexagonal faces | $\Sigma : \Gamma \to K$ | $[\zeta\zeta 0], [110]$ |
| | $L$ | Center of a hexagonal face | $\Lambda : \Gamma \to L$ | $[\zeta\zeta\zeta], [111]$ |
| | $W$ | Corner point | $Z : X \to W$ | $[\zeta 00], [100]$ |
| | $X$ | Center of a square face | $\Delta : \Gamma \to X$ | $[0\zeta 0], [010]$ |
| bcc | $H$ | Corner point joining four edges | $\Delta : \Gamma \to H$ | $[0\zeta 0], [010]$ |
| | $N$ | Center of a face | $\Sigma : \Gamma \to N$ | $[\zeta\zeta 0], [110]$ |
| | $P$ | Center point joining three edges | $\Lambda : \Gamma \to P$ | $[\zeta\zeta\zeta], [111]$ |
| hcp | $A$ | Center of a hexagonal face | $\Delta : \Gamma \to A$ | $[00\zeta], [001]$ |
| | $H$ | Corner point | $\Gamma \to H$ | $[\zeta\zeta\zeta], [111]$ |
| | $K$ | Middle of an edge joining two rectangular faces | $T : \Gamma \to K$ | $[\zeta\zeta 0], [110]$ |
| | $L$ | Middle of an edge joining a hexagonal and a rectangular face | $\Gamma \to L$ | $[\zeta 0\zeta], [101]$ |
| | $M$ | Center of a rectangular face | $\Sigma : \Gamma \to M$ | $[\zeta 00], [100]$ |

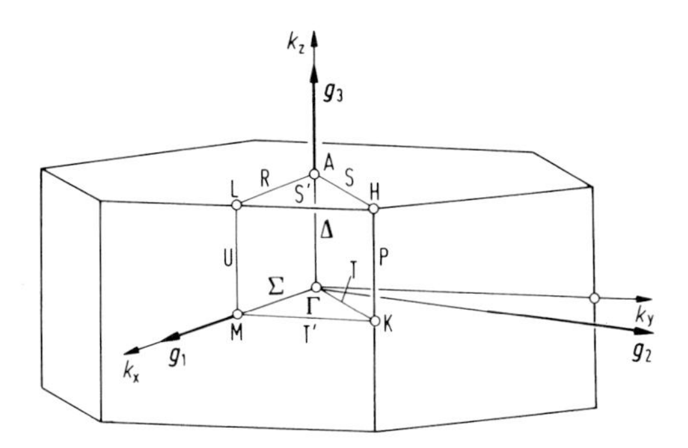


**Figure 4.15** First Brillouin zone of the hexagonal close-packed crystal lattice. $k_x, k_y, k_z$ are the axes of the Cartesian coordinate system in $\boldsymbol{k}$-space. The symmetry points and symmetry lines are indicated. See Table 4.1 for details.

solids. These data are especially useful for study of acoustic spectrum of single crystals.

# 5
# Homogeneous Electron Gas and Simple Metals

In this chapter we consider a gas that consists of free electrons. The homogeneous gas of interacting electrons, in which the positive charges are assumed to be uniformly distributed in space, is the simplest model representing condensed matter. This model is completely specified by the density $n$ of the electron gas or by the average distance $r_s$ between electrons.

Drude [5] applied the classical kinetic theory of gases to a metal, considering it as a gas of electrons. This model treats electrons as particles, which move in straight lines until they collide with one another. The compensated electrical charge is attached to much heavier particles, which are considered as immobile. The classical free-electron gas model explains very well the electrical and thermal conductivities of metals. Drude's work formed the basis for the purely classical theory that remains highly successful today, reinterpreted in the light of quantum physics. Moreover, the contemporary model allows one to calculate the strength of interatomic bonding for simple metals. Simple metals contain no d electrons in outer electronic shells. One refers to them as s-valent metals and sp-valent metals.

Let us consider a cubic box of size $L \times L \times L$ of a monovalent metal. Electrons are trapped by the positive charge that is spread inside the cubic box uniformly through the volume $V$ with an average charge density $\varrho_+ = n|e|$, where $n = N/V$ is the number of ions per unit volume and $e$ is the charge of the electron. This positive charge maintains the overall charge neutrality of the system. Initially, we make the assumption that the gas of electrons is not scattered by the underlying ionic medium. The homogeneous electron gas in the positive uniform medium is also called the jellium model of a solid.

The second assumption is that the electrons move independently of each other, so that each electron feels the average repulsive electrostatic field from all other electrons. This field would be completely canceled by the attractive electrostatic potential from smeared-out ionic background. Thus, one treats the s- and sp-valent metals as a metallic jellium within the independent particle approximation.

As a matter of fact, the state of electrons is correlated. The energy of electrons depends also on their spin. We shall recur to the interaction of electrons later.

*Interatomic Bonding in Solids: Fundamentals,Simulation,andApplications*, First Edition. Valim Levitin.

## 5.1
## Gas of Free Electrons

Thus, $N$ electrons are confined in a large volume $V = L^3$, with an average density $n = N/V$. We have considered a similar problem for one electron in Section 2.6. We are now going to determine the ground state properties of these electrons. In addition, we should apply the Pauli exclusion principle and choose boundary conditions. It is clear that the eigenfunction $\psi(\boldsymbol{r})$ must vanish whenever $\boldsymbol{r}$ is on the surfaces of the cube. This however, is often unsatisfactory, because it leads to the standing wave solutions of the Schrödinger equation. The transport of energy and charge by electrons is far more convenient to describe in terms of traveling waves. We can do this by imagining each face of the Brillouin zone to be joined to the face opposite it, so that an electron coming to the surface is not reflected back in, but leaves the Brillouin zone, simultaneously reentering it at the corresponding point on the opposite surface [7].

The Schrödinger equation for a free-electron gas takes the form

$$-\frac{\hbar^2}{2m}\left[\frac{\partial^2\psi(\boldsymbol{r})}{\partial x^2}+\frac{\partial^2\psi(\boldsymbol{r})}{\partial y^2}+\frac{\partial^2\psi(\boldsymbol{r})}{\partial z^2}\right]=E\psi(\boldsymbol{r})\,, \tag{5.1}$$

where $E$ is the total energy of the system. Within the free-electron approximation no potential energy term appears in the Schrödinger equation.

The boundary conditions for the three-dimensional cube are[1]:

$$\psi(x+L,y,z)=\psi(x,y,z)\,; \tag{5.2}$$

$$\psi(x,y+L,z)=\psi(x,y,z)\,; \tag{5.3}$$

$$\psi(x,y,z+L)=\psi(x,y,z)\,. \tag{5.4}$$

The eigenfunctions for (5.1) can be expressed as equations of the plane waves,

$$\psi_{\boldsymbol{k}}(\boldsymbol{r})=L^{-\frac{3}{2}}\exp[i(\boldsymbol{k}\cdot\boldsymbol{r})]\,, \tag{5.5}$$

where $\boldsymbol{r}$ is the radius-vector, $\boldsymbol{k}$ is the wave vector. $|\boldsymbol{k}| = 2\pi/\lambda$. We encounter here the Bloch waves.

One chooses the normalization constant in (5.5) so that the probability of finding the electron somewhere in the whole volume $L^3$ is unity:

$$\int_V [\psi_{\boldsymbol{k}}(\boldsymbol{r})]^2 d^3\boldsymbol{r}=1\,. \tag{5.6}$$

Substituting (5.5) and $(\boldsymbol{k}\cdot\boldsymbol{r}) = k_x x + k_y y + k_z z$ into (5.1) we arrive at the equation for the corresponding eigenvalue, that is, for energy $E$ of the electron as a function of wave vector $\boldsymbol{k}$,

$$E=\frac{\hbar^2}{2m}\left(k_x^2+k_y^2+k_z^2\right)=\frac{\hbar^2}{2m}k^2\,. \tag{5.7}$$

1) Considering an electron in the rectangular box (Section 2.6) we supposed that $\psi = 0$ at $x = L, y = L, z = L$. As opposed to that case, (5.2)–(5.4) are the periodicity conditions.

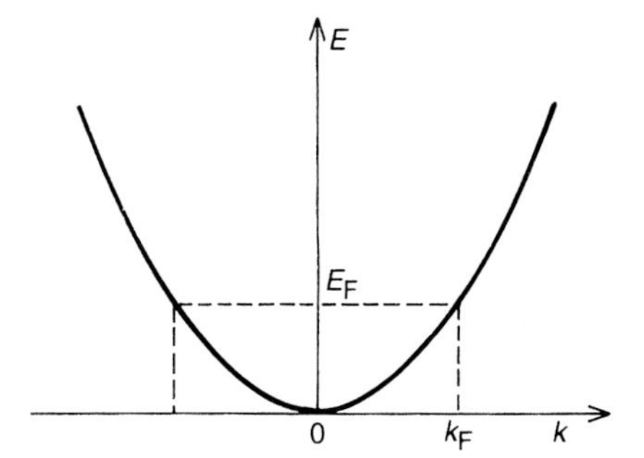


**Figure 5.1** Dependence of energy on the wave vector magnitude for the free electrons in the $\boldsymbol{k}$-space. Electrons are in states (that is, the electrons have the energy) from zero to the Fermi energy $E_F$.

The parabolic dependence of the total energy $E$ on $k$ is shown in Figure 5.1. For a free-electron gas this energy is purely kinetic, so that

$$E = \frac{\hbar^2}{2m} k^2 = \frac{p^2}{2m} . \tag{5.8}$$

Hence, from (5.7) we have momentum $\boldsymbol{p} = \hbar \boldsymbol{k}$. The plane wave $L^{-(3/2)} \exp[i(\boldsymbol{k} \cdot \boldsymbol{r})]$ is periodic in the direction of $\boldsymbol{k}$. The wave length $\lambda = h/p$ as we have already found experimentally for free particles, namely (2.3).

The periodicity conditions mean that

$$\exp(ik_x x) = \exp[ik_x(x+L)] = \exp(ik_x x) \cdot \exp(ik_x L) ;$$
$$\Longrightarrow \exp(ik_x L) = 1 . \tag{5.9}$$

And by analogy

$$\exp(ik_y L) = 1; \quad \exp(ik_z L) = 1 . \tag{5.10}$$

Since the function $\exp w = 1$ only if $w = 2\pi i n$, where $n$ is an integer, the components of the wave vector $\boldsymbol{k}$ are given by

$$k_x = \frac{2\pi}{L} n_x; \quad k_y = \frac{2\pi}{L} n_y; \quad k_z = \frac{2\pi}{L} n_z \tag{5.11}$$

and the vector $\boldsymbol{k}$ that determines the state of each electron in the $\boldsymbol{k}$-space is expressed as

$$\boldsymbol{k} = \frac{2\pi}{L}(n_x \boldsymbol{i} + n_y \boldsymbol{j} + n_z \boldsymbol{l}) , \tag{5.12}$$

where $\boldsymbol{i}$, $\boldsymbol{j}$, $\boldsymbol{l}$ are the basis vectors and $n_x$, $n_y$, $n_z$ are integers. It follows from (5.8) and (5.12) that there are only discrete levels of energy for the electrons. In practice we deal with enormous number of $k$ points, hereupon the energy levels differ to a very little degree.

It is seen from (5.12) that the minimal distance between points in $\boldsymbol{k}$-space (at $n_x = n_y = n_z = 1$) equals to $2\pi/L$, and the volume per the $\boldsymbol{k}$-point is $(2\pi/L)^3$.

## 5.2
## Parameters of the Free-Electron Gas

If a region of $\boldsymbol{k}$-space is of volume $\Omega$ then it contains

$$\frac{\Omega}{(2\pi/L)^3} = \frac{\Omega L^3}{8\pi^3} = \frac{\Omega V}{8\pi^3} \tag{5.13}$$

allowed values of $\boldsymbol{k}$. This means that the density of levels (number of allowed $\boldsymbol{k}$-points per unit volume of $\boldsymbol{k}$-space) is given by

$$N_k = \frac{\Omega V/(8\pi^3)}{\Omega} = \frac{V}{8\pi^3} . \tag{5.14}$$

Thus, the allowed values of the wave vector $\boldsymbol{k}$ in the $\boldsymbol{k}$-space are discrete and fall on a fine mesh. Each state corresponding to a given $\boldsymbol{k}$ value can contain not more than two electrons of the opposite $s$ signs according to the Pauli principle. At the temperature of absolute zero all the states of lowest energy will be occupied within a sphere of radius $k_F$, the so-called Fermi sphere. These electron states are shown in Figure 5.2. The volume occupied by electrons in $\boldsymbol{k}$-space at $T = 0$ equals to $\Omega_F = 4\pi k_F^3/3$.

From (5.7) we have for the Fermi energy

$$E_F = \frac{\hbar^2}{2m} k_F^2 . \tag{5.15}$$

The volume of unit cell that contains two electrons in $\boldsymbol{k}$-space is $(2\pi/L)^3$. Consequently, the total number of electrons $N$ can be expressed as

$$N = \frac{\frac{4}{3}\pi k_F^3}{(2\pi/L)^3} \cdot 2 = \frac{k_F^3 L^3}{3\pi^2} \tag{5.16}$$

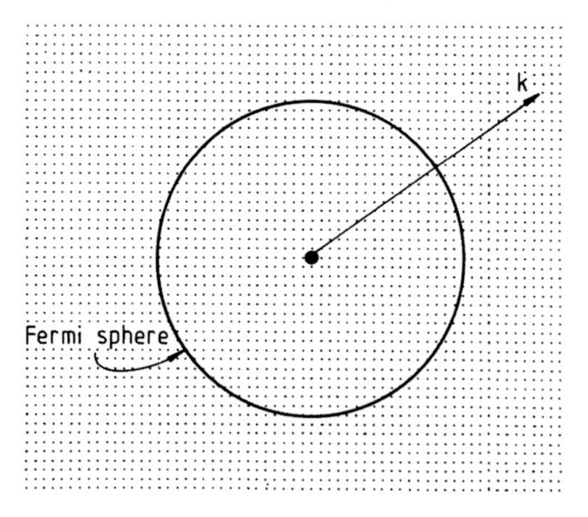


**Figure 5.2** The fine mesh of allowed $\boldsymbol{k}$-values. Only the $\boldsymbol{k}$-states within the Fermi sphere of the radius $\boldsymbol{k}_F$ are occupied at zero temperature.

or

$$k_F = (3\pi^2)^{\frac{1}{3}} N^{\frac{1}{3}} V^{-\frac{1}{3}} = (3\pi^2)^{\frac{1}{3}} n^{\frac{1}{3}} . \tag{5.17}$$

The value $k_F$ is called the Fermi momentum. It depends on the electron density. It is seen from (5.17) that for large $N$ the occupied region in $\boldsymbol{k}$-space is indistinguishable from a sphere.

Substituting (5.17) into (5.15) we arrive at the next equation for the Fermi energy

$$E_F = \frac{\hbar^2}{2m}(3\pi^2)^{\frac{2}{3}} n^{\frac{2}{3}} . \tag{5.18}$$

Thus, the Fermi energy of the free-electron gas is dependent on the density of electrons as $n^{2/3}$.

The values in (5.15)–(5.18) are in SI units.

A dimensionless parameter $r_s$ is universally used to describe properties of the electron gas. By definition, $r_s$ is the radius in atomic units of the sphere, which encloses one unit of the electron charge. It may be found from the equality

$$\frac{4}{3}\pi(r_s \cdot a_0)^3 = \frac{V}{N} = \frac{1}{n} , \tag{5.19}$$

and

$$r_s = \frac{1}{a_0}\left(\frac{3}{4\pi n}\right)^{\frac{1}{3}} , \tag{5.20}$$

where $a_0$ is the Bohr radius. Thus $r_s$ is a measure of the average distance between electrons in the uniform electron gas.

The parameter $r_s$ is small for a high-density electron gas and large for a low-density one. For metals, the $r_s$ parameter lies between 1.88 for beryllium and 5.63 for cesium. For semiconductors silicon and germanium the $r_s$ parameters equal to 2.00 and 2.08, respectively.

It is convenient to express properties of the electron gas in terms of $r_s$. We obtain from (5.17) and (5.20)

$$r_s = \frac{1}{k_F \cdot a_0}\left(\frac{9\pi}{4}\right)^{\frac{1}{3}} \tag{5.21}$$

or in SI system of units,

$$k_F = \frac{1.9192}{r_s a_0} . \tag{5.22}$$

The Fermi energy is given by

$$E_F = \frac{\hbar^2}{2m_e} k_F^2 = \frac{\hbar^2}{2m_e} \frac{3.6832}{(r_s a_0)^2} . \tag{5.23}$$

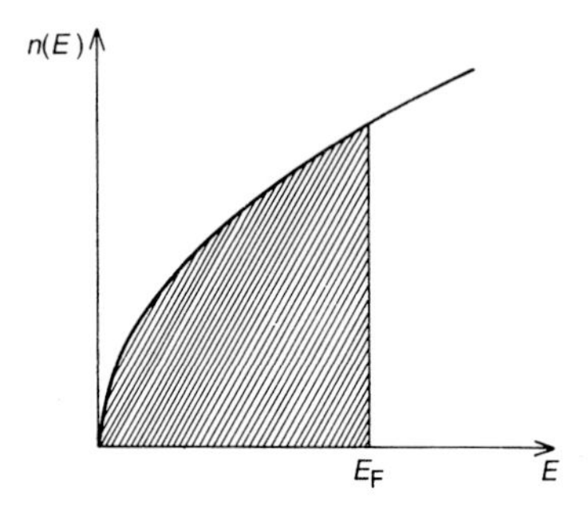


**Figure 5.3** Dependence of the number of states on the energy of electrons at the absolute zero of temperature. All the states up to the Fermi energy are occupied.

In atomic units,

$$E_F = \frac{3.6832}{r_s^2} . \tag{5.24}$$

Multiplying (5.24) by $E_{Ry} = 13.605$ eV we obtain the Fermi energy in electron volts.

The number of electron states of both spins with the energy less than $E$ may be obtained directly from (5.16) by writing it in the form (since $k^2 = 2m/\hbar^2 E$)

$$N(E) = \left(\frac{L^3}{3\pi^2}\right)\left(\frac{2m}{\hbar^2}\right)^{\frac{3}{2}} E^{\frac{3}{2}} . \tag{5.25}$$

Differentiating (5.25) with respect to energy gives the density of states in the volume unit per the unit interval of energy

$$n(E) = \left(\frac{2m}{\hbar^2}\right)^{\frac{3}{2}}\left(\frac{L^3}{3\pi^2}\right) E^{\frac{1}{2}} . \tag{5.26}$$

The density of states $n(E)$ is the number of electron states per the energy interval from $E$ to $E + dE$. Figure 5.3 illustrates the filling of states by electrons at temperature of the absolute zero. The temperature of electrons $T_F$ that corresponds to the Fermi energy can be estimated as

$$T_F = \frac{E_F}{k_B} . \tag{5.27}$$

For aluminum the Fermi temperature is of the order of $10^5$ K.

## 5.3
## Notions Related to the Electron Gas

Some definitions connected with the electron gas enumerated.

- *Homogeneous electron gas* is a collection of electrons uniformly distributed in space. Properties of this gas are completely specified by its density.
- *Jellium model* is a quantum mechanical model of interacting electrons in a solid where the positive charge is replaced by an uniform background.
- *Independent electron approximation.* The interactions of the electrons with one another are ignored. Electrons must only obey the Pauli exclusion principle.
- *The Hartree–Fock approximation.* The independent electron approximation, which includes the electron–electron Coulomb interaction and the exchange interaction between electrons with parallel spins.
- *Free-electron approximation.* The ions have no effect on the motion of electrons. The ions maintain overall charge neutrality only.
- *Nearly free-electron model of metals.* A gas of free electrons into which a lattice of positive ions is immersed.
- *The Wigner crystal.* The electrons are localized and form a regular lattice. The positive charge is uniformly spread through the system.

## 5.4
## Bulk Modulus

Let us consider the pressure of electron gas within the framework of the independent electron approximation.

The average energy of free-electron gas containing $N$ electrons is given by

$$E = \frac{\int_0^{E_\mathrm{F}} E n(E) d E}{\int_0^{E_\mathrm{F}} n(E) d E} N . \tag{5.28}$$

Substitution of the density of states $n(E)$ from (5.26) to (5.28) leads to

$$E = \frac{\int_0^{E_\mathrm{F}} E \left(2m/\hbar^2\right)^{\frac{3}{2}} \left(L^3/(3\pi^2)\right) E^{\frac{1}{2}}}{\int_0^{E_\mathrm{F}} \left(2m/\hbar^2\right)^{\frac{3}{2}} \left(L^3/(3\pi^2)\right) E^{\frac{1}{2}}} N = \frac{3}{5} E_\mathrm{F} N . \tag{5.29}$$

Because the electrons do not interact with one another the energy $\frac{3}{5} N E_\mathrm{F}$ is a pure kinetic. The pressure of this gas can be found as $p = -(\partial E/\partial V)_N$. Bulk modulus can be calculated taking the derivative of the pressure, $B = -V\,(\partial p/\partial V)_N$. Note that the number of electrons is constant.

Substituting (5.18) into (5.29) we arrive at

$$E = \frac{3}{5} N (3\pi^2)^{\frac{2}{3}} N^{\frac{2}{3}} V^{-\frac{2}{3}} . \tag{5.30}$$

**Table 5.1** Comparison of bulk moduli calculated for the free-electron gas and experimental data for alkali metals.

| Metal | $r_s$ | $E_F$ (eV) | $B$ (GPa) Theory | $B$ (GPa) Experiment |
|---|---|---|---|---|
| Li | 3.23 | 4.80 | 24.57 | 11.50 |
| Na | 3.93 | 3.24 | 9.20 | 6.42 |
| K | 4.86 | 2.12 | 3.18 | 2.81 |
| Rb | 5.20 | 1.85 | 2.27 | 1.92 |
| Cs | 5.63 | 1.58 | 1.52 | 1.43 |

$$p = -\left(\frac{\partial E}{\partial V}\right)_N = \frac{2}{5}\frac{N}{V}\frac{\hbar^2}{2m}(3\pi^2)^{\frac{2}{3}} N^{\frac{2}{3}} V^{-\frac{2}{3}} = \frac{2}{5}\frac{N}{V} E_F \,. \tag{5.31}$$

Now we can find bulk modulus,

$$\begin{aligned} B = -V\left(\frac{\partial p}{\partial V}\right)_N &= -VN\frac{\partial}{\partial V}\left[\frac{2}{5}\frac{\hbar^2}{2m}(3\pi^2 N)^{\frac{2}{3}} V^{-\frac{5}{3}}\right] \\ &= -VN\frac{2}{5}\frac{\hbar^2}{2m}(3\pi^2 N)^{\frac{2}{3}}\left(-\frac{5}{3}\right)(V)^{-\frac{8}{3}} \\ &= \frac{2}{3}N\frac{\hbar^2}{2m}(3\pi^2 N)^{\frac{2}{3}}(V)^{-\frac{5}{3}} \\ &= \frac{2}{3}\frac{N}{V}\frac{\hbar^2}{2m}(3\pi^2 N)^{\frac{2}{3}}(V)^{-\frac{2}{3}} = \frac{2}{3}\frac{N}{V}E_F = \frac{2}{3}n\,E_F. \end{aligned} \tag{5.32}$$

Substitution of the values to the end expression of (5.32) and calculation leads to a design formula

$$B = \frac{2}{3}n\,E_F = \frac{2}{3}\frac{n}{4\pi r_s^2 a_0^2}\frac{\hbar^2}{2m_e}(3\pi^2)^{\frac{2}{3}} n^{\frac{2}{3}} = \frac{3.578\,17 \times 10^{-39}}{r_s^5 a_0^5}\,\text{GPa}\,. \tag{5.33}$$

The theory predicts results close to the experiment for potassium, rubidium and cesium and fails in the case of lithium and sodium (Table 5.1). Note that the model of independent electrons fails for elements with a high electron density ($r_s = 3.23$ and 3.93).

## 5.5 Energy of Electrons

Here we describe kinetic energy of electrons, the energy of electron–ion interaction, the exchange energy, and the correlation energy.

The first term in an expression for the total energy is the kinetic energy. All terms except the first one in the total energy come from the Coulomb interaction between the particles and from a quantum interaction between electrons. The potential energy terms cannot be found exactly. Instead, approximate expressions for them are obtained by different authors.

**Kinetic Energy.** The foregoing model of the free-electron gas would exhibit no bonding because the only contribution to the energy is the repulsive kinetic energy. The kinetic energy is given by (5.29), $E_{\text{kin}} = (3/5) E_{\text{F}} N$. In terms of $r_{\text{s}}$ the average kinetic energy per electron in atomic units can be expressed as

$$E_{\text{kin}} = \frac{2.2099}{r_{\text{s}}^2} . \tag{5.34}$$

The electrons are free particles, which interact pairwise according to the Coulomb law. Potential energy of interaction per electron can be expressed as

$$E_{\text{pot}} = -\int_{\infty}^{r} \frac{1}{4\pi\varepsilon_0} \frac{e^2}{r^2} dr = \frac{1}{4\pi\varepsilon_0} \cdot \frac{e^2}{r} \tag{5.35}$$

In the homogeneous electron gas the average kinetic energy per electron is inversely proportional to the square of the characteristic length of the system, that is $E_{\text{kin}} \sim 1/r_{\text{s}}^2$, see (5.34). The average potential energy per particle is proportional to $1/r_{\text{s}}$. When the electron gas has sufficiently high density ($r_{\text{s}}$ is small) the kinetic energy term will be larger than the potential energy term, $E_{\text{kin}} > E_{\text{pot}}$. In this case electrons behave like free particles, since the potential energy is a perturbation on the dominant kinetic energy. The metals lithium and sodium differ from all the rest in a relatively small $r_{\text{s}}$ and high kinetic energy of electrons. That is why the theory overestimates the predicted values of bulk moduli (Table 5.1).

**The Coulomb Interaction.** The Coulomb interaction occurs between the electrons and the uniform positive background. In the model of the homogeneous electron gas, the time-average electron density is uniform throughout the system, as is the positive background. These equal and opposite charged densities cancel each other, so that the net system is charge neutral. Consequently, the direct Coulomb energy is zero. The term that describes the direct Coulomb interaction between electrons is omitted because it is canceled by the direct interaction with the positive background.

## 5.6 Exchange Energy and Correlation Energy

Until now, we have considered a system of noninteracting electrons that obey the Pauli exclusion principle. The next step is to take into account the interaction between electrons.

The exchange interaction between electrons is a purely quantum effect. It is the result of the fact that the eigenfunction of indistinguishable electrons is subject to exchange symmetry. The eigenfunctions describing two electrons must be inverted in sign (antisymmetric) if the two electrons are swapped (Section 3.3).

The motion of the electrons is correlated: parallel spin electrons keep apart from each other according to the Pauli exclusion principle. This leads to a decrease in energy by an amount called the exchange energy. The electron exchange term describing the correlation between electrons with parallel spins is found within the Hartree–Fock approximation. The exchange energy $E_{ex}$ prevents two parallel-spin electrons from being found at the same point in space. The exchange contribution to the ground state energy per electron in terms of parameter $r_s$ is

$$E_{ex} = -\frac{3}{2\pi}(k_F a_0)\left(\frac{e^2}{2a_0}\right) = -\frac{0.9163}{r_s} . \quad (5.36)$$

The two terms that correspond kinetic energy and exchange energy for the ground state are

$$E_{kin} + E_{ex} = \frac{2.2099}{r_s^2} - \frac{0.9163}{r_s} + \ldots \quad (5.37)$$

The energy in (5.37) is given in atomic units per the electron. It is seen that the exchange energy is negative whereas the kinetic energy is positive. As $E_{ex}$ is negative, the interaction favors electrons with parallel spins, and it is necessary to spend work to separate the electrons.

The energy terms (5.37) comprise the Hartree–Fock theory. This is defined to be the kinetic energy, the direct Coulomb energy, which is zero, and the exchange energy. Within the Hartree–Fock method the antisymmetric eigenfunction is approximated by a single Slater determinant (3.26). Exact eigenfunctions, however, cannot generally be expressed as single determinants. The single-determinant approximation does not take into account the correlation energy, leading to a total electronic energy different from the exact solution of the nonrelativistic Schrödinger equation. Therefore the Hartree–Fock limit is always above this exact energy. One uses the term “correlation energy” to mean the difference between the true value of the ground state energy and the Hartree–Fock computational result.

The electrons with anti-parallel spins keep apart to lower their mutual Coulomb repulsion. This leads to a lowering of energy by an amount called correlation energy. Figure 5.4 illustrates the dependence of the correlation energy on parameter $r_s$. A contribution of the correlation energy is considerable for the electron gas of high density.

A zone around each electron turns out to be depleted of other electrons as if it is surrounded by an additional positive potential proportional to $1/r_s$. It is illustrated in Figure 5.5. The electron and its positive hole move together through the electron gas.

The dominant term in the expression for the exchange-correlation energy can be expressed as

$$E_{ex} + E_{corr} = -\frac{0.9163}{r_s} + (-0.115 + 0.0313 \ln r_s) , \quad (5.38)$$

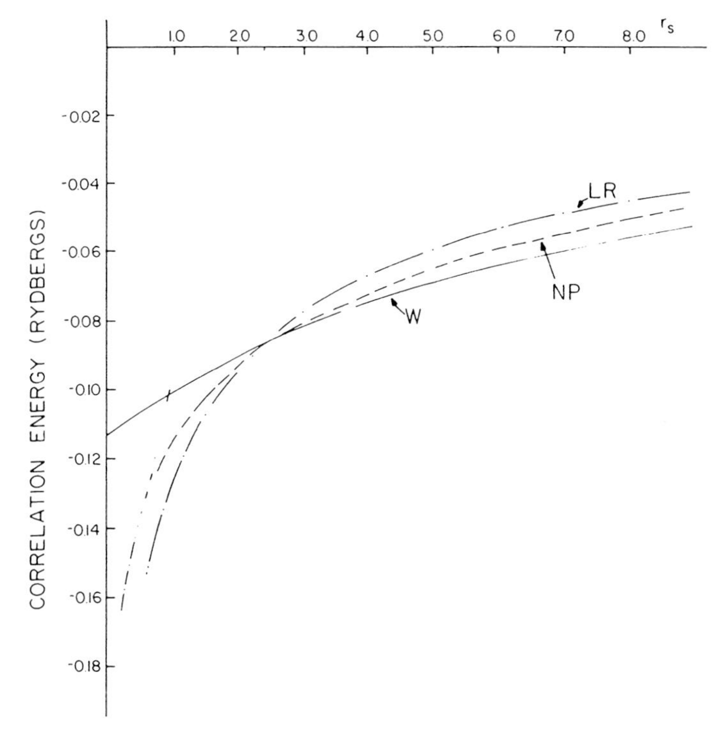


**Figure 5.4** The correlation energy of the electron gas as a function of parameter $r_S$. The curves are interpolation-schemes according to Wigner (W), Noziers and Pines (NP), and Lindgren and Rosen (LR) (after [6]).

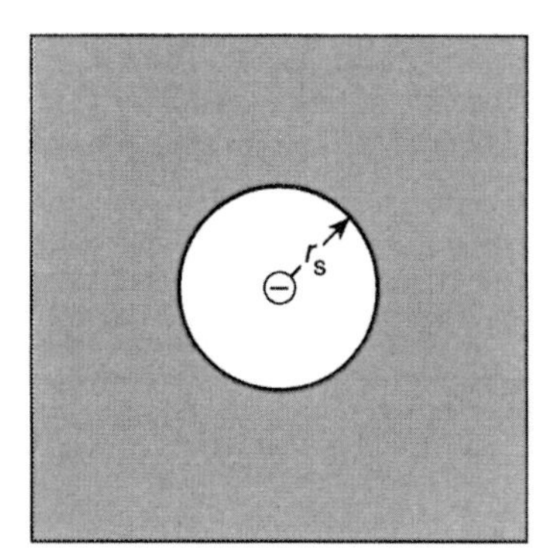


**Figure 5.5** The exchange-correlation hole of a given electron within a free-electron gas.

where the first and the second term represent the exchange energy and correlation energy, respectively. The both energies are negative. This means that interactions of electrons gain the total energy.

The ground state energy is expressed as a power series, in increasing powers of $r_S$. Although it is usually unsafe to extrapolate from just two terms, in fact this is a series in order $(r_s)^0$. The zeroth power could be interpreted as either a constant or as $\ln r_s$.

## 5.7
## Low-Density Electron Gas: Wigner Lattice

The calculation of the correlation energy is based on a Seitz theorem which relates the ground state energy to the chemical potential. The chemical potential is defined as the energy needed to add an electron to the material. Of course, this is just the Fermi energy of the metal. The chemical potential is the negative of the work function. The work function is the energy required to remove an electron from the solid and take it to infinity with zero kinetic energy.

Some interpolation-schemes have been proposed for calculating the correlation energy as a smooth function between limits of high and low density of electrons. The high-density system implies free-electron states and the low-density system implies localized electrons.

Some formulas have been proposed for the correlation energy $E_{corr}$. The Wigner formula

$$E_{corr} = -\frac{0.88}{r_s + 7.8} \tag{5.39}$$

is widely used.

At low densities, Wigner supposed that the electrons would become localized and form a regular lattice. The electron lattice is a close-packed structure such as fcc or hcp, in which electrons vibrate around their equilibrium positions. This lattice is immersed in a positive charged background.

At very low density the potential energy becomes more important than the kinetic energy. Results of the Wigner calculations are presented by a curve W in Figure 5.4.

The equation proposed by Noziers and Pines is

$$E_{corr} = -0.115 + 0.0313 \ln r_s \ . \tag{5.40}$$

In metals $r_s$ value varies from 1.88 to 5.63. This means that the correlation energy ranges approximately from −0.10 to −0.06 Ry. One may compare this with the exchange energy ranging from −0.57 to −0.23 Ry in the same interval of $r_s$.

The total ground state energy per electron is given by

$$E_g = \frac{2.2099}{r_s^2} - \frac{0.9163}{r_s} + (-0.115 + 0.0313 \ln r_s) \ . \tag{5.41}$$

## 5.8
## Near-Free Electron Approximation: Pseudopotentials

A near-free electron model (NFE) of simple metals is described as a gas of free electrons into which a lattice of positive ions is immersed.

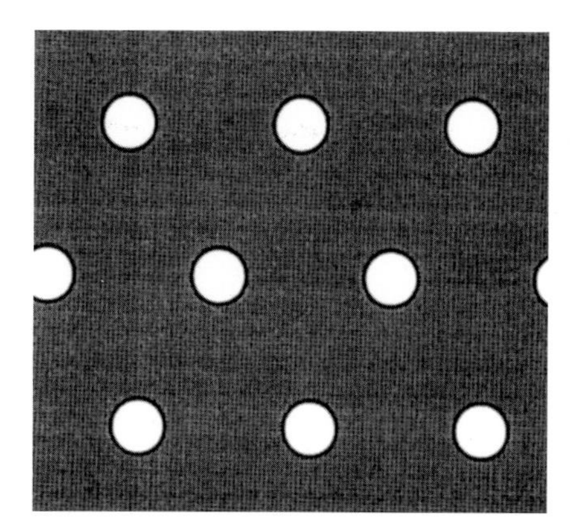

**Figure 5.6** Scheme of the near-free electron model of simple metals. The white circles represent the Wigner–Seitz cells, in which the point-positive charges are located. The lattice of the cells is immersed in a free-electron gas.

The Wigner–Seitz cells containing positively charged points at their centers form a lattice in the real-space (Figure 5.6). (The Wigner–Seitz cell is constructed in a real-space around a given lattice-site by bisecting the near-neighbor position vectors, see Section 4.3). This lattice is immersed into a free-electron gas. The ionic lattice would presumably be a close-packed structure as bcc, fcc, or hcp.

For a homogeneous electron gas, the electronic contribution to the ground state energy has been presented by (5.41). The calculations of electron–ion interactions poses some difficulties. The other contributions can be evaluated using formulas for the free-electron model. One needs expression for the ionic potential which acts upon the electron. According to the Coulomb law, one has $E_{\text{pot}} \rightarrow -\infty$ if $r \rightarrow 0$. For simple metals, one easy procedure is to construct a pseudopotential which gives the correct electron behavior for motion outside of the ion. This is usually done by fitting the potential to atomic energy levels. The primary application in study of electronic structure is to replace the strong Coulomb potential of the nucleus and the effects of the tightly bound core electrons by an effective ionic potential acting on the valence electrons. A pseudopotential can be generated in an atomic calculation and then used to compute properties of valence electrons in molecules or solids, since the core states remain almost unchanged. Furthermore, the fact that pseudopotentials are not unique gives one freedom to choose such a form of a dependence that simplify the calculations and the interpretation of the resulting electronic structure. The fundamental idea of a "pseudopotential" is the replacement of one problem with another [1].

Figure 5.7 illustrates the Ashkroft pseudopotential. This ionic potential has an empty core and can be described as

$$V(\mathbf{r}) = \begin{cases} 0 \quad \text{for} \quad r < R_c \\ \frac{-Ze^2}{4\pi\varepsilon_0 r} \quad \text{for} \quad r > R_c \,. \end{cases}$$

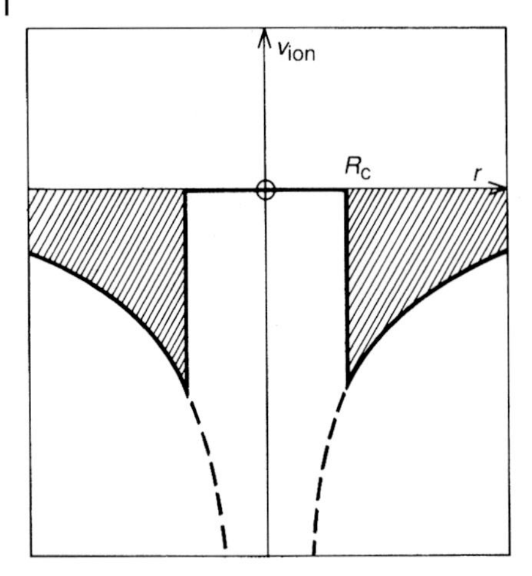


**Figure 5.7** The Ashkroft pseudopotential. This is an "empty core" model potential, in which the potential is zero inside a radius $R_C(l)$, which is different for each $l$. $l$ is the azimuthal quantum number.

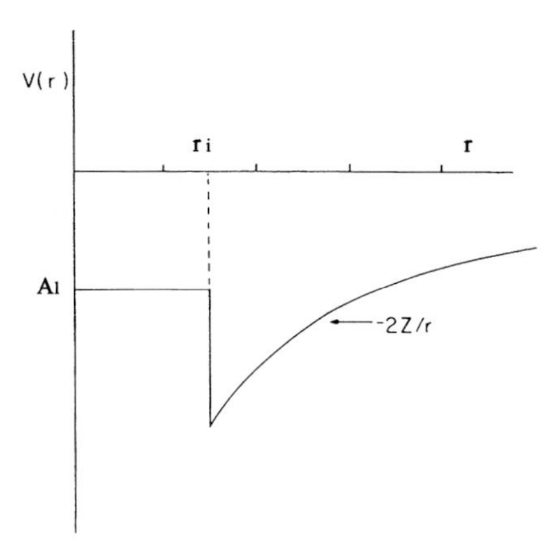


**Figure 5.8** The pseudopotential proposed by Heine and Abarenkov that is a square-well model potential with value $A_l$ inside a cutoff radius $\mathbf{r}_i$.

The model pseudopotential of Heine and Abarenkov is presented in Figure 5.8. The distance $r_i$ is interpreted as the radius of the ion core. One may at first select $r_i$ from a range of reasonable values. After a value for $r_i$ is chosen, it remains fixed throughout the rest of the calculations. The parameter $A_l$ (Figure 5.8) is then varied to give the proper energy state. For each eigenvalue $E_{ns}$ the parameter $A_l$ has a different value. $r_i$ was found to lie between 1.5 and 3.0 atomic units.

## 5.9
## Cohesive Energy of Simple Metals

The cohesive energy of a solid is the energy required to disassemble it into atoms. This is in fact the bonding energy.[2] The units of the cohesive energy are Ry/atom or eV/atom or kJ/mol or kcal/mol (1 eV/atom = 96.35 kJ/mol = 23.05 kcal/mol).

A most important test of the theory is to compare their data with the bonding energy of real metals. This comparison has been performed. It turned out that the cohesive energy of simple metals could be explained by the theory described in Sections 5.2–5.4. Simple metals are those without d electrons in the electron shells.

The Wigner–Seitz model of a spherical unit cell is used for calculation. This is a sphere of radius $r_s a_0$ with one nucleus at the center. Each sphere has overall neutrality, since one-electron charge at the center is canceled by the positive charge inside the volume of the sphere. In this model the spheres exert no electrical forces on each other. Of course this is only an approximate model, since the unit cells are not truly spheres – spheres cannot be packed together to cover all volume. However, the error made by the approximation is remarkably small.

For $Z > 1$ the Wigner–Seitz radius $r_{WS} = Z^{1/3} r_s$. The total electrostatic energy from electron–electron interactions $E_{e-e}$ and electron–ion interactions $E_{e-i}$ in atomic units is given by

$$E_{e-e} = \frac{6}{5}\frac{Z^2}{r_{WS}} = 1.2\frac{Z^{\frac{5}{3}}}{r_s} , \tag{5.42}$$

$$E_{i-e} = -\frac{3Z^2}{r_{WS}} . \tag{5.43}$$

These values must be divided by $Z$ to obtain the contribution per electron. The total electron–electron interaction energy, which includes the influence of correlation and exchange is given by

$$E_{e-e} = -\frac{0.9163}{r_s} + 1.2\frac{Z^{\frac{2}{3}}}{r_s} + E_{corr}(r_s) . \tag{5.44}$$

For a finite nucleus the electron–ion term can be expressed as

$$E_{i-e} = -\frac{3Z}{r_{WS}}(1-\delta) , \tag{5.45}$$

where the quantity $\delta$ accounts for the finite size of the nucleus. $\delta$ would be zero for a point ion. For simple metals it ranges from 0.2 to 0.3. One concludes that the quantity $E_{i-e}$ is mostly potential energy.

For real metals, the kinetic energy term reflects the fact that the effective electron mass $m_{eff}$ may not be unity (in atomic units). It is defined as $m_{eff} = \hbar^2 k^2/(2E)$, see (5.7).

2) The energy of an aggregate of atoms in a solid is less than the energy of separate atoms. Thus, the cohesive energy is negative.

**Table 5.2** Cohesive energies and their components (in rydberg per atom) of alkali metals. $r_s$ and $m_{eff}$ are given in atomic units.

| Metal | $r_s$ | $m_{eff}$ | $E_{ion}$ | $E_{kin}$ | $E_{e-e}$ | $E_{i-e}$ | $E_{coh}$ (theory) | $E_{coh}$ (exper.) |
|---|---|---|---|---|---|---|---|---|
| Li | 3.24 | 1.45 | 0.396 | 0.145 | 0.008 | −0.687 | −0.138 | −0.117 |
| Na | 3.96 | 0.98 | 0.378 | 0.144 | −0.003 | −0.609 | −0.090 | −0.084 |
| K | 4.86 | 0.93 | 0.319 | 0.101 | −0.011 | −0.486 | −0.077 | −0.073 |
| Rb | 5.23 | 0.89 | 0.307 | 0.091 | −0.013 | −0.459 | −0.075 | −0.061 |
| Cs | 5.63 | 0.83 | 0.286 | 0.084 | −0.015 | −0.421 | −0.066 | −0.061 |

The electron–electron effects are included in the term $E_{e-e}$. A theoretical value of the cohesive energy can now be calculated as

$$E_{coh} = E_{ion} + E_{kin} + E_{e-e} + E_{i-e} \,. \tag{5.46}$$

In Table 5.2, the results of calculations of the cohesive energy and the experimental data are presented. The agreement is satisfactory for all alkali metals except for lithium (the discrepancy between calculated and measured values is 18%) and rubidium (23%). For other metals the fit is good (from 5 to 8%). The cohesive energies are small varying from 0.83 to 1.59 eV/atom, so that the alkali metals are not strongly bound. Indeed, the metals are mechanically soft.

Of the four numbers on the right-hand side of (5.46) $E_{ion}$ is certainly the most accurate, and $E_{i-e}$ is probably the second best. The smallness of the electron–electron interaction term $E_{e-e}$ confirms the notion of the exchange and correlation hole. The interactions are small because the electrons tend to avoid each other.

The level of agreement between the near-free electron theory and experiments for metals with one electron in s shell may generally be considered as satisfactory.

# 6
# Electrons in Crystals and the Bloch Waves in Crystals

In Chapter 3 we have considered the probability density functions describing the electron locations in the atom. It is generally accepted to use the term "atomic orbital" as a synonym of the probability density function.

In a single isolated atom the electrons occupy atomic orbitals, which form a discrete set of energy levels (see Figure 3.2 and (3.6)). If several atoms are brought together into a molecule or a solid, their atomic orbitals split. More than two electrons cannot have the same energy according to the Pauli principle. Each subsequent couple of electrons in a system must have another value of energy than the prior one. Therefore the number of orbitals grows with the number of atoms. When a large number of atoms (of the order of $10^{23}$) are brought together to form a solid, the number of orbitals becomes exceedingly large. The difference between energy levels becomes very small, so the levels may be considered to form continuous bands of energy rather than the discrete energy levels.

However, there are orbitals in some solids that never have specific values of energy, no matter how many atoms are aggregated. These values are grouped in continued intervals thus forming band gaps. One names these gaps as "forbidden" bands of energy.

The delocalized electrons in a crystal lattice of a solid have properties of waves. The electron waves interact with atoms of the lattice and scattered waves diffract with each other. The allowed and forbidden ranges of the electron energy arise as a result. The electronic band structure of solids implies certain intervals of energy of electrons in the crystal lattice.

## 6.1
## The Bloch Waves

The ions in a perfect crystal are arranged in a regular array. The electron moving freely inside a crystal is affected by a potential $V(\boldsymbol{r})$ with the periodicity of the unit cell, that is

$$V(\boldsymbol{r}) = V(\boldsymbol{r} + \boldsymbol{R}) \,, \tag{6.1}$$

*Interatomic Bonding in Solids: Fundamentals,Simulation,andApplications*, First Edition. Valim Levitin.

where $\boldsymbol{R}$ is the lattice translation vector. Any periodic function $\rho(\boldsymbol{r})$ defined for the crystal, such as the density of electrons, which has the period of the unit cell, is given by

$$\rho(\boldsymbol{r}) = \rho(\boldsymbol{r} + \boldsymbol{R}) \,. \tag{6.2}$$

The period of the potential $V$ is of the order of $10^{-10}$ m. The de Broglie wavelength of electrons has the same order of magnitude. The electronic waves interact with the periodic potential of the crystal lattice.

We know (see Section 3.2) that the expression

$$-\frac{\hbar^2}{2m_\mathrm{e}}\nabla^2 + V(\boldsymbol{r}) \tag{6.3}$$

is called Hamiltonian $H$. In quantum physics, the Hamiltonian $H$ is an operator corresponding to the total energy of the system. It is of fundamental importance in quantum theory.

Thus, the state of an electron in the periodic potential $V(\boldsymbol{r})$ of a crystal lattice is characterized by the equation

$$H\psi(\boldsymbol{r}) \equiv \left[-\frac{\hbar^2}{2m_\mathrm{e}}\nabla^2 + V(\boldsymbol{r})\right]\psi(\boldsymbol{r}) = E\psi(\boldsymbol{r}) \,, \tag{6.4}$$

where $E$ is the eigenvalue. It equals to the total energy of the electron.

The question arises concerning the types of the eigenfunctions $\psi(\boldsymbol{r})$ that are solutions of (6.4). The Bloch theorem asserts that plane waves represent the electron eigenfunctions in the optimal way.

Consequently, the solution of (6.4) can be chosen to have the form of a plane wave times a periodic function with the period equal to the $\boldsymbol{r}$ vector of the direct crystal lattice:

$$\psi_{n,\boldsymbol{k}}(\boldsymbol{r}) = u_{n,\boldsymbol{k}}(\boldsymbol{r}) \exp i(\boldsymbol{k} \cdot \boldsymbol{r}) \,, \tag{6.5}$$

where

$$u_{n,\boldsymbol{k}}(\boldsymbol{r}) = u_{n,\boldsymbol{k}}(\boldsymbol{r} + \boldsymbol{R}) \,. \tag{6.6}$$

The eigenfunction consists of the product of a plane wave envelope function and a periodic function which has the same period as the potential.

For the given $n$ the eigenfunctions and eigenvalues are periodic functions in the reciprocal lattice,

$$\psi_{n,\boldsymbol{k}}(\boldsymbol{r}) = \psi_{n,\boldsymbol{k}+\boldsymbol{g}}(\boldsymbol{r}) \,. \tag{6.7}$$

The corresponding energy is periodic with periodicity of the reciprocal lattice vector

$$E_{n,\boldsymbol{k}} = E_{n,\boldsymbol{k}+\boldsymbol{g}} \,. \tag{6.8}$$

As the energies associated with the index $n$ vary continuously with wave vector $\boldsymbol{k}$ we speak about an energy band with the index $n$. Because the eigenvalues for the given $n$ are periodic in $\boldsymbol{k}$, all distinct values of $E_n(\boldsymbol{k})$ occur for $\boldsymbol{k}$ values within the first Brillouin zone (BZ) of the reciprocal lattice.

In (6.5) the subscript $n$ indicates the band index and $\boldsymbol{k}$ is a continuous wave vector that is confined to the first Brillouin zone of the reciprocal lattice. The index $n$ appears in the Bloch theorem because for a given $\boldsymbol{k}$ there are many solutions to the Schrödinger equation. Because the eigenvalue problem is set in a fixed finite volume, we generally expect to find an infinite family of solutions with discretely-spaced eigenvalues which we label with the band index $n$. The wave vector $\boldsymbol{k}$ can always be confined to the first Brillouin zone. The vector $\boldsymbol{k}$ takes on values within the Brillouin zone corresponding to the crystal lattice, and particular directions like $\Gamma, \Delta, \Lambda, \Sigma$ (see Figures 4.13–4.15).

The Bloch theorem (6.5) also holds for any vector that does not lie in the first zone. The Bloch function is periodic in the reciprocal space.

The periodic potential $V(\boldsymbol{r})$ determines the properties of the Bloch function. It includes the interaction between all electrons and ions. One applies the Hartree–Fock procedure in self-consistent approximation in order to find the potential $V(\boldsymbol{r})$. Consequently, the many-particle problem is reduced to the one-electron problem by means of the averaged field $V(\boldsymbol{r})$.

An example of the Bloch function is presented in Figure 6.1.

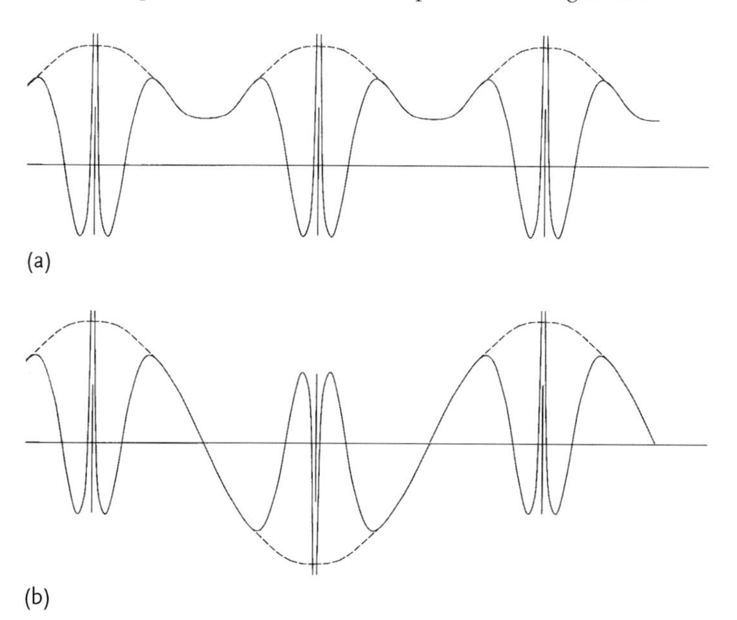


**Figure 6.1** Schematic illustration of the Bloch waves in one dimension: (a) at $k = 0$; (b) at the Brillouin zone boundary $k = \pi/a$. The envelope is the smooth function that implies a periodic array of atomic-like 3s functions.

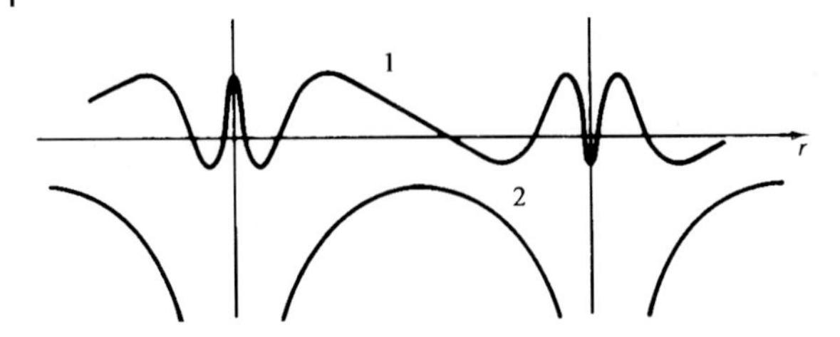


**Figure 6.2** The eigenfunction $\psi(r)$ (1) and the potential (2) versus the interatomic distance for the crystal lattice of sodium. The eigenfunction changes its sign when entering a neighboring cell.

An alternative form of the Bloch theorem is given by

$$\psi_k(\boldsymbol{r} + \boldsymbol{R}) = \exp(i\boldsymbol{k} \cdot \boldsymbol{R})\psi_k(\boldsymbol{r}) \,. \tag{6.9}$$

Equation (6.9) asserts that the Bloch function has the same form in each cell (the Brillouin zone) but the phase of the wave varies from cell to cell in a linear manner. A Bloch function that is a superposition of atomic eigenfunction such as 3s-orbitals in metallic sodium is shown in Figure 6.2.

## 6.2
## The One-Dimensional Kronig–Penney Model

The influence of the periodic potential of the crystal lattice on the electronic structure is introduced through the one-dimensional Kronig–Penney model that illustrates the essential features of band theory of solids.

Let us consider this simple but significant enough physical model of the electron movement in a field of periodic potential. Figure 6.3 illustrates the distribution of the one-dimensional potential $V(x)$ in the Kronig–Penney model. This potential comprises square wells that are separated by barriers of height $V_0$ and thickness $b$. The potential is periodic with the period $a$ so that

$$V(x) = V(x + na) \,, \tag{6.10}$$

where $n$ is an integer. The probability of locating the electron in any given well must be the same for all wells, so that

$$\psi^2(x) = \psi^2(x + na) \,. \tag{6.11}$$

Consequently, one can write

$$\psi_k(x)\exp(ikna) = \psi_k(x + na) \,. \tag{6.12}$$

It is easy to see that the left-hand side of (6.12) is the expression of the plane wave, and $\boldsymbol{k}$ is the wave vector, $|\boldsymbol{k}| = 2\pi/\lambda$. Expression $\exp(ikna)$ is a phase multiplier.

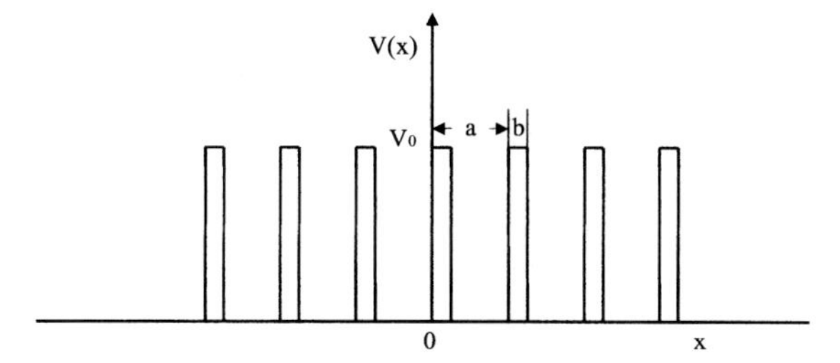


**Figure 6.3** The one-dimensional Kronig–Penney potential. The rectangular wells model of the interionic distances, whereas the barriers of a $V_0$ height model the one-dimensional ion lattice.

Equation (6.12) is the usual statement of the Bloch theorem in one dimension, see (6.5).

Thus, the translational symmetry of the potential leads to the eigenfunctions being characterized by a wave vector $\boldsymbol{k}$ (the Bloch vector). It is only defined modulo $2\pi/a$ since $k + p(2\pi/a)$ results in the same phase vector in (6.12) as $k$ alone ($p$ is an integer). It is, therefore, customary to label the eigenfunction $\psi_k(x)$ by restricting $\boldsymbol{k}$ to lie within the first Brillouin zone that is defined by

$$-\frac{\pi}{a} \leq k \leq +\frac{\pi}{a} . \tag{6.13}$$

In a one-dimensional case $na$ is the magnitude of a direct lattice vector, whereas a value of $p(2\pi/a)$ is the magnitude of a reciprocal lattice vector. In the first Brillouin zone one has $n = p = 1$, and the product of the direct and reciprocal vectors equals to $2\pi$.

Inside the first square well $b < x < a$, the potential $V = 0$, and the solution of the Schrödinger equation

$$-\frac{\hbar^2}{2m_e}\frac{d^2\psi_k(x)}{dx^2} = E\psi_k(x) \tag{6.14}$$

can be written as

$$\psi_k(x) = A\exp(iKx) + B\exp(-iKx) , \tag{6.15}$$

where

$$K = \left(\frac{2m_e E}{\hbar^2}\right)^{\frac{1}{2}} . \tag{6.16}$$

We arrive at the value of the coefficient $K$ substituting (6.15) into (6.14). The energy $E$ is the eigenvalue corresponding to eigenfunction $\psi_k$. The value of $K \sim E^{1/2}$ is a direct measure of the energy of the electron.

For the interval $0 \leq x \leq b$ under the potential barrier $V = V_0$ the solution of the Schrödinger equation

$$-\frac{\hbar^2}{2m_e}\frac{d^2\psi_k(x)}{dx^2} + V_0\psi_k(x) = E\psi_k(x) \tag{6.17}$$

can be written as a linear combination of exponential functions, namely

$$\psi_k(x) = C \exp(\gamma x) + D \exp(-\gamma x) , \tag{6.18}$$

where

$$\gamma = \left[ \frac{2m_e(V_0 - E)}{\hbar^2} \right]^{\frac{1}{2}} . \tag{6.19}$$

The coefficients $A, B, C, D$ in (6.15) and (6.18) are yet to be determined.

The desired solution must be continuous. There is tunneling of the electron between wells. Because of this, the function $\psi_k(x)$ and its first derivative $d\psi_k(x)/dx$ must be matching at the boundaries at $x = b$ and $x = a$, respectively.

One arrives after transformations at a transcendental equation (see Appendix B)

$$\mu \frac{\sin Ka}{Ka} + \cos Ka = \cos ka , \tag{6.20}$$

where a dimensionless parameter $\mu$ measures the strength of the Kronig–Penney barrier. The parameter $\mu$ is given by

$$\mu = \frac{m_e}{\hbar^2} \cdot abV_0 . \tag{6.21}$$

Equation (6.20) links the energy of the electron $E = \hbar^2/(2m)K^2$ to the wave vector $k$. It is obvious that it has solutions for which the magnitude of the left-hand side is less or equal to unity, since $|\cos ka| \leq 1$. Correspondingly, there are energy gaps for the electron. The spectrum of energies has a band structure.

The allowed values of the electron energy alternate the forbidden values. As illustrated in Figure 6.4, energy gaps that open up in the spectrum as traveling solutions are only found between the segments a–b; c–d; e–f; g–h.

The one-dimensional Kronig–Penney model is sufficiently general as it works for a wide range of electron states, from the nearly-free electron (NFE) regime (in which the band gaps are small and the electronic states are free-electron-like) to the tight-binding (TB) regime (in which the band gaps of energy are large).

Figure 6.5 presents the change in behavior of the energy bands which takes place as the strength of the barrier $\mu$ varies. For free electrons $\mu = 0$; any energy $E = \hbar^2 k^2/(2m_e)$ is allowed. For free atom $\mu = \infty$; wave functions are standing waves; only discrete values of energy $E$ are allowed. If electrons move in a periodic potential a number of bands of energies are allowed. The wave functions are the plane Bloch waves $\psi_k(x) = u(x)\exp(ikx)$. If the potential barrier is strong enough, $\mu > 1$, energy bands are narrowed and space far apart. This corresponds to crystals in which electrons are tightly bound to ion cores, and wave functions do not overlap much with adjacent cores. d-valent transition metals, sp-valent semiconductors are well described by the tight-binding (TB) approximation.

If potential barriers between wells are weak, $\mu < 1$, energy bands are wide and spaced close together. This is typically for metals with weakly-bound electrons, that is for alkali metals. Here the model of nearly-free electrons (NFE) works well. The nearly-free electron approximation describes well s- and sp-valent metals.

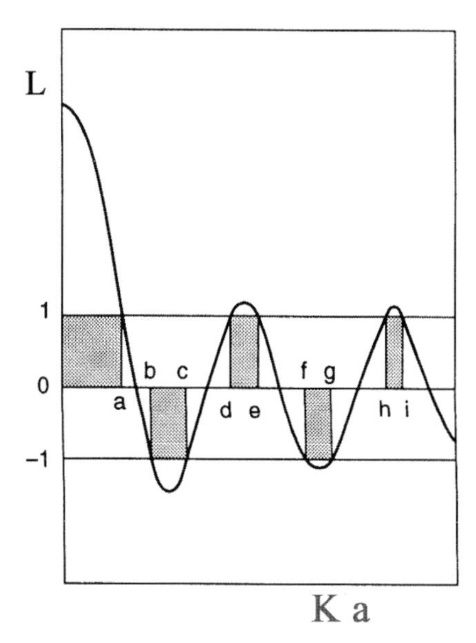


**Figure 6.4** The function $L = \mu(\sin Ka/a) + \cos Ka$ that appears on the left-hand side of (6.20) versus value of $K \cdot a$. Traveling electron wave is forbidden for those ranges of $K \cdot a$ for which the magnitude of the function is greater than unity as shown by shaded regions.

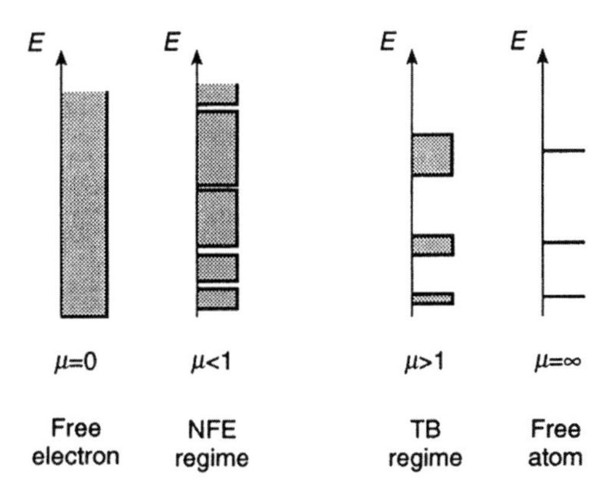


**Figure 6.5** The effect of the Kronig–Penney barrier $\mu$ on the energy bands. $\mu = m_e/\hbar^2 abV_0$ (after [3] by permission of Oxford University Press).

## 6.3 Band Theory

For simplicity, let us consider a process of an approach of several atoms that eventually form an united atomic system. When $N$ atoms are brought together the individual eigenfunctions of electrons begin to overlap. A given energy level of the system is split into $N$ distinct energy levels.

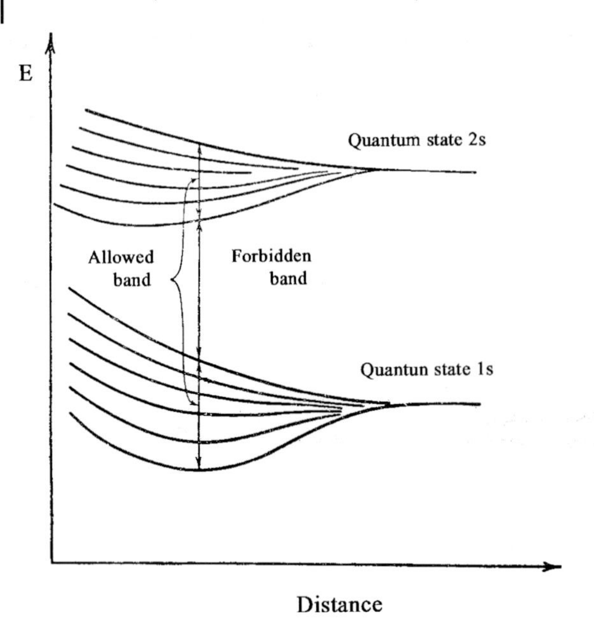


**Figure 6.6** Schematic drawing of the splitting of two energetic levels and formation of energetic bands in a system of six atoms, as a function of the separation distance between adjacent atoms.

Figure 6.6 illustrates the formation of quantum bands of energy from six energy levels of six atoms at their approaching. According to the Pauli principle it is impossible for more than two electrons to be in the same quantum state, say 1s or 2s. That is why the atomic energy levels split and transform into bands. The splitting of the energy levels begins when the center-to-center atomic separation becomes small enough for the atoms to start overlapping. As one can see from Figure 6.6, the "forbidden" energetic zone is situated between two "allowed" zones. The bands will be also formed by p, d, ... levels.

As we go to a system containing a growing number of atoms, the distance between the energy levels decreases. As more and more atoms are added to the system, each set of split levels contains more and more levels spread over the same energy range at a particular distance. At the atom–atom distances found in solids, a few tenths of nanometer, the energy spread is of the order of a few electron volts (Figure 3.17). Consider then that a solid that contains of the order of $10^{23}$ atoms per mole and we have to conclude that the levels of each set in a solid are so extremely closely spaced in energy that they practically form a continuous energy band.

Figure 6.7 shows schematically the energy levels for one-dimensional models that consist of two isolated atoms, a two-atom molecule and a four-atom crystal. We see that electrons in inner shells of the atoms are not sufficiently influenced by nearby atoms. These electrons are localized at particular atoms because the potential barriers between the atoms are relatively high and wide for them. Their energy levels do not split appreciably because their eigenfunctions do not overlap. On the

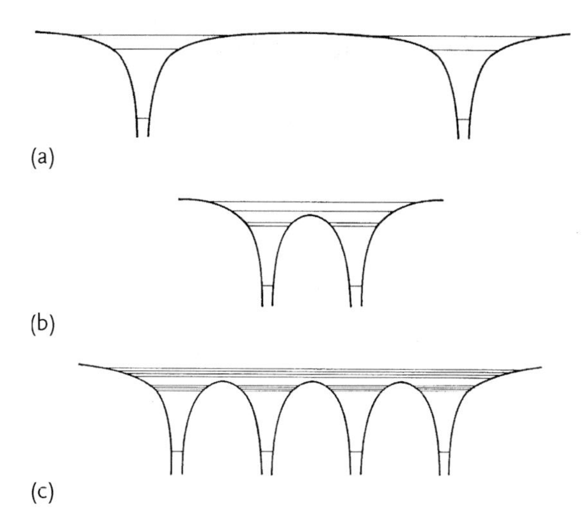


**Figure 6.7** (a) Energy-level scheme for two isolated atoms; (b) energy-level scheme for the same two atoms in a diatomic molecule; (c) the analogous scheme for four atoms in a modeling one-dimensional crystal (after [2]).

other hand, the valence electrons are not localized. They become a part of the whole system (Figure 6.7c). That is to say, these electrons are collectivized.

## 6.4 General Band Structure: Energy Gaps

In first approximation, it can be said that electron waves are diffracted by a crystal plane that is located at the boundary of the Brillouin zone. However, only electrons of a specified wavelength undergo the diffraction. Imagine an electron wave of length $\lambda$ that propagates in a direction perpendicular to atom planes. The parallel planes are at a distance $a = d$ from each other. According to the Bragg equation (4.15), the reflection of waves in the direction $\theta = \pi/2$ occurs on condition

$$n\lambda = 2a \sin\frac{\pi}{2}\,, \tag{6.22}$$

where $n = 1, 2, 3\ldots$ As $\lambda = 2\pi/k$ the reflection from planes will occur on condition $\lambda = 2\pi/k_{\mathrm{BZ}}$, where $k_{\mathrm{BZ}}$ denotes the wave vector on the zone boundary. Then it follows from (6.22)

$$k_{\mathrm{BZ}} = n\frac{\pi}{a} = n\frac{\pi}{d}\,. \tag{6.23}$$

The first Brillouin zone is of a primary importance because all the solutions can be completely characterized by their behavior in the first Brillouin zone. Usually a band structure is only plotted in the first Brillouin zone.

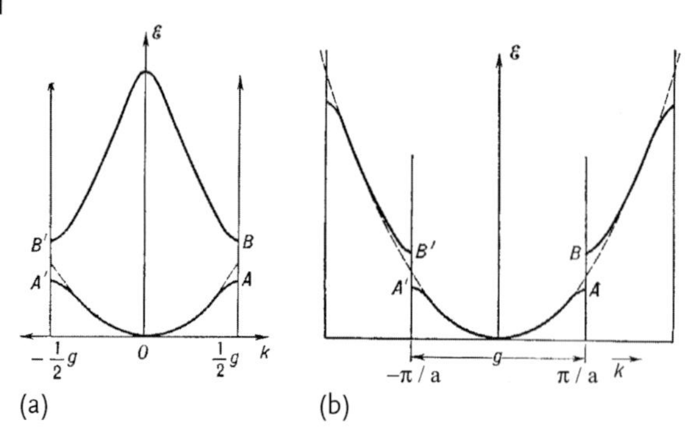


**Figure 6.8** Energies of an electron in an one-dimensional lattice of a lattice spacing $a$. (a) The function $E(k)$ is plotted in the first Brillouin zone. (b) Energies as a function of wave vector. Electrons can not have energies indicated by segments $AB$ and $A'B'$.

So electrons with wave vectors on or near the boundary of Brillouin zone are diffracted; all others ("normal") are not. We expect that "normal" electrons will not feel any diffraction. They obey the relation for the total energy $E = \hbar^2 k^2/(2m_e)$.

The condition $\boldsymbol{k} - \boldsymbol{k}_0 = \mathbf{g}$, see (4.16), is satisfied at the boundary of the Brillouin zone. For electrons with the $\boldsymbol{k}_{BZ}$ vectors we must expect major modifications. The energy value of an electron with the $\boldsymbol{k}_{BZ}$ wave vector turns out to be split into two quantities,

$$E_{k_{BZ}} = \frac{\hbar^2}{2m_e} k^2 \pm \Delta E \,. \tag{6.24}$$

Thus, the electrons at the Brillouin zone boundary can have two different energies for the same wave vector and thus the same states. One energy value is somewhat lower than the free-electron gas value, the other one is somewhat higher. Energies between these values are unattainable for any electrons – there is now an energy gap in the $E = E(\boldsymbol{k})$ relation for all $\boldsymbol{k}_{BF}$ vectors ending on the Brillouin zone.

Figure 6.8 illustrates dependence $E(k)$ for an one-dimensional case.

It is appropriate to recognize two tractable limits of the Bloch theorem: a very weak periodic potential and a very strong periodic potential (so strong that electrons can hardly move from atom to atom). The both extreme limits give rise to bands with gaps between them. In both cases, the bands are qualitatively very similar, that is, real potentials, which must lie somewhere between the two extremes.

Figure 6.9a–f illustrates a variety of the accepted band structure representations for nearly-free electron model. The Figure introduces the repeated-zone, extended-zone and reduced-zone images. The original free-electron parabola $E = \hbar^2 k^2/(2m_e)$ is shown in Figure 6.9a. To leading order in the weak one-dimension periodic potential this curve remains correct except the value of $\boldsymbol{k}$ near the reciprocal lattice vector $\mathbf{g}$. One can imagine that in this point "the Bragg plane" reflects the electron wave since the Bragg condition holds. Another free-electron parabola is centered at $\boldsymbol{k} = \mathbf{g}$, and two parabolas are crossed each other at the

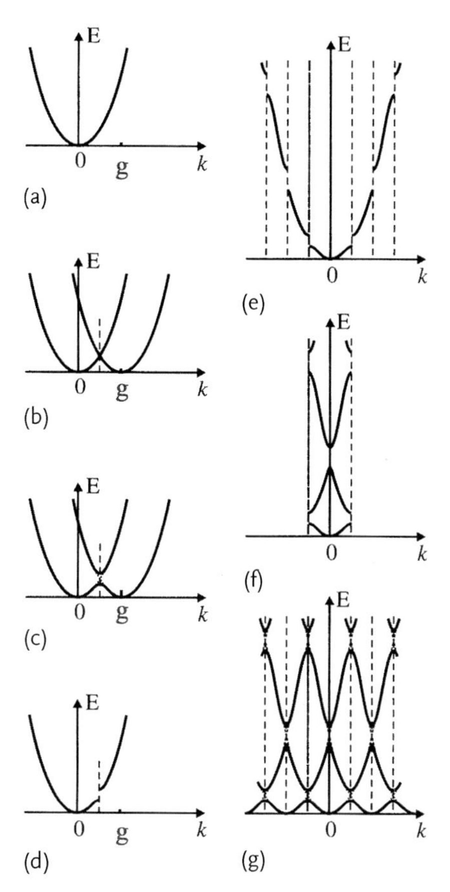


**Figure 6.9** The evolution of nearly-free electron bands: (a) energy of a free electron as a function of the wave vector is the parabola, $E = \hbar^2 k^2/(2m_e)$; (b) replication of the free-electron dispersion curves due to periodicity of $k$-space: a second parabola is drawn centered on the reciprocal lattice vector g; (c) the distortion of two parabolas at the point on the Brillouin zone boundary where two dispersion curves cross, the degeneracy of the two parabolas is split; (d) the effect on the electronic dispersion of a band gap opening at each successive Brillouin zone boundary; the energy gap takes place in the neighborhood of a Bragg "plane;" the periodicity of $k$-space means that the curves (d) must be present in every Brillouin zone; (e) is the extended zone scheme, in which each Brillouin has one band shown; (f) is the reduced zone scheme whereby all bands are shown within the first Brillouin zone; (g) is the repeated-zone or periodic-zone scheme (after [7]).

point $\frac{1}{2}$g, Figure 6.9b. Both curves have zero slope at this point. Figure 6.9b allows one to get Figure 6.9c. The original free-electron curve is therefore modified as in Figure 6.9d. When all the reciprocal vectors ("Bragg planes") are included, we end up with a set of curves such as those shown in Figure 6.9e. This particular way of depicting the energy levels is known as the extended-zone scheme. If we went on specifying all the levels by a wave vector $\boldsymbol{k}$ in the first Brillouin zone, then we must translate the pieces of Figure 6.9e, through reciprocal lattice vectors, into the first

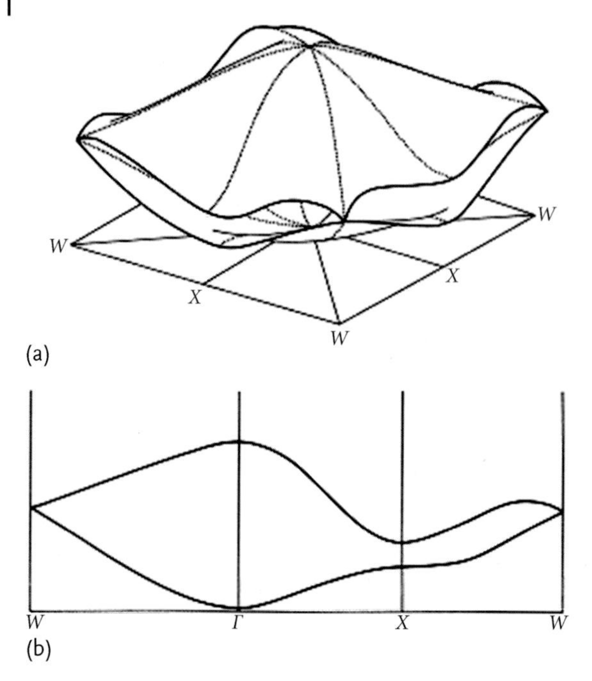


**Figure 6.10** The scheme of construction of a band structure graph: the three-dimensional image (a) and a section (b). Energy of electrons is plotted along $Oy$ axis.

Brillouin zone. The result is shown in Figure 6.9f. The representation is that of the reduced-zone scheme. One can also emphasize the periodicity of the labeling in $k$-space by periodically extending Figure 6.9f throughout all $k$-space to arrive at Figure 6.9g. This representation is the repeated-zone scheme. The repeated-zone scheme is the most general representation used in literature, but it is highly redundant, since the same level is shown many times, for all equivalent vectors $k, k \pm g, k \pm 2g, \ldots$ [7].

Construction of a graph for obtaining the band structure using the slightly adjusted parabolas of the free-electron gas model is simple. One constructs the parabolas along major directions of the reciprocal lattice, interpolate in between, and fold them back into the first Brillouin zone. How this can be done for the free-electron gas is shown in Figure 6.10. The lower part of Figure 6.10a,b (the "cup," $\Gamma \rightarrow W, \Gamma \rightarrow X$,) is contained in the first Brillouin zone. The upper parts (the "top," $W \rightarrow \Gamma, X \rightarrow W$) come from the second Brillouin zone, but is now folded back into the first one. It thus would carry a different band index. This could be continued ad infinitum; but Brillouin zones with energies well above the Fermi energy are of no real interest.

Figure 6.10b shows tracings along major directions. The directions are chosen that lead from the center of the Brillouin zone in the more generalized picture – to special symmetry points. These points are labeled in Table 4.1. Evidently, they contain most of the relevant information in condensed form. It would be sufficient for most purposes to know the $E_n(k)$ curves – the dispersion relations – along the major directions of the reciprocal lattice ($n$ is the band index). This is exactly what is done when real band diagrams of crystals are shown.

## 6.5 Conductors, Semiconductors, and Insulators

Figure 6.11a shows the dependence of energy of the electron on the magnitude of the wave vector $k$. It is often convenient to consider a simplified energy diagram (Figure 6.11b).

One distinguishes the conductivity band and the valency band.

The empty bands do not contain electrons and therefore do not contribute to the electrical conductivity of the solid. Partially filled bands do contain electrons as well as unoccupied energy levels which have a slightly higher energy. These unoccupied levels enable carriers of the electrical charge to gain energy when moving in an applied electric field. Electrons in a partially filled band therefore do contribute to the electrical conductivity of the solid.

The widely used classification of materials based upon the filling of the bands is illustrated in Figure 6.11b.

- Conductors (metals) have partially filled bands with no excitation gaps or the conductivity band is overlapped with the valency band, so that electrons can conduct electricity. In solids from the monovalent alkali atom like sodium, the band containing valent electrons behaves like a conductor. Only half of the levels of the isolated 3s allowed band of sodium are filled because a sodium atom has a single electron in the 3s level, whereas the excursion principle allows such a level to accommodate two electrons. Hence, electrons in this solid can easily acquire a small amount of additional energy. Thus any applied electric field will be effective in giving energy to electrons, and the solid will be a conductor. Conductors are also found in cases where valence band and conduction band overlap.
- Semiconductors have only a small gap, so that thermal energies are sufficient to excite the electrons to a degree that allows electric conduction. The completely filled band in a semiconductor is close to the next empty band so that electrons

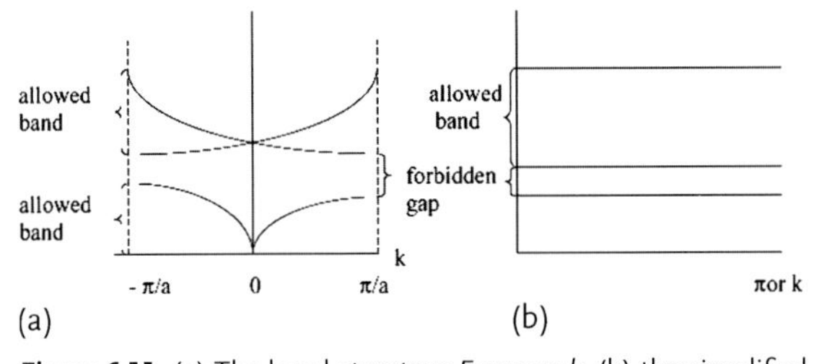


**Figure 6.11** (a) The band structure $E$ versus $k$; (b) the simplified energy diagram.

can pass into the next band yielding an almost full band below an almost empty band.

- Insulators have filled bands with a large energy gap of forbidden energies separating the ground state from all excited states of the electrons.

## 6.6 Classes of Solids

The most convenient classification scheme of solids is based on the physical character of the interatomic binding forces in various classes of crystalline materials. According to this classification, all solids fall into one of five general categories: metallic, covalent, ionic, molecular, and hydrogen-bonded crystals. Some materials may belong to more than one category, thus, the distinction is in many cases not a sharp one.

**Metallic bond** In metals the collectivized electrons move in the lattice potential of the positive ions and are shared by all atoms in the crystal. The ions provide the rigid framework, that is, the crystal lattice, through which the valence electrons move about freely like the particles of a gas. The metallic bond is not a directional one.

Metallic systems are conductors because there is no energy gap for electronic excitation. The energy bands of metals are partially filled. Then the bands can easily accept different numbers of electrons, leading to the ability of metals to form alloys with different valency, and to the tendency for metals to adopt close-packed structures, such as fcc, hcp, and bcc.

Due to presence of free electrons the metals and the metallic solid solutions are excellent conductors of electricity and heat. The electrons easily absorb energy of incident radiation or the lattice vibration.

We have seen in previous chapters that the good examples of metallic crystals are the alkali metals, which can be correctly described by the near-free electron model. The valence electrons in these metals are completely separated from their ion cores and form a nearly uniform gas.

The properties of transition metals are to a considerable degree determined by their not completely filled d shell. We consider transition metals in Chapter 13.

**Covalent bond** If the outermost electronic shells are only partly filled, then the electrons in outer shells rearrange themselves when the eigenfunctions of neighboring atoms start to overlap. This overlap of the outermost shells generally leads to a lowering of the total electronic energy.

The covalent bond is composed of sharing between the atoms of an even number of electrons, two per single bond. The most striking feature of the covalent bond is its strong directional property. The electron states are changed from those of isolated atoms or ions to well-defined bond states in solids.

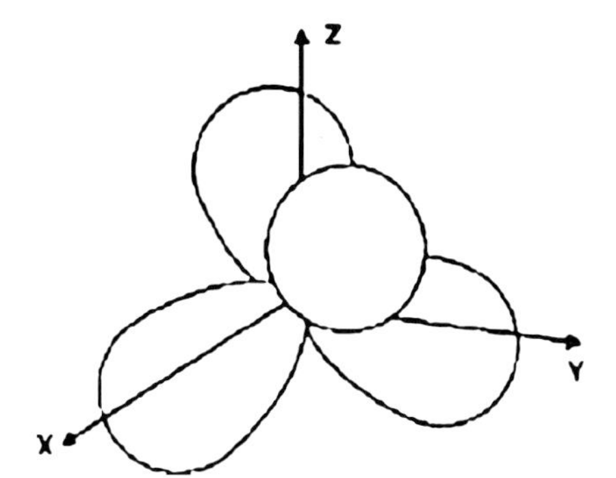


**Figure 6.12** The electron shells in a carbon atom.

Because there are no free electrons, the covalent solids are poor conductors of electricity and heat. Some of them are insulators (diamond) or semiconductors (silicon, germanium, compounds of elements III and V group of the Mendeleev Table). The covalent elements have a high melting temperature and are hard on account of their rigid electronic structure.

The electronic configuration of a neutral carbon atom is $^6$C: $1s^2 2s^2 2p^2$. In the valence state the carbon atom has tetrahedral bond angels and the configuration $^6$C: $1s^2 2s^1 2p^3$. If the nucleus of a C-atom is situated at the origin, the electron shells are stretched along the directions $[111], [\bar{1}1\bar{1}], [\bar{1}\bar{1}1], [1\bar{1}\bar{1}]$ (Figure 6.12). The angle between any two directions is $\alpha = 109°28'$ since $\cos\alpha = -(-1-1+1)/(\sqrt{3}\sqrt{3}) = -0.3333$.

**Ionic bond** An ionic crystal is built from positive and negative ions. The ionic bond results from the electrostatic interaction of the alternating anions and cations.

The electronic configuration of the ions is essentially an inert gas configuration, the charge distribution of each ion being spherically symmetric. Consequently, ionic solids crystallize in close-packed structures. Apparently, the anions and cations do not behave like hard (nonoverlapping) spheres. Experimentally, the observed interatomic distance in solid ionic compositions is less than the sum of the ionic radii.

**The van der Waals bond** occurs due to forces between polar or polarisable molecules. The dipole–dipole, dipole–induced dipole and dispersion interaction between molecules are referred to as van der Waals bonds.

**Hydrogen bond** An atom of hydrogen can be attracted by rather appreciable forces to two atoms, instead of only one. The hydrogen atom with 1s orbital forms the covalent bond and gains a partial positive charge. This charge can attract another electronegative atom or an atomic group.

# 7
# Criteria of Strength of Interatomic Bonding

In this chapter we consider properties of solids that are related to strength of interatomic bonding. We both bear in mind such macroscopic characteristics as elastic constants, melting temperature, and microscopic parameters as amplitudes of atom vibrations, energy of vacancy formation, the Debye temperature. All these properties can be measured experimentally with a high precision. Therefore they are factors that characterize the interatomic bonding in solids.

## 7.1
## Elastic Constants

The Hooke law states that the strain of a body is proportional to the applied stress. The law is applicable at the macroscopic level of strength of a solid (see Table 1.1, the characteristic length is greater than 0.01 m). Practically, the Hooke law holds within the elastic range of a material for relative small values of strain $\varepsilon$ ($\varepsilon \ll 1$).

Cauchy generalized the Hooke law to a three-dimensional elastic body and stated that the six components of stress are linearly related to the six components of strain. The law can be written in a tensor form as

$$\sigma_{ij} = C_{ijkl}\varepsilon_{kl} , \tag{7.1}$$

where $\sigma_{ij}$ are components of the stress, $\varepsilon_{kl}$ are components of the strain. $C_{ijkl}$ are modules of elasticity (or elastic stiffness constants). The components of the stress are linear functions of the elastic strain.

One component of stress is normal to the surface and represents the normal stress. The other two components are tangential to the surface and represent the shear stresses. In Figure 7.1 the stress tensor $\sigma_{ij}$ for a body of a cubic shape is shown. The first subscript $i$ indicates direction of the surface normal, along which the stress acts. The second subscript $j$ indicates direction of the stress component. Thus, $\sigma_{xy}$ means the component of stress applied along the normal to $yOz$ plane and acting along $y$ axis. $\sigma_{xx}, \sigma_{yy}, \sigma_{zz}$ are the normal components of the stress, the rest components are the shear ones.

*Interatomic Bonding in Solids: Fundamentals,Simulation,andApplications*, First Edition. Valim Levitin.

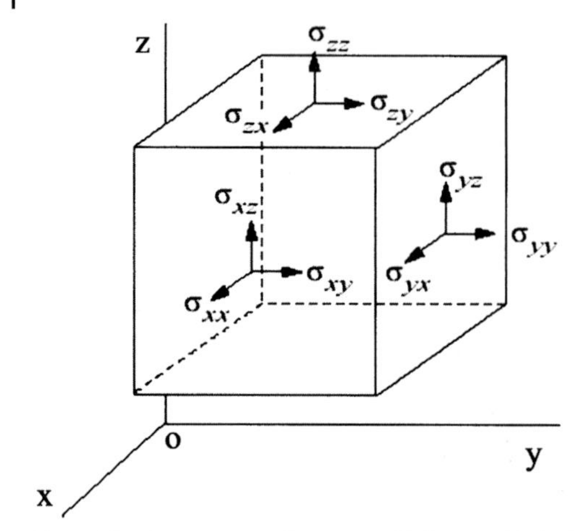


**Figure 7.1** Components of stress applied to a cubic body.

The relationship between stress and strain written in a matrix form is given by

$$\begin{bmatrix} \sigma_{xx} \\ \sigma_{yy} \\ \sigma_{zz} \\ \sigma_{yz} \\ \sigma_{zx} \\ \sigma_{xy} \end{bmatrix} = \begin{bmatrix} C_{11} + C_{12} + C_{13} + C_{14} + C_{15} + C_{16} \\ C_{21} + C_{22} + C_{23} + C_{24} + C_{25} + C_{26} \\ C_{31} + C_{32} + C_{33} + C_{34} + C_{35} + C_{36} \\ C_{41} + C_{42} + C_{43} + C_{44} + C_{45} + C_{46} \\ C_{51} + C_{52} + C_{53} + C_{54} + C_{55} + C_{56} \\ C_{61} + C_{62} + C_{63} + C_{64} + C_{65} + C_{66} \end{bmatrix} \begin{bmatrix} \varepsilon_{xx} \\ \varepsilon_{yy} \\ \varepsilon_{zz} \\ \varepsilon_{yz} \\ \varepsilon_{zx} \\ \varepsilon_{xy} \end{bmatrix} . \tag{7.2}$$

Remind that this notation means

$$\begin{aligned} \sigma_{xx} &= C_{11}\varepsilon_{xx} + C_{12}\varepsilon_{yy} + C_{13}\varepsilon_{zz} + C_{14}\varepsilon_{yz} + C_{15}\varepsilon_{zx} + C_{16}\varepsilon_{xy} , \\ \sigma_{yy} &= C_{21}\varepsilon_{xx} + C_{22}\varepsilon_{yy} + C_{23}\varepsilon_{zz} + C_{24}\varepsilon_{yz} + C_{25}\varepsilon_{zx} + C_{26}\varepsilon_{xy} , \\ &\vdots \\ \sigma_{xy} &= C_{61}\varepsilon_{xx} + C_{62}\varepsilon_{yy} + C_{63}\varepsilon_{zz} + C_{64}\varepsilon_{yz} + C_{65}\varepsilon_{zx} + C_{66}\varepsilon_{xy} . \end{aligned}$$

In expression (7.2) the notations for subscripts of $C_{ijkl}$ are simplified according to the rule:

$$xx \to 1 , \quad yy \to 2 , \quad zz \to 3 , \quad yz \to 4 , \quad zx \to 5 , \quad xy \to 6 .$$

Thus, each component of the stress depends upon all six components of the strain and vice versa. It follows from conditions of balance that $C_{ij} = C_{ji}$. As a result there is only 21 independent $C_{ij}$ values instead of 36.

The number of the elastic constants decreases for crystals with the cubic lattice due to the symmetry to three independent components: $C_{11}$, $C_{12}$ and $C_{44}$.

The $C_{ij}$ matrix in case of the cubic crystal lattice is written as

$$\begin{bmatrix} C_{11} & C_{12} & C_{12} & 0 & 0 & 0 \\ C_{12} & C_{11} & C_{12} & 0 & 0 & 0 \\ C_{12} & C_{12} & C_{11} & 0 & 0 & 0 \\ 0 & 0 & 0 & C_{44} & 0 & 0 \\ 0 & 0 & 0 & 0 & C_{44} & 0 \\ 0 & 0 & 0 & 0 & 0 & C_{44} \end{bmatrix} . \tag{7.3}$$

The elastic constants are of importance in theory of solid state. The point is that their values can be calculated from first principles or from other models of solids and be compared with the experimental data.

Most polycrystalline solids are considered to be isotropic, where, by definition, the material properties are independent of direction. Such materials have only two independent variables (that is elastic constants) in matrix (7.3), as opposed to the 21 elastic constants in the general anisotropic case. The two elastic constants are the Young modulus $E$ and the Poisson ratio $\nu$. The alternative elastic constants bulk modulus $B$ and shear modulus $\mu$ can also be used. For isotropic materials, $\mu$ and $B$ can be found from $E$ and $\nu$ by a set of equations, and on the contrary.

The shear modulus $\mu$ is defined as the ratio of shear stress to engineering shear strain on the loading plane,

$$\mu = \frac{\sigma_{xy}}{\varepsilon_{xy} + \varepsilon_{yx}} = \frac{\sigma_{xy}}{2\varepsilon_{xy}} = \frac{\sigma_{xy}}{\gamma_{xy}} = \frac{E}{2(1+\nu)} . \tag{7.4}$$

The engineering shear strain $\gamma_{xy} = \varepsilon_{xy} + \varepsilon_{yx} = 2\varepsilon_{xy}$ is a total measure of shear strain in the $yOz$ plane. In contrast, the shear strain $\varepsilon_{xy}$ is the average of the shear strain on the $yOz$ face along the $y$ direction, and on the $xOz$ face along the $x$ direction.

The Hooke law in form of stiffness matrix is given by

$$\begin{bmatrix} \sigma_{xx} \\ \sigma_{yy} \\ \sigma_{zz} \\ \sigma_{yz} \\ \sigma_{zx} \\ \sigma_{xy} \end{bmatrix} = \frac{E}{(1+\nu)(1-2\nu)} \times \begin{bmatrix} 1-\nu & \nu & \nu & 0 & 0 & 0 \\ \nu & 1-\nu & \nu & 0 & 0 & 0 \\ \nu & \nu & 1-\nu & 0 & 0 & 0 \\ 0 & 0 & 0 & 1-2\nu & 0 & 0 \\ 0 & 0 & 0 & 0 & 1-2\nu & 0 \\ 0 & 0 & 0 & 0 & 0 & 1-2\nu \end{bmatrix} \begin{bmatrix} \varepsilon_{xx} \\ \varepsilon_{yy} \\ \varepsilon_{zz} \\ \frac{\gamma_{yz}}{2} \\ \frac{\gamma_{zx}}{2} \\ \frac{\gamma_{xy}}{2} \end{bmatrix} , \tag{7.5}$$

where $\gamma_{kl}$ is the shear strain.

## 7.2
## Volume and Pressure as Fundamental Variables: Bulk Modulus

The equation of state as a function of pressure and temperature describes perhaps the most fundamental property of condensed matter [1]. The stable structure of a solid at given pressure $p$ and temperature $T$ determines all other properties of a material. The total energy $E$ at $T = 0$ as a function of volume $V$ is the most convenient quantity for the theoretical analysis because it is the straightforward procedure to carry out the structure calculations at fixed volume. Comparison of theory and experiment is one of the touchstones of the *ab initio* research of interatomic forces.

The fundamental quantities in these researches are pressure $p$, volume $V$, total energy of a solid $E$ and bulk modulus $B$. Energy depends upon volume of the solid,

$$E = E(V) . \tag{7.6}$$

In a crystal, $E$ is the energy taken per cell of volume $V = V_{\text{cell}}$. The pressure $p$ is given by

$$p = -\frac{dE}{dV} . \tag{7.7}$$

The definition of bulk modulus is expressed as

$$B = -V\frac{dp}{dV} = V\frac{d^2E}{dV^2} . \tag{7.8}$$

All quantities are determined for a fixed number of particles.

The experimental methods have made it possible to study materials over large ranges of pressures, which can change the properties of materials completely. Figure 7.2 shows the dependence of energy on volume for silicon calculated using the plane wave pseudopotential method and the local density approximation (these methods are described in Chapter 8). The calculated $E(V)$ curve fits with experiment.

In general, as the distance between the atoms decreases there is a tendency for all materials to transform to metallic structure, which is closely packed at the highest pressures. Thus, many interesting examples involve materials that have large-volume open structures at ordinary pressure, and which transform to more close-packed structures under pressure.

## 7.3
## Amplitude of Lattice Vibration

In a crystal lattice, vibrations of different frequencies appear simultaneously, forming the vibrational spectrum. This spectrum is characterized by a function of distribution of vibration by the frequencies $g(\nu)$. The physical meaning of the function $g(\nu)$ is that being multiplied by a small interval $d\nu$, it gives the number of vibrations

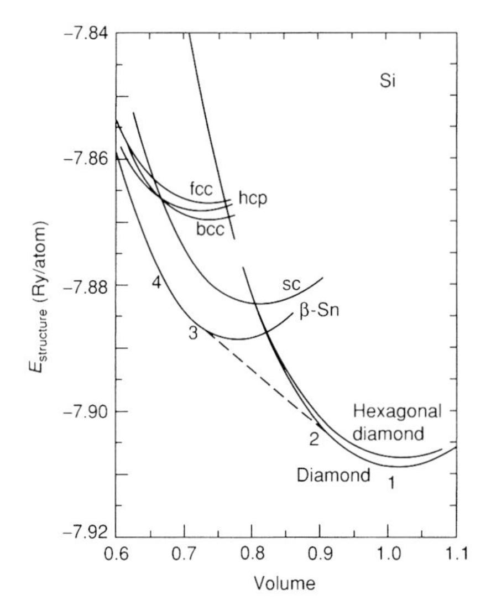


**Figure 7.2** Energy versus volume for various structures of silicon found by means of *ab initio* calculations. Transition pressures are given by the slopes of the tangent lines to the phases (after [8]).

whose frequencies cover all frequencies from $\nu$ to $\nu + d\nu$. The quantity $g(\nu)d\nu$ is assumed to be equal to the part of the total number of vibrations belonging to the same interval of frequencies $[\nu, \nu + d\nu]$. Similarly, the number of vibrations in an interval $[\nu_1, \nu_2]$ is represented by the integral of the function of the distribution density taken between the corresponding limits,

$$N_{[\nu_1,\nu_2]} = \int_{\nu_1}^{\nu_2} g(\nu)d\nu \,. \tag{7.9}$$

The method of neutron spectroscopy is the most efficient tool for study frequencies of the crystal lattice vibrations. This method is based on the scattering of the low-energy, so-called heat neutrons by the nuclei of solids. The wavelength of the normal vibrations and of the heat neutrons are values of the same order as the energies are. As a result of the interaction of the low-energy neutrons with solids, quanta of normal vibrations of the crystal lattice (phonons) are created or, conversely, annihilated. The collision neutron–phonon changes the state of the neutron essentially and this change can be detected experimentally.

Figure 7.3 presents the frequency spectra for potassium, aluminum, niobium and molybdenum. The curves have basically the same shape: a relatively diffuse low-frequency part and a narrow high-frequency peak. Obviously, the peak corresponds to the optical vibrational modes and is caused by displacements of the atoms toward each other. The range of the frequencies depends on the strength of the interatomic bonding in the crystal lattice. The higher strength, the higher frequencies of the normal vibrations. The frequency interval for simple metal

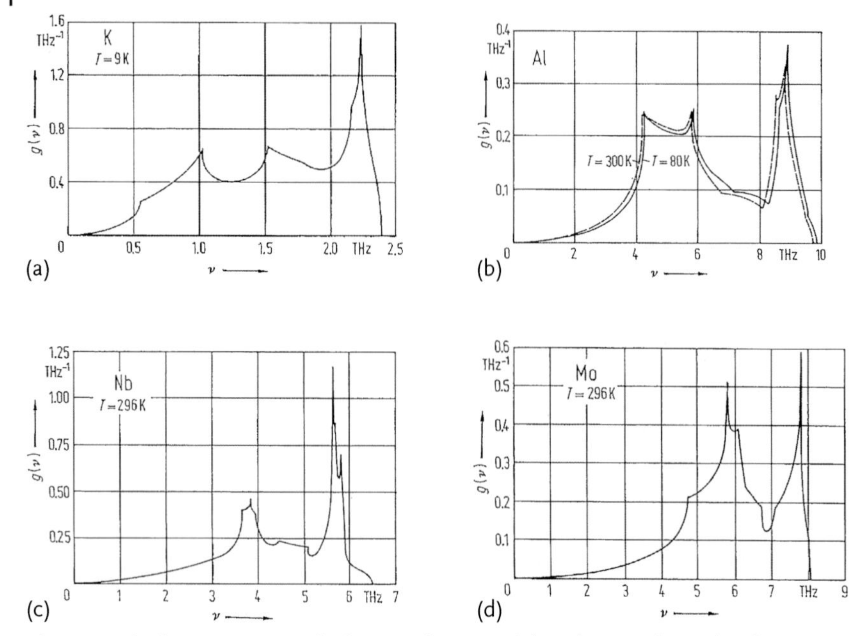


**Figure 7.3** The frequency spectra for four metals: K (a), Al (b), Nb (c), and Mo (d) (after [9]).

potassium is from 0 to 2.3 THz, whereas for aluminum it is from 0 to 9.8 THz. The frequency interval and the coordinates of the peaks for molybdenum are shifted to the right compared with those of niobium.

The atom mass is the second factor that has an effect on the frequencies. The greater the atom mass, the less the frequency.

The vibrating displacements of the atoms from the equilibrium position occur in different directions. The arithmetic mean of atomic displacements is zero, because all directions of displacements of atoms from the equilibrium position in a crystal lattice are equiprobable. The mean-square amplitude of the atom vibrations $\sqrt{\overline{u^2}}$ is a measure of the average heat displacements of atoms in the crystal.

The displacement vector of an atom can be decomposed along three coordinate axes. Each of $3N$ lattice waves has its frequency $\nu_i$ and its amplitude $u_i$. The value of the mean-square amplitude can be calculated if the frequency distribution function $g_i(\nu)$ is known. According to the definition of the mean value of a function, we have

$$\overline{u^2} = \frac{\sum_{i=1}^{N}\sum_{\alpha=1}^{3} u_{i\alpha}^2(\nu) g_i(\nu)}{\sum_{i=1}^{N} g_i(\nu)}, \tag{7.10}$$

where the sum is taken over all waves of the crystal lattice and over three directions of displacements.

Let us choose the Cartesian coordinate system in the lattice. Then, the squared amplitude of the atomic displacements in some direction $\alpha$ is given by

$$\overline{u_{i\alpha}^2} = \overline{u_{ix}^2} + \overline{u_{iy}^2} + \overline{u_{iz}^2}. \tag{7.11}$$

Since the heat displacements along all axes are equiprobable the terms in the right-hand side are equal to each other. Therefore, the mean-square amplitude of the $i$th oscillatory mode along one of the axes is

$$\overline{u_{ix}^2} = \frac{1}{3}\overline{u_{i\alpha}^2} . \tag{7.12}$$

In a three-dimensional crystal lattice, $3N$ waves propagate in three independent directions. The total kinetic energy of the crystal lattice consisting of $N$ atoms of the same mass $m$ can be found by the obvious formula,

$$E_{\text{kin}} = \frac{1}{2}\sum_{n=1}^{N} m\dot{u}_n^2 . \tag{7.13}$$

The mean value of the total energy of vibrating lattice is equal to the doubled value of $E_{\text{kin}}$. The total energy of an $i$-mode propagating in a direction $\alpha$ is given by [10]

$$\overline{E_{i\alpha}} = \frac{1}{2}m\omega_{i\alpha}^2\overline{u_{i\alpha}^2} , \tag{7.14}$$

where the angular frequency $\omega_{i\alpha} = 2\pi\nu_{i\alpha}$. Consequently,

$$\sqrt{\overline{u_{i\alpha}^2}} = \frac{\sqrt{2\overline{E_{i\alpha}}/m}}{\omega_{i\alpha}} . \tag{7.15}$$

The mean-square amplitude of an oscillatory mode is inversely proportional to its frequency. Smaller amplitudes correspond to the vibrations of higher frequencies. One can see from the same equation (7.15) that larger amplitudes are typical for atoms of a smaller mass than for those of more massive atoms.

According to the Planck formula, the mean energy of a quantum oscillator is given by

$$\overline{E_{i\alpha}} = \left[\frac{\hbar\omega_{i\alpha}}{\exp\left(\frac{\hbar\omega_{i\alpha}}{kT}\right) - 1} + \frac{\hbar\omega_{i\alpha}}{2}\right] , \tag{7.16}$$

where $\hbar = h/2\pi$; $h$ is the Planck constant; $T$ is the temperature.

Suppose that the temperature is sufficiently high, so that the condition $\hbar\omega \ll kT$ holds. Then one can neglect the second term in the right-hand side of (7.16) and expand the exponent term into a series retaining only first two terms. As a result, the formula takes the form $\overline{E_{i\alpha}} = kT$. Hence, under this assumption we obtain

$$\overline{u_{i\alpha}^2} = \frac{2kT}{m\omega_{i\alpha}^2} . \tag{7.17}$$

That is, at relatively high temperatures the square of the mean-square amplitude is proportional to the temperature.

The basic tool for the measurement of the mean-square amplitude of vibrations $\overline{u^2}$ is X-ray technique. The energy of the diffracted X-rays is reduced due to independent heat atom vibrations.

## 7.4
## The Debye Temperature

According to the Debye model, acoustic spectrum of a solid consists of $3N$ independent waves. A maximal frequency $\nu_m$ of lattice vibration exists. The characteristic Debye temperature $\theta$ corresponds to this frequency:

$$h\nu_m = k_B\theta \ , \tag{7.18}$$

where $h$ is the Planck constant, $k_B$ is the Boltzmann constant. The Debye theory was developed in order to estimate the phonon contribution to heat capacity of solids. The Debye temperature $\theta$ is a boundary value. If the temperature $T > \theta$ the energy of a crystal is uniformly distributed over all vibrations of the crystal lattice. Under this condition, the heat capacity of solids is a constant value. At temperatures which are small compared to the Debye temperature, $(T < \theta)$, the dependence of heat capacity of solids on the absolute temperature $T$ is cubic.

The Debye temperature is an useful characteristic of the strength of a material. A strong material has stiffer interatomic bonding and therefore higher frequencies of the atom vibration. Accordingly, the stronger the material the higher the Debye temperature. Different measurement methods result in slightly different values of the Debye temperature.

Diamond has the highest Debye temperature of 1860 K. The semiconductors silicon and germanium have 625 and 290 K, respectively. For metals, the Debye temperature ranged from 100 K for potassium to 470 K for iron.

## 7.5
## Melting Temperature

Melting is a fundamental process in which a crystal undergoes a phase transition from a solid to a melt.

It is intuitively obvious that the higher melting temperature of a solid the stronger is the interatomic bonding. It is interesting to note that the using of the melting temperature as a primary characteristic of the interatomic bonding strength is due to theories dealing with elastic constants of a solid and with the amplitude of atomic vibration.

The first theory explaining the mechanism of melting in the bulk was proposed by Lindemann, who linked the melting transition to the increased vibration of atoms in the crystal as follows. The average amplitude of thermal vibrations increases when the temperature of the solid increases. At some point the amplitude of vibration becomes so large that the atoms start to invade the space of their nearest neighbors and disturb them and the melting process initiates. Because of quantitative calculation difficulties based on the present model Lindemann offered a simple criterion: melting might be expected when the mean-square amplitude of vibrations $\sqrt{\overline{u^2}}$ exceeds a certain threshold value (namely when the amplitude reaches at least 10% of the nearest neighbor distance). Assuming that all atoms vibrate about their equilibrium positions with the same frequency (the Einstein approximation) the average thermal vibration energy can be estimated based on the equipartition theorem as

$$E = 4\pi^2 m \nu_E^2 \overline{u^2} = k_B T \,, \tag{7.19}$$

where $m$ is atom mass, $\nu_E$ is the Einstein frequency, $k_B$ is the Boltzmann constant.

According to the Lindemann criterion, melting temperature is expressed as

$$T_m = \frac{4\pi^2 c_L m a^2 \nu^2}{k_B} \,, \tag{7.20}$$

where $c_L$ is the Lindemann constant, $a$ is the parameter of unit cell. The Lindemann constant is assumed to be the same for crystals with similar structure.

Born proposed another theory that is based on the fact that a liquid differs from a crystal in having zero resistance to the shear stress. The distances between the atoms are increased due to the thermal expansion, hence the restoring forces between the atoms are reduced, and therefore the shear elastic moduli decrease with rising temperature. The softening of the shear moduli leads to a mechanical instability of the solid structure and finally to a collapse of the crystal lattice at some temperature. Born derived the general conditions for stability of a crystal lattice. He analyzed the free energy of a solid with a cubic crystalline lattice. For a lattice to be stable, the free energy must be represented by a positively defined quadratic form, and this is satisfied when the following inequalities for the shear elastic coefficients hold true:

$$C_{11} - C_{12} > 0 \,; \quad C_{11} + 2C_{12} > 0 \,; \quad C_{44} > 0 \,.$$

According to Born, $C_{44}$ goes to zero and the melting temperature can be found from the condition

$$(C_{44})_{T=T_m} = 0 \,.$$

The melting temperatures of transition metals are shown in Figure 13.1.

## 7.6 Cohesive Energy

The cohesive energy is the energy gained by arranging the atoms in a crystalline state, compared with energy of the isolated atoms. The energy of a solid is less than the total energy of atoms, which compose this solid. Consequently, we should follow the rule of signs: cohesive energy is a negative value.

Insulators and semiconductors have large cohesive energies; these solids are bound together strongly and have good mechanical strength. Metals with the sp bonding have relatively small cohesive energies. This type of bonding is weak.

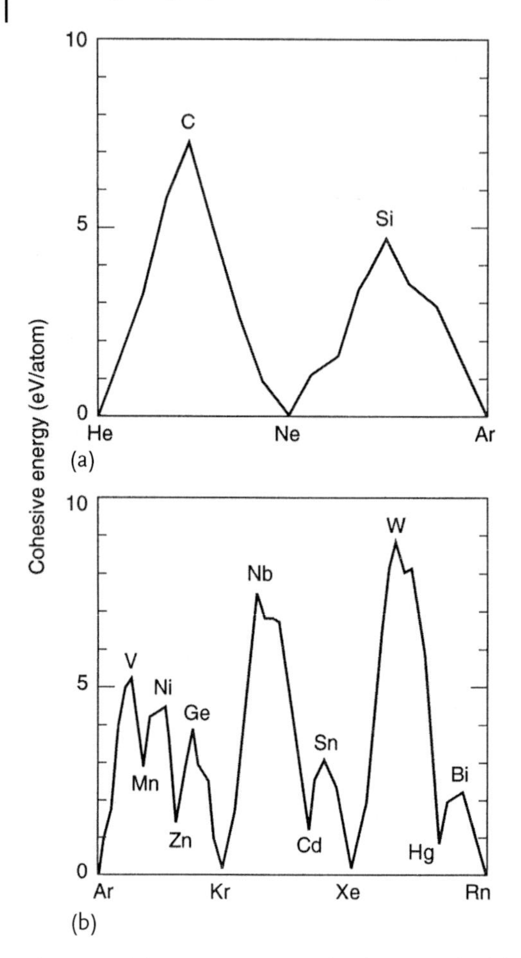


**Figure 7.4** Cohesive energy across the short periods of the Mendeleev periodic table (a) and the long periods (b) (after [3]).

The variation of the cohesive energy across the periodic table is illustrated by Figure 7.4. We see a very marked trend across the short periods of the cohesion increasing until the electron valence shell is half-full at carbon or silicon, thereafter decreasing to almost zero when the valence shell is full for the noble gas solids neon or argon. A similar trend is exhibited across the 4d and 5d transition metal series where the cohesion peaks at niobium and tungsten, respectively, in the vicinity of a half-full valence d shell. The minima at cadmium and mercury corresponding to the d shell being full and nearly core-like.

## 7.7
## Energy of Vacancy Formation and Surface Energy

The formation energy of a vacancy is related to the strength of interatomic bonding of solids. According to a simple phenomenological model of vacancy formation a central atom is surrounded by $Z$ nearest neighbors. Creation of a vacancy by removing this central atom will mean breaking $Z$ bondings and then replacing the atom on a free surface thereby restoring on the average $Z/2$ bondings. Thus, in the first approximation, the formation energy of a vacancy is equal to $Z/2$ times the bonding strength between a pair of atoms. An empirical relationship holds between energy of the vacancy formation $E_{\mathrm{v}}$ and cohesive energy $E_{\mathrm{coh}}$,

$$E_{\mathrm{v}} \approx \frac{1}{2} E_{\mathrm{coh}} \,. \tag{7.21}$$

The vacancy formation scales to the melting point as it is seen in Figure 7.5 for metals with the body-centered crystal (bcc) lattice.

In Figure 13.2 the dependence of energy of the vacancy formation on number of electrons in 4d shell is shown.

The reversible work $d\,W$ required for an external force to create an infinitesimal area $d\,A$ of the surface is directly proportional to this area,

$$-d\,W = \gamma\, d\,A \,, \tag{7.22}$$

where $\gamma$ is a proportionality factor. It is dependent on temperature, volume, and the particle number.

The minus in (7.22) implies that the work performed by an external force is expended for increase in the Helmholtz free energy of $d\,F$. Since for a reversible and equilibrium process $d\,F = -d\,W$ the surface contribution to the free energy must be proportional to the increment in the surface area,

$$d\,F = \gamma\, d\,A \,. \tag{7.23}$$

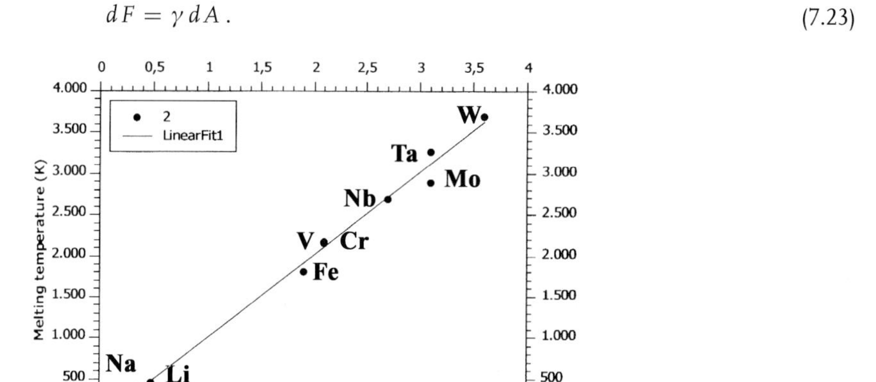


Figure 7.5 The close fit between energy of the vacancy formation and melting temperature for bcc metals. Data on the vacancy formation energy of Schulz [11].

The coefficient of proportionality $\gamma$ has a certain physical meaning, namely its value is equal to the excess of the free energy per unit area. The surface energy $\gamma$ is also defined as

$$\gamma = \left(\frac{dF}{dA}\right)_{T,V,N} . \tag{7.24}$$

The temperature, the volume of crystal, and the number of atoms are assumed to be constant. The reversibility requirement in the definition of $\gamma$ means that the composition in the surface region is also in thermodynamic equilibrium. The distances between atoms remain unchanged.

The unit of surface energy is $\mathrm{J\,m^{-2}}$ or $\mathrm{N\,m^{-1}}$.

The surface energy depends on a crystal plane orientation. The experimental measurement of the surface energy for solids is extremely difficult. There are few data about values of $\gamma$ in the literature.

## 7.8 The Stress–Strain Properties in Engineering

Figure 7.6 illustrates a typical curve of tensile strain of a material. From the origin O to the point called proportional limit, the stress–strain curve is a straight line. This linear relation between elongation and the axial force causing it was first noticed by Hooke in 1678 and is called Hooke's Law that states that within the proportional limit the stress is directly proportional to the strain or

$$\sigma = E\varepsilon , \tag{7.25}$$

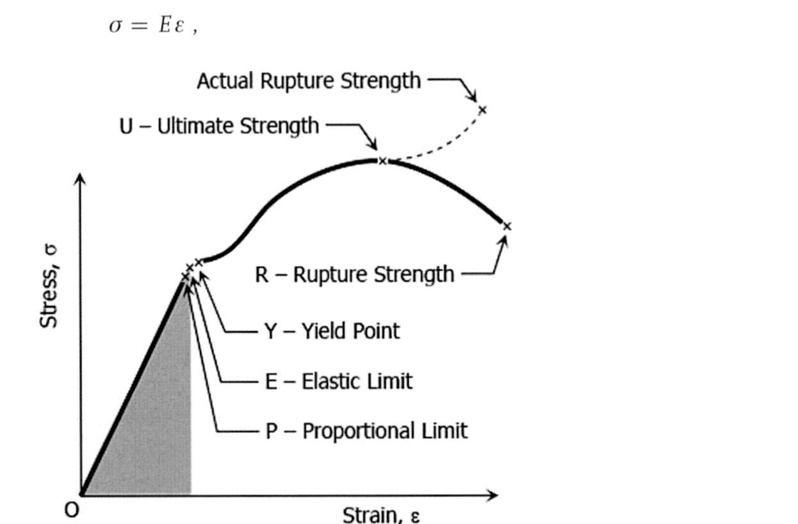


**Figure 7.6** The diagram stress–strain in engineering.

where $\sigma$ is the stress, $\varepsilon$ is the strain. The constant of proportionality $E$ is called the modulus of elasticity or the Young modulus and it is equal to the slope of the stress–strain diagram from point O to point P.

# 8
# Simulation of Solids Starting from the First Principles ("*ab initio*" Models)

Computer simulation of solids has gradually developed from mere aids in supporting experimental data to powerful tools for predicting new materials and their structures and properties.

Different approaches to simulation and modeling of solids were evolved, the most important of which are:

- The computational method starting from first principles (the so-called "*ab initio* method").
- The tight-binding approximation.
- The embedded atom method.
- Application of empirical potentials in study of the interatomic interaction.
- The computer technique of modeling of the kinetic processes.

## 8.1
## Many-Body Problem: Fundamentals

The structure and energy of isolated atoms of chemical elements are well-known (Figure 3.13). The atomic emission spectroscopy allows one to determine the composition of electronic shells and the energy of electrons in an atom with a desired high precision. The structure of atoms has specifically been investigated by measurements of the frequency, intensity, and broadening of their spectral lines.

The particles which a solid consists of (nuclei, ions, electrons) are charged and have quantum properties. These particles constitute both isolated atoms and solids. However, an atom which interacts with other atoms and constitutes a solid, ceases to be oneself. An atom of a chemical element in a solid has different properties than an individual atom of the same element.

Nevertheless, there is a logical trend to determine properties of a solid and explain them based on properties of the "free" atoms that constitute this solid.

The approach that attempt to make use of "first principles" or "*ab initio*" for prediction material properties distinguishes itself, evidently, among other approaches. This method requires only a specification of the atoms, which enter into the composition of a solid, by their atomic numbers. Many of the *ab initio* methods that

*Interatomic Bonding in Solids: Fundamentals,Simulation,andApplications*, First Edition. Valim Levitin.

physicists and chemists use have existed for more than a decade. All of the *ab initio* methods were continuously refined over these years and all benefited from the use of increasingly powerful computers.

Which characteristics of solids would we like to predict? Among these are first of all:

- The crystal structure.
- The energy of a solid per atom.
- The electronic structure.
- The forces between atoms.
- The strength of interatomic bonding, that is, properties such as cohesive energy, elastic constants, phonon spectrum, energy of the vacancy formation, the surface energy.

The fundamental difficulty in condensed matter theory as well as in materials science is a problem of interaction of enormous number of particles. This problem is called the many-body problem. The electrons in a solid interact strongly both with ions or nuclei and with each other making up the crystal structure.

The most difficult problem in any electronic structure calculation is the need to take into consideration the effects of the electron–electron interaction.

Because of the large difference in mass between the electrons and nuclei and the fact that the forces acting on the particles are the same, the electrons respond practically instantaneously to the motion of the nuclei. The nuclei may be considered almost stationary on the time-scale of the motion of the electrons. Thus, the nuclei can be treated adiabatically, leading to a separation of electronic and nuclear coordinates in the many-body eigenfunction. This is the Born–Oppenheimer approximation. This "adiabatic principle" reduces the many-body problem to an easier problem of studying of the dynamics of the electrons in some quasi-frozen-in configuration of the nuclei. Quantum physics simulations deal with approximate solutions to the simplified Schrödinger equation which only accounts the dynamics of the light electrons with the nuclei positions remaining unchanged. However, even with this simplification, the many-body problem remains formidable, which means that further simplifications have to be made.

The theory of independent electrons (so-called the independent electron approximation) assumes that the electrons are uncorrelated except that they must obey the exclusion Pauli principle. All electron calculations involve finding a solution to a Schrödinger-like equation for an electron of a spin $\sigma$,

$$\left[-\frac{\hbar^2}{2m_e}\nabla^2 + V^{\sigma}_{\text{eff}}(\boldsymbol{r})\right]\psi^{\sigma}_i(\boldsymbol{r}) = E^{\sigma}_i\psi^{\sigma}_i(\boldsymbol{r}) , \tag{8.1}$$

where $V^{\sigma}_{\text{eff}}(\boldsymbol{r})$ is an effective potential that acts on the electron at a point $\boldsymbol{r}$, $E^{\sigma}_i$ is eigenvalue (that is total energy), and $\psi^{\sigma}_i$ is eigenfunction of the electron.

The solution to the equation having the form (8.1) is at the heart of the computer simulation of solids. However, the modern way of calculations follows the Kohn–Sham method (see Section 8.5), which also involves potentials chosen to incorporate the exchange energy and correlation energy.

One writes (8.1) in the form

$$H\psi_i^\sigma(\mathbf{r}) = E_i^\sigma \psi_i^\sigma(\mathbf{r}) . \tag{8.2}$$

We deal here again with a basic concept of the Hamiltonian $H$.

$$H = -\frac{\hbar}{2m_e}\nabla^2 + V_{\text{eff}}^\sigma(\mathbf{r}) . \tag{8.3}$$

Hamiltonian is an operator corresponding to the total energy of the system.[1]

The electrons in the ground states occupy the lowest eigenstates of (8.1) obeying the exclusion principle. Excited states involve occupation of higher energy eigenstates. If the Hamiltonian is spin independent, then up and down spin states are the degenerate ones and one can simply consider spin as a factor of two in the calculation.

The Hamiltonian for a system consisting of electrons and nuclei is given by [1]

$$\begin{aligned} H = &-\frac{\hbar^2}{2m_e}\sum_i \nabla_i^2 - \frac{1}{4\pi\varepsilon_0}\sum_{i,I}\frac{Z_I e^2}{|\mathbf{r}_i - \mathbf{R}_I|} + \frac{1}{2}\frac{1}{4\pi\varepsilon_0}\sum_{i\neq j}\frac{e^2}{|\mathbf{r}_i - \mathbf{r}_j|} \\ &-\sum_I \frac{\hbar^2}{2M_I}\nabla_I^2 + \frac{1}{2}\frac{1}{4\pi\varepsilon_0}\sum_{I\neq J}\frac{Z_I Z_J e^2}{|R_I - R_J|} , \end{aligned} \tag{8.4}$$

where electrons are denoted by lower case subscriptions $i$ and $j$, and nuclei, with charge $Z_I$ and mass $M_I$, are denoted by uppercase subscriptions $I$ and $J$. Hamiltonian (8.4) is similar to Hamiltonian (3.27) for multi-electron atom. Remember that the first term corresponds to kinetic energy of electrons, the second term describes interaction electrons and nuclei, the third term characterizes interaction of electrons with each other. The fourth term in (8.4) describes kinetic energy of nuclei and can be regarded as a small one because of the inverse mass $1/M_I$ factor. The last term corresponds to interactions between nuclei. Ignoring the nuclear kinetic energy, the fundamental Hamiltonian for the theory of electronic structure can be compactly written as

$$H = T + V_{\text{ext}} + V_{\text{int}} + E_{\text{II}} , \tag{8.5}$$

where $T$ is the kinetic energy operator for electrons, $V_{\text{ext}}$ is the potential acting on the electrons due to the nuclei, $V_{\text{int}}$ is potential of the electron–electron interaction, and $E_{\text{II}}$ is the interaction of nuclei with one another and any other terms that contribute to the total energy of the system.

1) An operator is a rule that transforms a given function into another function. The operator is a symbol that tells us to do something with whatever follows the symbol. The differentiation $d/dx$ is an example of the operator, it transforms a differentiable function $f(x)$ into another function $f'(x)$.

## 8.2
## Milestones in Solution of the Many-Body Problem

Let us briefly consider the basic stages of advance in the many-body problem. It is very interesting and instructive to learn which scientists were involved and how they overcame, step-by-step, various difficulties and obstacles.

- It is remarkable that as early as 1927, Douglas Hartree pioneered a self-consistent field method (SCF) for theory of atoms. Hartree proposed the numerical iterative technique for solution to quantum equations.
- In 1937, John Slater proposed to partition a crystal into regions of atomic nucleus and an interstitial space. He used different functions for approximation of these two areas of a many-body system. He replaced the trial eigenfunction for many particles by a determinant (3.26).
- Vladimir Fock developed a technique known as the Hartree–Fock theory, which takes into account the exchange energy of electrons.
- In 1964, Pierre Hohenberg and Walter Kohn published a theory of an inhomogeneous interacting electron gas. These authors put forward an idea that the electronic density plays a central role in properties of this gas. The authors in fact are the founders of the density functional theory (DFT).
- Walter Kohn and Lu Jeu Sham proved in 1965 that it is possible to replace the many-electron problem by an equivalent set of auxiliary self-consistent one-electron equations. They derived such a set of equations.
- Several authors advanced a pseudopotential theory to model the electron–ion interactions.
- Theoretical work concerning structure and energy of solids could not represent major achievements until the advent of high power computers (supercomputers).
- Scientists made steady progress in the iterative minimization technique. They use this technique in order to solve the supercomputer-aided set of equations arisen in the many-body problem.
- The linearized augmented plane wave method (LAPW) was developed and widely used to investigate materials.
- The technique of computer simulation makes rapid progress. One may assume that we are now at the threshold where the simulation is starting to make a major impact in development of real materials.

## 8.3
## More of the Hartree and Hartree–Fock Approximations

To describe completely the quantum behavior of electrons in solids it is necessary to calculate the eigenfunction for the many-electron system. In principle, a solution may be obtained from the time-independent Schrödinger equation, but in practice the potential experienced by each electron is dictated by the behavior of all other electrons in the solid. Of course, the influence of nearby electrons is much stronger than that of far-away electrons since the interaction is electrostatic in nature, but the fact remains that the motion of any one electron is strongly coupled to the motion of the other electrons in the system. To solve the Schrödinger equation directly for all these electrons would thus require us to solve a system of around $10^{23}$ differential equations. Such a calculation is beyond the capabilities of present-day computers, and it is likely to remain so for the foreseeable future.

One of the earliest attempts to solve the problem was made by Hartree. He simplified the problem by making an assumption about the form of the many-electron eigenfunction, namely that it was just the product of a set of single-electron eigenfunctions. In an uniform system these eigenfunctions would take the form of simple plane waves. Having made this assumption it was possible to proceed using the variational principle.

This principle states that if a given system may be described by a set of unknown parameters then the set of parameter values, which minimize the total energy, most correctly describe the ground state of the system. The system exists in the ground state when it is not perturbed by outside influences.

In the equation (8.1), $\psi_i^\sigma$ is the eigenfunction, which is a function of each of the spatial coordinates and spins of each of the $N$ electrons in a solid, so

$$\psi_i^\sigma = \psi_i^\sigma(\boldsymbol{r}_1, \boldsymbol{r}_2, \ldots, \boldsymbol{r}_N) . \tag{8.6}$$

Each electron is defined by three spatial variables and its spin. Although $\psi_i^\sigma$ is a function of all coordinates of all electrons, Hartree supposed that it was possible to approximate the eigenfunction of a system $\psi$ by a product of individual electron eigenfunctions,

$$\psi = \psi_1^\sigma(\boldsymbol{r})\psi_2^\sigma(\boldsymbol{r}) \ldots \psi_N^\sigma(\boldsymbol{r}) . \tag{8.7}$$

This expression for the eigenfunction is known as the Hartree product. The product of individual one-electron eigenfunctions (8.7) seemed to be a good approximation for the full eigenfunction of the system to its author.

By using the variational method, Hartree found the Hamiltonian equation of the many-electron system. The Hartree potential was no longer coupled to the individual motions of all the other electrons, instead it was simply dependent upon the time-averaged electron distribution of the system. This was an important simplification in the many-body problem.

However, the Pauli exclusion principle states that it is not possible for two electrons of the same system to possess the same set of quantum numbers. Mathematically, the Pauli exclusion principle can be accounted for by ensuring that the eigenfunction of a set of electrons $\psi$ is antisymmetric under swap of any two electrons.

That is to say that the process of swapping any one of the electrons for any other of the electrons leaves the eigenfunction unaltered except for a change of the sign. Any eigenfunction possessing such a property will tend to zero (indicating zero probability) as any pair of electrons with the same quantum numbers approach each other. The Hartree product eigenfunction is symmetric (that is, stays precisely the same after interchange of two electrons), so the Hartree approach effectively ignores the Pauli exclusion principle.

However, in addition to the Hartree potential (which describes the direct Coulomb interaction between an electron and the average electron distribution) there is a second type of potential influencing the electrons, namely the so-called exchange potential (see Section 3.3). In 1930 Fock published the first method of calculations using properly antisymmetric determinant eigenfunctions. Fock included exchange in a total-energy calculation. The Hartree–Fock approach is to minimize the total energy with respect to all degrees of freedom of the eigenfunction with the restriction that it has the form of the Slater determinant (3.26). This is known as the Hartree–Fock method.

If we look at an electron with spin-up, then the Pauli exclusion principle means that other nearby spin-up electrons will be repelled. Spin-down electrons will not be affected since they have a different spin quantum number. Thus, our spin-up electron is surrounded by a region which has been depleted of other spin-up electrons. Consequently, this region is positively charged (remember that the average electron distribution exactly balances the positive charge of the ion cores, and that this region is relatively depleted of electrons). Similarly, if we had considered a spin-down electron from the start, then we would have found a region depleted of other spin-down electrons. The edge of the electron-depleted region is not clearly defined, but nevertheless we call this region the exchange hole. Notably, the exchange potential contributes a binding energy for electrons in a neutral uniform system, so correcting one of the major failings of the Hartree theory.

The exchange hole of an electron involves only electrons of the same spin and the probability vanishes, as it must, for finding two electrons of the same spin at the same point $\boldsymbol{r} = \boldsymbol{r}'$. The exchange energy can be interpreted as the lowering of the Coulomb energy because of interaction of each electron with a positive exchange hole.

On the other hand, the Hartree–Fock approach includes the exchange effect but completely ignores the other effect, which somewhat balances it. This missing effect is the electrostatic correlation of electrons. In general, correlation is the most important for electrons of opposite spin, since electrons of the same spin are automatically kept apart by the exclusion principle. For the ground state, the correlation energy is always negative and any approximation of it should be negative.

We can also visualize a second type of hole in the electron distribution caused by simple electrostatic processes. If we consider a region immediately surrounding any electron (regardless of spin) then we expect to see fewer electrons than in the average simply because of their electrostatic repulsion. Consequently, each electron is surrounded by an electron-depleted region known either as the Coulomb hole or the correlation hole (because of its origin in the correlated motion of the electrons). Just as in the case of the exchange hole the electron-depleted region is slightly positively charged. The effect of the correlation hole is twofold. The first is obviously that the negatively charged electron and its positively charged hole experience a binding force due to simple electrostatics. The second effect is more subtle and arises because any entities which interact with the electron over a distance an order larger than the diameter of the correlation hole will not interact with the bare electron but rather with the whole pair electron plus correlation hole (which of course has a smaller magnitude charge than that of the electron alone). Thus any other interaction effects, such as exchange, will tend to be reduced (that is, screened) by the correlation hole.

One can now clearly see why the Hartree–Fock approach fails for solids: firstly the exchange interaction should be screened by the correlation hole rather than acting in full, and secondly the binding between the correlation hole and electron is ignored. Nevertheless she Hartree–Fock approach gives quite creditable results for small molecules. This is because there are far fewer electrons involved than in a solid, and so correlation effects are minimal compared to exchange effects.

## 8.4 Density Functional Theory

The breakthrough was made by Hohenberg and Kohn [12]. They proved two significant theorems. The first theorem is: the total ground-state energy of a many-electron system is a unique functional.[2] of the electron density $n(\boldsymbol{r})$. Value of $n$ is measured in electrons per unit of volume and depends on coordinates.

The density functional theory that has been formulated by Hohenberg and Kohn is an exact theory of many-body systems. The theory asserts that the ground-state electron density $n_0(\boldsymbol{r})$ determines all properties of a system in the ground state and in excited states.

The probability density that the $N$ electrons are at a particular set of coordinates $\boldsymbol{r}_1, \boldsymbol{r}_2, \ldots, \boldsymbol{r}_N$ equals to the product $\psi(\boldsymbol{r}_1, \boldsymbol{r}_2, \ldots, \boldsymbol{r}_N) \cdot \psi^*(\boldsymbol{r}_1, \boldsymbol{r}_2, \ldots, \boldsymbol{r}_N)$. The eigenfunction for $N$ electrons cannot be experimentally observed. A quantity closely related to it is the density of electrons at a particular position $n(\boldsymbol{r})$. This can be written in terms of individual electron eigenfunctions as

$$n(\boldsymbol{r}) = 2 \sum_i \psi_i(\boldsymbol{r}) \cdot \psi_i^*(\boldsymbol{r}) \,. \tag{8.8}$$

Here, the term inside the summation is the probability that an electron with an eigenfunction $\psi_i(\boldsymbol{r})$ is located at position $\boldsymbol{r}$. The summation goes over all the individual electron eigenfunctions. The factor of 2 appears because electrons can have two different spins. The point is that the electron density $n(\boldsymbol{r})$, which is a function of only three coordinates, contains a great amount of information that is actually physically observable.

This seemingly simple result, by focusing on the electron density rather than the many-body eigenfunction, allowed the authors to derive an effective one-electron-

2) A functional is a function of a function, the electron density being the function of coordinates, see below.

type Schrödinger equation, namely

$$-\frac{\hbar^2}{2m}\nabla^2\psi(\boldsymbol{r}) + [V_{\text{H}}(\boldsymbol{r}) + V_{\text{xc}}(\boldsymbol{r})]\psi(\boldsymbol{r}) = E\psi(\boldsymbol{r}) . \tag{8.9}$$

It is directly analogous to the Hartree equation (8.1), except for an additional term $V_{\text{xc}}(\boldsymbol{r})$ to the average Hartree electrostatic potential $V_{\text{H}}(\boldsymbol{r})$ (denoted by $V_{\text{eff}}^{\sigma}(\boldsymbol{r})$ in (8.1)). This means that each electron feels an extra attractive potential, the exchange-correlation potential.

The formulation of the Hohenberg–Kohn theorems applies to any system of interacting particles in an external potential $V_{\text{ext}}(\boldsymbol{r})$, including any problem of electrons and fixed nuclei, where the Hamiltonian is written as

$$H = -\frac{\hbar^2}{2m_{\text{e}}}\sum_i \nabla_i^2 + \sum_i V_{\text{ext}}(\boldsymbol{r}) + \frac{1}{2}\frac{1}{4\pi\varepsilon_0}\sum_{i\neq j}\frac{e^2}{|r_i - r_j|} . \tag{8.10}$$

In atomic units ($\hbar = m_{\text{e}} = e = 1/(4\pi\varepsilon_0) = 1$, Section 3.1) the Hohenberg–Kohn Hamiltonian is given by

$$H = -\frac{1}{2}\sum_i \nabla_i^2 + \sum_i V_{\text{ext}}(\boldsymbol{r}) + \frac{1}{2}\sum_{i\neq j}\frac{1}{|r_i - r_j|} . \tag{8.11}$$

The starting point of the local electron density functional theory is the fundamental theorem stating that the total ground-state energy of an electron many-body system is a functional of the electronic charge density.

The difference between a function $f(x)$ and a functional $F[f]$ is that the function is defined to be a mapping of a variable $x$ to a result (a number) $f(x)$; whereas a functional is a mapping of an entire function $f$ to a result (a number) $F[f]$. The functional $F[f]$, denoted by square brackets, depends upon the function $f$ over its range of definition $f(x)$ in terms of its argument $x$.

Consequently, the difficult problem for a system interacting electrons has been mapped onto that of a system of noninteracting electrons moving in an effective potential. It is essential that the electron "feels" the summarized field of all other electrons. The effective potential for an electron in a point is a sum of the nucleus potential and potentials of all other electrons.

The fundamental tenet of the density functional theory is that any property of a system of many interacting particles can be viewed as a functional of the ground-state density $n_0(\boldsymbol{r})$; that is, one scalar function of position, in principle, contains all the information about the many-body eigenfunctions for the ground state and all exited states.

The proof for such functionals given in the original work of Hohenberg and Kohn [12] is simple. In the main, the density functional theory has become the primary tool for calculation of electronic structure in condensed matter. However, the authors provide no guidance for constructing the functionals whatsoever, and no exact functionals are known for any system of more than one electron.

The second Hohenberg–Kohn theorem defines an important property of the functional: the electron density that minimizes the energy $E[n]$ is the true electron density corresponding to the full solution of the Schrödinger equation.

We follow the author of [1] stating the density functional theory.

The relations established by Hohenberg and Kohn can be stated as follows:

- The total energy $E$ of a system of interacting electrons in the Coulomb potential due to the nuclei in a solid is given exactly as a functional of the ground-state electronic density $n$.

$$E = E[n] \,. \tag{8.12}$$

- For spin-polarized systems $E$ and the other ground-state properties become functionals of the spin densities,

$$E = E[n_\uparrow, n_\downarrow] \,. \tag{8.13}$$

- For any system of interacting particles in an external potential $V_{\text{ext}}(\boldsymbol{r})$ the potential $V_{\text{ext}}(\boldsymbol{r})$ is determined uniquely, up to a constant, by the ground-state particle density $n_0(\boldsymbol{r})$.
- The total energy is variational, and this is the key to its usefulness. The ground-state density is that density which minimizes the energy. A universal functional for the energy $E[n]$ in terms of the electronic density $n(\boldsymbol{r})$ can be defined, valid for any external potential $V_{\text{ext}}(\boldsymbol{r})$. For any particular $V_{\text{ext}}(\boldsymbol{r})$, the exact ground-state energy of the system is the global minimum value of this functional, and the density $n(\boldsymbol{r})$ that minimizes the functional is exactly the ground-state density $n_0(\boldsymbol{r})$.
- All properties of the system are completely determined by the given ground-state density $n_0(\boldsymbol{r})$ only.
- Density of electrons in the ground state can be considered as a basic variable, and all properties of the system can be considered as unique functionals of the ground-state density. The functional $E[n]$ alone is sufficient to determine the exact ground-state energy. The total energy functional according to Hohenberg–Kohn theory $E_{\text{HK}}$ is given by

$$E_{\text{HK}}[n] = T[n] + E_{\text{int}}[n] + \int d^3r\, V_{\text{ext}}(\boldsymbol{r}) n(\boldsymbol{r}) + E_{\text{II}}[n] \,, \tag{8.14}$$

where $T[n]$ is the electron kinetic energy, $E_{\text{int}}[n]$ arises from the interaction of electrons with each other, $V_{\text{ext}}$ is the potential due to interaction electrons with nuclei (or ions), $E_{\text{II}}$ is the interaction of the nuclei with one another.

These statements may look promising but it is clear that no prescription has been given to solve the problem. It only follows that if the functional is known, then by minimizing the total energy of the system with respect to variations in the density function $n(\boldsymbol{r})$ one can find the exact ground-state density and energy.

Figure 8.1 illustrates the relations that were established by Hohenberg and Kohn.

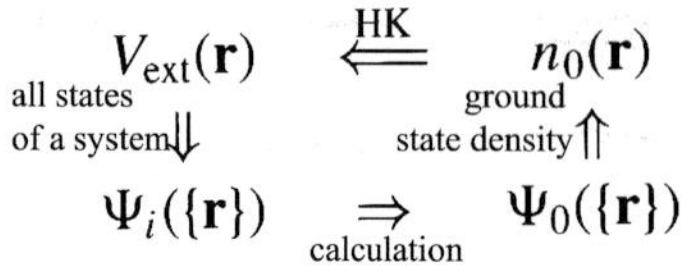

$$\begin{array}{ccc} V_{\text{ext}}(\mathbf{r}) & \stackrel{\text{HK}}{\Longleftarrow} & n_0(\mathbf{r}) \\ \text{all states of a system} \Downarrow & & \text{ground state density} \Uparrow \\ \Psi_i(\{\mathbf{r}\}) & \underset{\text{calculation}}{\Rightarrow} & \Psi_0(\{\mathbf{r}\}) \end{array}$$

**Figure 8.1** Schematic representation of the Hohenberg–Kohn theory. The short arrows denote [starting from $V_{\text{ext}}(r)$] the usual solution of the Schrödinger equation. The potential $V_{\text{ext}}(r)$ determines all states including the ground-state $\psi_0(\{r\})$ and the ground-state density $n_0(r)$. The long arrow labeled HK denotes the Hohenberg–Kohn theorem, which completes the cycle (after [1] by permission of Cambridge University Press).

## 8.5 The Kohn–Sham Auxiliary System of Equations

The Hohenberg–Kohn theorem says nothing specific about the form of $E_{\text{HK}}[n]$, and therefore the utility of the density functional theory depends on the choice of sufficiently accurate approximations for it. In order to do this, the unknown functional, $E[n]$, is rewritten as a sum of the Hartree total energy and another but presumably smaller unknown functional called the exchange-correlation functional, $E_{\text{xc}}[n]$.

Kohn and Shahm published the second excellent work in 1965 [13] whose formulation of density functional theory has become the basis of much of present-day methods for treating electrons in atoms, molecules, and solids. The approach proposed by Kohn and Sham replaced the original many-body problem by an auxiliary independent-particle problem.

Kohn and Sham assumed that the ground-state density of the original interacting system is equal to that of another yet unknown noninteracting system.[3] They showed how it is possible to replace the many-body problem by an equivalent set of self-consistent *one-electron* equations.

This leads to independent-particle equations for the noninteracting system that can be considered exactly soluble (in practice by numerical means) with all the difficult many-body terms incorporated into an exchange-correlation functional of the electron density. The Kohn–Sham approach has indeed led to very useful approximations that are now the basis of most calculations that attempt to make "*ab initio*" predictions for the properties of solids and large molecular system. The approach is remarkably accurate, most notably for "wide-band" systems, such as the group IV and II–V semiconductors, the sp bonded metals like sodium and aluminum, insulators like diamond, sodium chloride, and molecules with covalent and ionic bonds. It also appears to be successful in many cases in which the electrons have stronger effects of correlation, such as in transition metals.

The Kohn–Sham construction of an auxiliary system rests upon two assumptions:

3) Walter Kohn was awarded in 1998 the Nobel Prize in chemistry for his development of the density functional theory. The future Nobel laureate arrived in England at the age of 15 along with approximately 10 000 other children rescued in the famous Kindertransport operation immediately after the annexation of Austria by Hitler. Walter Kohn is from a Jewish family, both of his parents were later killed in the Holocaust.

$$\begin{array}{ccccccc} V_{\text{ext}}(\mathbf{r}) & \stackrel{\text{HK}}{\Longleftarrow} & n_0(\mathbf{r}) & \stackrel{\text{KS}}{\Longleftrightarrow} & n_0(\mathbf{r}) & \stackrel{\text{HK}_0}{\Longrightarrow} & V_{\text{KS}}(\mathbf{r}) \\ \Downarrow & & \Uparrow & & \Uparrow & & \Downarrow \\ \Psi_i(\{\mathbf{r}\}) & \Rightarrow & \Psi_0(\{\mathbf{r}\}) & & \psi_{i=1,N_e}(\mathbf{r}) & \Leftarrow & \psi_i(\mathbf{r}) \end{array}$$

**Figure 8.2** Schematic representation of the Kohn–Sham approach (compare to Figure 8.1). The notation $\text{HK}_0$ denotes the Hohenberg–Kohn theorem applied to the noninteracting system. The arrow labeled KS provides the connection in both directions between the many-body and independent-particle system. Note that any two values are "computationally" connected and therefore a solution of the independent-particle problem determines all properties of the many-body system (after [1] by permission of Cambridge University Press).

- The exact ground-state density can be represented by the ground-state density of an auxiliary system of noninteracting electrons.
- The auxiliary Hamiltonian is chosen to have the usual kinetic operator and an effective local potential $V_{\text{eff}}^{\sigma}$ acting on an electron of spin $\sigma$ at point $\boldsymbol{r}$.

According to the Kohn–Sham approach the ground-state energy functional $E_{\text{KS}}$ is rewritten as

$$E_{\text{KS}} = T[n] + \int d\boldsymbol{r}^3 V_{\text{ext}}(\boldsymbol{r})n(\boldsymbol{r}) + E_{\text{H}}[n] + E_{\text{II}} + E_{\text{xc}}[n] \,, \tag{8.15}$$

where $T[n]$ is the kinetic energy, $V_{\text{ext}}(\boldsymbol{r})$ is the external potential due to the nuclei and any other external field, $E_{\text{H}}[n]$ is the Hartree expression, $E_{\text{II}}$ is the interaction between the nuclei, and the term $E_{\text{xc}}[n]$ involves effects of exchange and correlation. It is an advantage of the Kohn–Sham approach that the authors incorporated orbitals to define the kinetic energy $T[n]$ of independent electrons. $E_{\text{H}}[n]$ is the self-interaction energy of the electron density $n(\boldsymbol{r})$ treated as a classical charge density. The Hartree component of the electron–electron interaction energy is expressed as

$$E_{\text{H}}[n] = \frac{e^2}{2}\int d^3\boldsymbol{r}d^3\boldsymbol{r}'\frac{n(\boldsymbol{r})n(\boldsymbol{r}')}{|\boldsymbol{r}-\boldsymbol{r}'|} \,. \tag{8.16}$$

Relation between the actually and auxiliary systems is shown in Figure 8.2.

## 8.6
## Exchange-Correlation Functional

The right side of the Kohn–Sham equation (8.15) contains five terms. Four of them can be written down in simple analytical forms. The contribution to the energy of the last one, that is, exchange-correlation term $E_{\text{xc}}[n]$, is unknown. However, there are several useful approximations to the exchange-correlation energy [14]. Local density approximation (LDA) is the simplest one. It approximates the energy in question by a local function that is equal to the known energy of the uniform electron gas. This known energy is then applied to realistic systems (molecules and solids). In the local density approximation, the local exchange-correlation potential in the Kohn–Sham equations is defined as the spatially uniform electron gas with the same electron density as the local electron density,

$$V_{XC}^{(LDA)} = V_{XC}^{\text{electron gas}}[n(\boldsymbol{r})] . \tag{8.17}$$

The exchange-correlation functional depending on $n$ for the uniform electron gas is known to a high precision. The local density approximation is successfully applied for solids because the valence electron density in many bulk materials varies slowly. This approximation fails for atoms and molecules because the electron density in this case cannot be satisfactorily approximated by the uniform electron gas.

The next approximation to the Kohn–Sham functional is the generalized gradient approximation (GGA). In this approximation, the local gradient of the electron density as well as the electron density are used in order to incorporate more information about electron gas. The physical idea of the generalized gradient approximation is simple; real electron densities are not uniform, so including information on the spatial variation in the electron density can create a functional that better describes real materials. The equations on which the GGA are based are valid for slowly varying densities. In the generalized gradient approximation, the exchange-correlation functional is expressed as

$$V_{XC}^{(GGA)}(\boldsymbol{r}) = V_{XC}[n(\boldsymbol{r}), \nabla n(\boldsymbol{r})] . \tag{8.18}$$

GGA functionals satisfy the uniform density limit. In addition, they satisfy several known, exact properties of the exchange-correlation hole. Two widely used nonempirical functionals that satisfy these properties are the Perdew–Wang (PW) functional [15] and the Perdew–Burke–Ernzerhof (PBE) functional [16]. Because GGA functionals include more physical ingredients than the LDA functional, it is often assumed that GGA functionals should be more accurate than the LDA. This is quite often true, but there are exceptions. One example of such a situation is in the calculation of the surface energy of transition metals and oxides.

The LDA and GGA functionals are both nonempirical functionals. Empirical functionals have also been proposed. Empirical functionals are fitted to selected experimental data and, therefore, contain a number of parameters that have been introduced and adjusted during the fitting process.

Weighted density approximation (WDA) incorporates more nonlocal information about electron gas by means of a model pair correlation function. Weighted density approximation greatly improves the calculated energies of atoms and bulk properties of solids. Nonetheless, the weighted density approximation is more computationally demanding than both LDA and CGA.

The relationship between various approximations can be understood using the exact expression for the exchange-correlation energy in terms of the pair correlation function [14],

$$E_{XC}[n] = \iint d^3r d^3r' \frac{n(\boldsymbol{r})n(\boldsymbol{r}')}{|\boldsymbol{r}-\boldsymbol{r}'|} \overline{g}[n, \boldsymbol{r}, \boldsymbol{r}'] = \iint d^3r d^3r' \frac{n(\boldsymbol{r})\overline{n}_{XC}[n, \boldsymbol{r}, \boldsymbol{r}']}{|\boldsymbol{r}-\boldsymbol{r}'|} \tag{8.19}$$

where $\overline{g}$ is the coupling constant average (from $e^2 = 0$ to $e^2 = 1$ in atomic units) of the pair correlation function of the electron gas in question, and $\overline{n}_{XC}$ is the coupling constant averaged exchange-correlation hole. Since the exchange-correlation hole must be a depletion containing exactly one electron charge, $E_{xc}$ is invariably negative. The physical meaning of this expression is that the exchange-correlation energy is given by the Coulomb interaction of each electron with its exchange-correlation hole, reduced in magnitude by a kinetic energy contribution, which corresponds to the energy required to dig out the hole. This reduction is accounted for by using the coupling constant average instead of the full strength pair correlation function, and induces contributions to the kinetic energy beyond the single particle level. The spherically symmetric Coulomb interaction in (8.19) means that only the spherical average of the exchange-correlation hole needs to be correct to obtain the correct energy. This fact is important for success of simple approximation like LDA.

From the above brief overview one can conclude that the development of density functional theory and the demonstration of the tractability and accuracy of the local density approximation to it constitutes an important milestone in solid-state physics and chemistry.

## 8.7 Plane Wave Pseudopotential Method

We have considered the electron–electron interaction in the previous sections. Here we discuss the electron–ion interaction.

The one-electron Schrödinger equation and consequently, the Kohn–Sham equations, still pose substantial calculational difficulties [17].

In the atomic region near the nucleus the kinetic energy of electrons is large resulting in rapid oscillations of the eigenfunction that requires fine grids for an accurate numerical representation. On the other hand, the large kinetic energy makes the Schrödinger equation stiff, so that a change in environment has little effect on the shape of the eigenfunction. Therefore, the eigenfunction in the atomic region can be represented well already by a small basis set.

In the bonding region between the atoms the situation is opposite. The kinetic energy is small and the eigenfunction is smooth. However, the eigenfunction is flexible and responds strongly to the environment. This requires large and nearly complete basis sets.

In most cases, the core electrons are quite strongly bounded with each other, and do not respond to motions of the valence electrons. Thus, they may be regarded as practically fixed. This is the essence of the pseudopotential approximation [14]. The strong core potential is replaced by a pseudopotential (compare dashed and solid lines in Figure 8.3). The corresponding electron eigenfunction is replaced by a smooth function. This pseudoeigenfunction represents the ground state of the electron and mimics the eigenfunction of valence electron outside a selected core radius. In this way, both the core states and the oscillations in the valence eigenfunctions are removed.

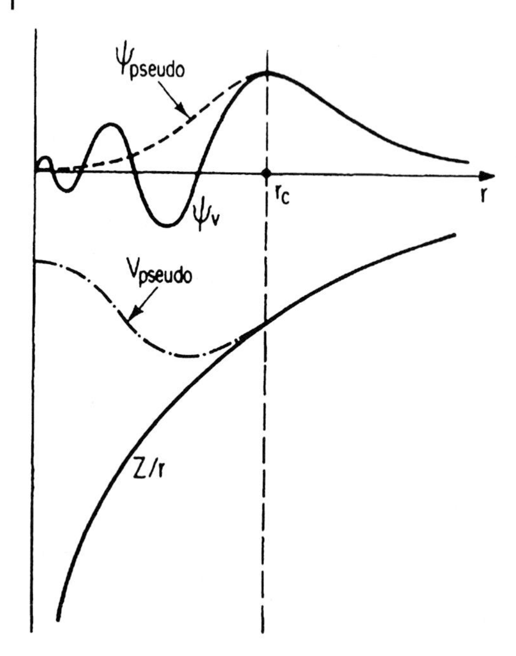


**Figure 8.3** Schematic illustration of the core potential $Z/r$ (in a.u.) and the electron eigenfunction $\psi_v$ (solid lines) and a pseudopotential, and a corresponding pseudoeigenfunction (dashed lines). At a distance of $r_c$ the values match.

Pseudopotential theory allows one to replace the strong electron–ion potential with a much weaker potential – a pseudopotential – that describes all the salient features of a valence electron moving through the solid, including relativistic effects. Thus the original solid is now replaced by pseudo valence electrons and pseudo ion cores. These pseudoelectrons experience exactly the same potential outside the core region as the original electrons but have a much weaker potential inside the core region. The fact that the potential is weaker is crucial, however, because it makes the solution of the Schrödinger equation much simpler by allowing expansion of the eigenfunctions in a relatively small set of plane waves. The use of plane waves as basis functions makes the accurate and systematic study of complex low-symmetry configurations of atoms much more tractable.

That is of the essence that the resulting pseudoeigenfunction $\psi_{\text{pseudo}}$ may be well represented by plane waves. The most general solution has to satisfy the Bloch theorem (Section 5.2) and boundary conditions. Each electronic wave function in a periodic crystal lattice can be written as the product of a cell-periodic part and a wave-like part,

$$\psi_i(\boldsymbol{r}) = u_i(\boldsymbol{r}) \exp(i\boldsymbol{k} \cdot \boldsymbol{r}) . \tag{8.20}$$

The cell-periodic part of the wave function can be expanded using a discrete set of plane waves whose wave vectors are reciprocal lattice vectors of the crystal,

$$u_i(\boldsymbol{r}) = \sum_{g} c_{i,g} \exp(i\boldsymbol{g} \cdot \boldsymbol{r}) \,, \tag{8.21}$$

where **g** vectors are defined by $\boldsymbol{g} \cdot \boldsymbol{r} = 2\pi M$, see Section 4.4, (4.13). Therefore each electronic wave function (eigenfunction) can be written as a sum of plane waves,

$$\psi_i(\boldsymbol{r}) = \sum_{g} c_{i,k+g} \exp i[(\boldsymbol{k} + \boldsymbol{g}) \cdot \boldsymbol{r}] \,, \tag{8.22}$$

where the summation is carried out over all the vectors of reciprocal lattice. The coefficients $c_{i,k+g}$ are to be determined.

Pseudopotentials regard an atom as a perturbation of the free electron gas [17]. The most natural basis functions are plane waves. Plane wave basis sets are, in principle, complete and suitable for sufficiently smooth eigenfunctions. The disadvantage of the comparably large basis sets is balanced by their extreme computational simplicity. Finite plane-wave expansions are however absolutely inadequate for description of strong oscillations of the eigenfunctions near the nucleus. In the pseudopotential approach the Pauli repulsion of the core electrons is therefore described by an effective potential that expels the valence electrons from the core electron. The resulting eigenfunctions are smooth and can be represented well by plane waves. The price to pay is that all information on the charge density and eigenfunctions near the nucleus is lost.

Figure 8.4 shows the iteration procedure for determination of a pseudopotential. All-electron calculations are performed for an isolated atom in its ground state and some excited states, using a given form for the exchange-correlation density functional. This provides valence electron eigenvalues and valency electron eigenfunctions for the atom, as one can see in Figure 8.4. A parameterized form of pseudopotential is chosen. The parameters are then adjusted, so that a pseudo-atom calculation using the same form for exchange-correlation as in the all-electron atom gives both pseudoeigenfunctions that match the valence eigenfunctions outside some cutoff radius $r_c$ and pseudoeigenvalues that are equal to the valence eigenvalues. The ionic pseudopotential obtained in this fashion is then used, without further modification, for any environment of the atom. The electronic density in any new environment of the atom is then determined using both the ionic pseudopotential obtained in this way and the same form of exchange-correlation functional as employed in the construction of the ionic pseudopotential.

Finally, it should be noted that ionic pseudopotentials are constructed with $r_c$ ranging typically from one to two times the value of the core radius. It should also be noted that, in general, the smaller the value of $r_c$ the more "transferable" is the potential.

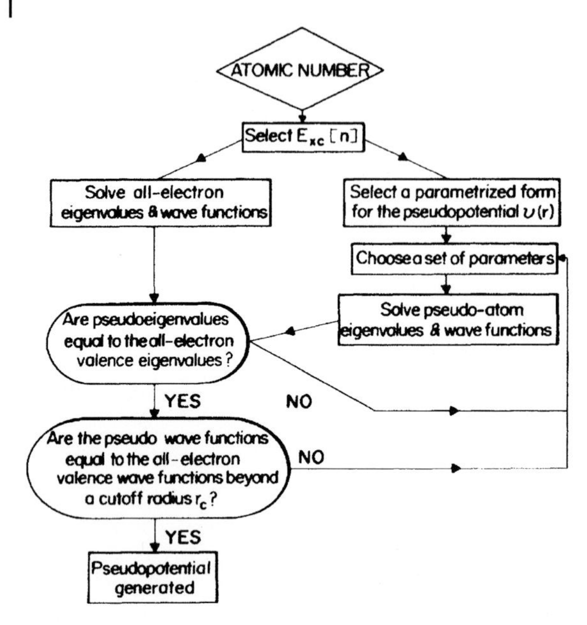


**Figure 8.4** Flow chart describing the construction of an ionic pseudopotential for an atom (after [18]).

## 8.8
## Iterative Minimization Technique for Total Energy Calculations

Only the minimum value of the Kohn–Sham energy functional has physical meaning. At the minimum, the Kohn–Sham energy functional is equal to the ground-state energy of the system of electrons with the ions in positions $\boldsymbol{R}_{\mathrm{I}}$.

It is necessary to determine the set of eigenfunctions $\psi_i$ that minimize the Kohn–Sham energy functional $E[\{\psi_i\}]$. These are given by self-consistent solutions of the Kohn–Sham equations:

$$\left[-\frac{\hbar^2}{2m}\nabla^2 + V_{\mathrm{ext}}(\boldsymbol{r}) + V_{\mathrm{H}}(\boldsymbol{r}) + V_{\mathrm{xc}}^{\sigma}(\boldsymbol{r})\right]\psi_i^{\sigma}(\boldsymbol{r}) = E_i^{\sigma}\psi_i^{\sigma}(\boldsymbol{r})\,, \tag{8.23}$$

where $\psi_i^{\sigma}$ is the eigenfunction of an electron of spin $\sigma$, $E_i^{\sigma}$ is the Kohn–Sham eigenvalue, and $V_{\mathrm{H}}$ is the Hartree potential of the electron given by (8.16).

We have already noted that the difficulties of the many-body problem were overcome by Kohn and Scham, who showed that the task of finding the right electron density can be expressed in a way that involves solving of a set of equations in which each equation only involves a *single* electron. The main difference is that the Kohn–Sham equations are missing the summations that appear inside the Hamiltonian (8.4) of the full Schrödinger equation. This is because the solutions of the Kohn–Sham equations are single-electron eigenfunctions that depend only on three spatial variables and spin.

The exchange-correlation potential $V_{xc}$ is given formally by the functional derivative

$$V_{xc}(\boldsymbol{r}) = \frac{\delta E_{xc}[n(\boldsymbol{r})]}{\delta n(\boldsymbol{r})} \tag{8.24}$$

The Kohn–Sham equations represent a mapping of the interacting many-electron system onto a system of noninteracting electrons moving in an effective potential due to all the other electrons. If the exchange-correlation energy functional was known exactly, then taking the functional derivative with respect to the electron density would produce an exchange-correlation potential that included the effects of exchange and correlation exactly.

The Kohn–Sham equations must be solved self-consistently so that the occupied electronic states generate a charge density that produces the electronic potential that was used to construct the equations. The sum of the single-particle Kohn–Sham eigenvalues does not give the total electronic energy because this overcounts the effects of the electron–electron interaction in the Hartree energy and in the exchange-correlation energy. The Kohn–Sham eigenvalues are not, strictly speaking, the energies of the single-particle electron states, but rather the derivatives of the total energy with respect to the occupation numbers of these states. Nevertheless, the highest occupied eigenvalue in an atomic or molecular calculation is nearly the unrelaxed ionization energy for that system.

The term within the brackets in (8.23) is the Hamiltonian. If an approximate expression for the exchange-correlation energy $V_{xc}^{\sigma}(\boldsymbol{r})$ is given the bulk of the work involved in a total-energy pseudopotential calculation is the solution of the eigenvalue problem.

As input data, the atomic numbers and positions of atoms in a solid must be given. The computational procedure requires an initial guess for the electron density $n^{\uparrow}(\boldsymbol{r}), n^{\downarrow}(\boldsymbol{r})$. From these densities the external potential, the Hartree potential, and exchange-correlation potential can be calculated. The effective potential that is the sum of these three potentials is obtained for each of the $\boldsymbol{k}$ points included in the calculation. Then the Kohn–Sham equation is solved to obtain the eigenfunctions. The eigenfunctions generate a different electron density. A new density is used to construct a new effective potential. The process is repeated until the solutions are self-consistent.

The numerical solution to Kohn–Sham equations is presented in Figure 8.5.

There are a set of Schrödinger-like independent-particle equations which must be solved subject to the condition that the effective potential $V_{eff}^{\sigma}(\boldsymbol{r}) = V_{ext}(\boldsymbol{r}) + V_{H}(\boldsymbol{r}) + V_{xc}^{\sigma}(\boldsymbol{r})$ and the density $n(\boldsymbol{r}, \sigma)$ are consistent. An actual calculation utilizes a numerical procedure that successively changes $V_{eff}^{\sigma}$ and $n$ to approach a self-consistent solution. The computationally intensive step in Figure 8.5 is "solve KS (that is Kohn–Sham) equation" for a given potential $V_{eff}$. This step is considered as a "black box" that uniquely solves the equation for a given input the potential $V_{eff}^{\sigma\,\text{input}}$ to determine an output electronic density $n^{\sigma\,\text{output}}(\boldsymbol{r})$. Except for the exact solution, the input and output potentials and densities do not agree. To arrive at the solution one defines a new potential operationally which can start a new cycle

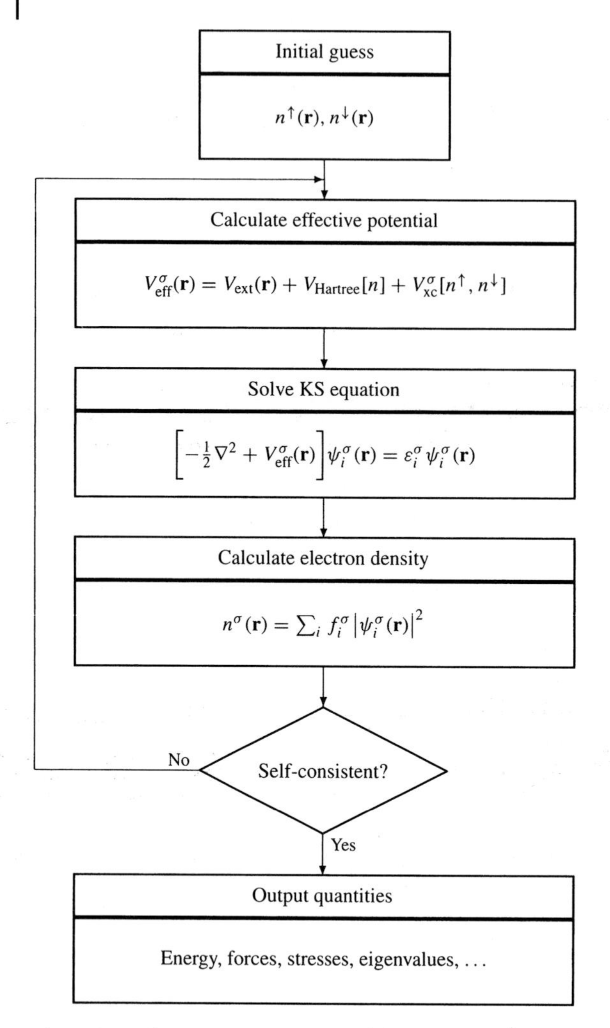


**Figure 8.5** Schematic representation of the self-consistent loop for solution to Kohn–Sham equations. Input values for two such loops are the electron densities with spin $n^{\uparrow}$ and $n^{\downarrow}$. An effective potential $V_{\text{eff}}^{\sigma}(r)$ is calculated for each spin as the first step. $\varepsilon_i^{\sigma}$ is energy of an $i$ electron of a spin $\sigma$ (after [1] by permission of Cambridge University Press).

with this new input potential. The iterative progression converges with a judicious choice of the new potential in terms of the potential or density comparing with the previous step (or steps).

One should carefully choose the number of $k$ points in order to ensure the convergence and a desired accuracy of calculations. Most density functional theory packages offer the option of choosing $k$ points.

The integrals over the Brillouin zone are given by

$$\overline{g} = \frac{V_0}{(2\pi)^3} \int_{BZ} g(\boldsymbol{k}) d\boldsymbol{k} , \tag{8.25}$$

where $V_0$ is the volume of the unit cell.

## 8.9 Linearized Augmented Plane Wave Method

Slater worked out the original augmented plane wave method (APW) as early as in 1937. Near an atomic nucleus the potential and eigenfunctions are similar to those in an atom. They are strongly varied but nearly spherical. Conversely, in the interstitial space between the atoms, both potential and eigenfunctions are smoother. Accordingly, space is divided into regions and different basis expansions are used in these regions.

Slater proposed the muffintin approximation of the potential field in a solid. In its simplest form, nonoverlapping spheres are centered at the atomic positions. Within these regions, the screened potential experienced by an electron is approximated to be spherically symmetric about the given nucleus. In the atom-centered regions, the wave functions can be expanded in terms of spherical harmonics and the eigenfunctions of a radial Schrödinger equation. In the interstitial region of constant potential, the single electron eigenfunctions can be expanded in terms of plane waves. Continuity of the potential between the atom-centered spheres and interstitial region is enforced.

The partitioning of space into two regions is shown in Figure 8.6.

The linearized augmented plane wave (LAPW) method is a modification of the APW method. It is a procedure for solving the Kohn–Sham equations for the total energy, ground-state density, and electronic band structure of solids. The adaptation to the problem is achieved by dividing the unit cell into two regions, too. The first region consists of the nonoverlapping atomic spheres, which are centered at atomic cites. The remaining interstitial space is the second region. In the LAPW method, the spheres must be nonoverlapping. This means that the sum of the sphere radii of two neighboring atoms must not exceed the distance between the

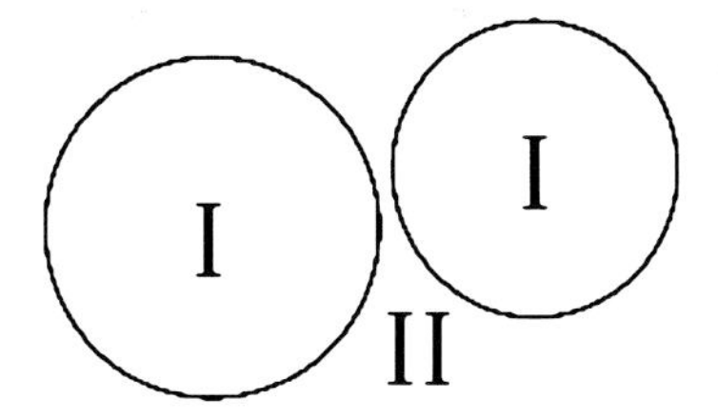


**Figure 8.6** The dual representation of space. I are the atomic spheres and II is the interstitial region.

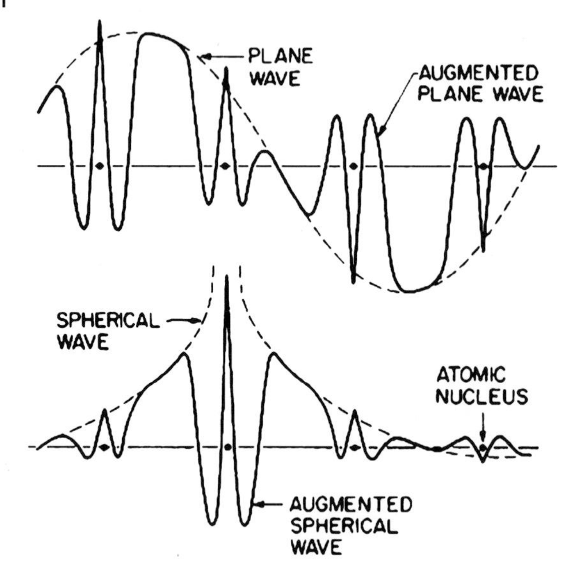


**Figure 8.7** The augmented plane wave and the augmented spherical wave. Both are used for representing the electron waves in the interatomic region.

two atoms. The spheres are chosen for computational convenience in the LAPW method. They are not atomic-sized representative.

Different basis equations are applied in these regions. Inside the spheres, one uses the radial solutions of equations. A linear combination of radial functions $u_l(r)\,Y_{lm}(\mathbf{r})$ and their derivatives with respect to the linearization parameters $E_l$ is utilized. These functions and derivatives are matched on the sphere boundaries.

In the interstitial region II the expansion of eigenfunction into plane waves is used. Figure 8.7 shows the augmented waves that represent the electronic waves in the interatomic region.

The program Wien2k "An Augmented Plane Wave and Local Orbitals Program for Calculating Crystal Properties [19]" is based on the linearized augmented plane wave (LAPW) . The method has proven to be one of the most accurate methods for the computation of the electronic structure of solids within the density functional theory. The program Wien2k is intended for calculation of different parameters of the energy, electronic and geometric structure of crystal solids. The program makes it possible to perform simulation, calculation and determination a whole number of characteristics:

- Plot of structure of the energy bands.
- Calculation of the X-ray spectra.
- Determination of the total and partial density of states.
- Determination of the electron density atoms in molecules.
- Optimization of the crystal lattice parameters.
- Calculation of energy of the crystal lattice and bulk modulus.
- Calculation of the interatomic forces.

- Computation of the magnetic dipolar anisotropy.
- Generation of wave functions and plotting on spatial grids.
- Calculation of phonon spectrum.

# 9
# First-Principle Simulation in Materials Science

> It is known that the human knowledge deserves the name of
> Science depending on the role played there by the number
>
> *Émile Borel*

We already know that the first-principle models involve no adjustable or experimentally-derived parameters. The simulation technique allows one to solve many actual problems in materials science, solid-state chemistry and metallurgy. Density functional theory was employed successfully in recent years because theorists performed total-energy calculations using the exchange-correlation potentials and showed that they reproduced a variety of ground-state properties only deviating a few percent from experimental data. Thus, the acceptance of local approximations to density functional theory has only emerged after many successful applications to many types of materials and systems.

The efficiency of these methods may be demonstrated by the following convincing examples.

## 9.1
## Strength Characteristics of Solids

The data calculated from first-principles and their comparison with experimental results for metals, semiconductors, oxides, and intermetallic compounds are presented in Table 9.1.

The results for pure metals are shown in eight rows of the Table 9.1. Potassium is a simple metal, aluminum and copper are described by the model of near-free electrons, molybdenum and tungsten are transition metals, and silver and gold are the noble metals.

The authors of [20] used the method of augmented spherical waves (ASW) for calculation of bulk moduli and cohesive energies of metals. Figure 8.7 (lower part) shows the augmented spherical waves that are applied for simulating of interatomic electron waves.

*Interatomic Bonding in Solids: Fundamentals,Simulation,andApplications,* First Edition. Valim Levitin.

**Table 9.1** Comparison of calculated from first-principles and experimental bulk moduli and cohesive energies of different materials. The heats of formation are presented for oxides in eV/atom. For intermetallic compounds the melting temperatures are given.

| Material | *B* (GPa) | | Cohesive energy (eV/atom) | | Ref. |
|---|---|---|---|---|---|
| | Calculated | Experimental | Calculated | Experimental | |
| K | 4 | 5 | −0.94 | −0.94 | [20] |
| Al | 87 | 88 | −4.01 | −3.39 | |
| Cu | 129 | 142 | −4.05 | −3.48 | |
| Mo | 268 | 271 | −6.72 | −6.82 | |
| W | 297 | 332 | −8.93 | −8.90 | [21] |
| Cu | 126 | 137 | −3.30 | −3.49 | [22] |
| Ag | 80 | 101 | −2.37 | −2.95 | |
| Au | 130 | 173 | −2.99 | −3.81 | |
| Si | 98 | 99 | −4.84 | −4.63 | [8] |
| Ge | 73 | 77 | −4.26 | −3.85 | |
| MgO | 163 | 160 | −10.5 | −10.3 | [23] |
| CaO | 111 | 111 | −11.8 | −11.0 | |
| SrO | 88 | 89 | −10.9 | −10.4 | |
| | | | **Melting temperature (K)** | | [24] |
| SbY | 67 | 66 | 1590 ± 310 | 2580 | |
| CoAl | 157 | 162 | 2070 ± 310 | 1700 | |
| NbIr | 320 | 301 | 2830 ± 390 | 2200 | |

The success of simulation is evident. The first-principle approach enables one to predict with a high accuracy the strength of interatomic bonding for metals with different electronic structures. The cohesive energy of metals varies from −0.94 eV/atom (0.905 kJ/mol) for potassium to −6.82 eV/atom (656.76 kJ/mol) for molybdenum and −8.90 eV/atom (857.07 kJ/mol) for tungsten.

A discrepancy between calculated and experimental values of 21–24% is obtained for the noble metals silver and gold.

The theory requires from a researcher to input only the composition of a solid and the supposed crystal structure. No fitting parameters are necessary. However, we should realize that the researcher must also choose preliminarily and validate an exchange-correlation functional, number of *k*-points, the cutoff parameter before he or she can expect to obtain a reasonable result. The convergence of the basis set solutions to the Kohn–Sham equations is controlled by a cutoff parameter $R_{ml} K_{\max} = 6 - 9$, where $R_{ml}$ is the smallest atomic sphere radius in the unit cell and $K_{\max}$ is a magnitude of the largest reciprocal lattice vector.

Calculated data for semiconductors silicon and germanium agree with the experimental data well. Discrepancy between theoretical and experimental data does not exceed 5% for Si and 10% for Ge.

**Table 9.2** Data of the first-principle simulation. Calculated and experimental lattice constants $a$ (a.u.) and bulk moduli $B$ (GPa) for metals are given. $n_a$ is a number of different investigations, in which the lattice constants were calculated; $n_B$ is a number of investigations, in which the bulk moduli were calculated.

| | Lattice | $n_a$ | $a_{calc}$ | $a_{exp}$ | $n_B$ | $B_{calc}$ | $B_{exp}$ |
|---|---|---|---|---|---|---|---|
| V | bcc | 6 | $5.59^{\pm 0.09}$ | 5.73 | 6 | $175^{\pm 37}$ | 162 |
| Cr | bcc | 4 | $5.34^{\pm 0.09}$ | 5.44 | 4 | $252^{\pm 22}$ | 190 |
| Ni | fcc | 4 | $6.57^{\pm 0.10}$ | 6.65 | 4 | $252^{\pm 22}$ | 186 |
| Nb | bcc | 6 | $6.22^{\pm 0.12}$ | 6.24 | 6 | $170^{\pm 11}$ | 170 |
| Mo | bcc | 6 | $5.95^{\pm 0.10}$ | 5.95 | 6 | $263^{\pm 19}$ | 273 |
| Pd | fcc | 6 | $7.38^{\pm 0.11}$ | 7.42 | 6 | $204^{\pm 27}$ | 181 |
| Ta | bcc | 5 | $6.20^{\pm 0.13}$ | 6.24 | 5 | $189^{\pm 11}$ | 200 |
| W | bcc | 5 | $5.99^{\pm 0.12}$ | 5.97 | 5 | $296^{\pm 25}$ | 323 |
| Pt | fcc | 5 | $7.48^{\pm 0.12}$ | 7.41 | 5 | $279^{\pm 32}$ | 278 |

The calculated binding energies in oxides (rows from 11 to 13 of Table 9.1) also agree well with experimental data. The first-principle simulation correctly predicts the greatest cohesive energy for CaO and similar values for MgO and SrO. Precise measurements of the bulk moduli for oxides is difficult. The reported values are in agreement with elastic constants measured by ultrasonic techniques.

The authors of [23] demonstrated the power of the linear augmented plane wave method (LAPW) for intermetallic compounds. They calculated lattice constants, electronic structure, and elastic moduli in SbY (the NaCl type structure), CoAl (the CsCl structure), and NbIr (the CuAu structure). The predicted bulk moduli are within 7% of experimental values (14–16th rows of Table 9.1).

The calculation, however, fails to predict the melting temperatures of these intermetallic compounds.

Table 9.2 presents the comparison of the data obtained by different authors in order to illustrate the simulation accuracy of lattice constants and bulk moduli of metals. The local density approximation and the generalized gradient approximations were used for every metal.[1)]

The results of calculations of the crystal lattice parameter $a$ are in a close fit with experiment. The coefficient of variation of the calculated lattice constants equals to 1.5–2.1%. All experimental values (except for vanadium) are within $a \pm \Delta a$ values computed by different investigators.

The variation coefficient of calculated bulk moduli $B$ for chromium, nickel, niobium, molybdenum, tantalum, and tungsten ranges from 6.0 to 8.7%; it equals to 21, 13 and 11% for vanadium, palladium and platinum, respectively. A minimal discrepancy with experiment takes place for niobium, molybdenum, tantalum, tungsten, and platinum. The maximal discrepancy (35%) occurs for chromium and nickel.

1) The references are given in the article [25].

**Table 9.3** Calculated and experimental values for energy of vacancy formation (eV/atom).

| Metal | V | Cr | Ni | Cu | Nb | Mo | Pd | Ag | Ta | W | Pt | Au |
|---|---|---|---|---|---|---|---|---|---|---|---|---|
| Calc. | 3.06 | 2.86 | 1.77 | 1.33 | 2.92 | 3.13 | 1.65 | 1.24 | 3.49 | 3.27 | 1.45 | 0.82 |
| Exper. | 2.20 | 2.27 | 1.79 | 1.28 | 2.85 | 3.00 | 1.70 | 1.11 | 3.10 | 3.95 | 1.35 | 0.93 |

In general, one may estimate as satisfactory the fit of strength data based on the first-principles simulation with experimental results.

## 9.2 Energy of Vacancy Formation

Vacancies in a crystal lattice act a significant part in the thermodynamic and kinetic behavior of solids. The energy of vacancy formation $E_v$ is a most important quantity, which determines the equilibrium vacancy concentration and contributes to processes of diffusion and strain. The activation energy of self-diffusion $Q_d$ is the sum of the energy of vacancy formation $E_v$ and the energy of vacancy diffusion $U_v$.

The first-principle calculation of point defects is much more difficult than the calculation of perfect crystals because the model looses the translational symmetry.

The self-consistent electronic structure calculations for the energy of vacancy formation are based on the local density approximation, equation (8.17). Authors of [25] used the supercells with 27 and 32 lattice sites for bcc and fcc metals, respectively. The atoms neighboring the vacancies are not allowed to relax from their perfect lattice position. It was found that the errors due to omission of the lattice relaxation around the vacancy were only of the order of one-tenth of an eV. The total energy of a supercell $E_{tot}$ depends on the number of atoms in the supercell $N$, the number of vacancies $v$ in a volume $\Omega$, $E_{tot} = E(N, v, \Omega)$. In the supercell approximation, the vacancy-formation energy is given by

$$E_v = E(N-1, 1, \Omega) - \frac{N-1}{N} E(N, 0, \Omega) . \tag{9.1}$$

The calculated values of the energy $E_v$ are given in Table 9.3 together with the experimental data.

The absolute values of the calculated vacancy-formation energy for fcc metals nickel, copper, and palladium are in good agreement with the experimental data. The difference between predicted and measured values is less than 4%. The theoretical vacancy-formation energies for bcc metals vanadium, chromium, tantalum show larger discrepancies, up to 25–36%. For niobium, molybdenum, and tungsten the theoretical values are closer to the experimental ones.

When considering along the rows of the Mendeleev periodic table, the calculated energies of the vacancy formation increase at the beginning and decrease at the end of the d series (see Figure 13.2). This means that the vacancy-formation energies follow the general trend known from melting temperature, cohesion, and surface energies. The parabolic trend in these properties arises from the fact that when going from left to right on transition metal series the number of electrons in the d shell increases. When adding electrons in the d states they first occupy the bonding states and after the band becomes half-full they proceed to occupy the antibonding states. One obtains the parabolic shape for the strength of d bonding, too.

## 9.3 Density of States

The electronic density of states (DOS) describes energy of electrons in solids. By definition, the density of states $\rho(E)$ is a value, which being multiplied by a small interval of energies $dE$ equals to a number of the electronic states with energies in an interval from $E$ to $E + dE$. In solids, the density distributions are not discrete like a spectral density but continuous. A high density of states at a specific energy level means that there are many states available for occupation. A density of states of zero means that no states can be occupied at that energy level. Local variations, most often due to distortions of the system, are often called local density of states (LDOS).

Calculations based on density functional theory and the plane waves approximation allow one to determine the energy distribution of electrons. Figure 9.1 shows the density of states for silver and platinum. One can see that the electron density is non-zero at the Fermi energy for both elements. An electric field applied to the material will accelerate electrons to higher energies than when there is no field. As a result both elements are conductors. Silver has a greater interval of states concentrated above the Fermi level than platinum. This seems to be a cause of difference in resistivity between silver and platinum. The resistivity of silver ($1.6 \times 10^{-8}\,\Omega$ m) is by the order of magnitude less than that of platinum ($10.7 \times 10^{-8}\,\Omega$ m).

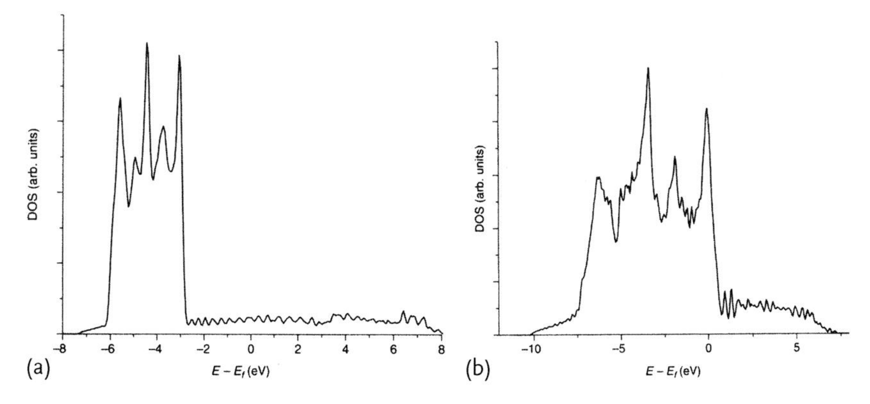


**Figure 9.1** Calculated electronic density of states for silver (a) and platinum (b). A difference between the energy of electrons and the Fermi energy $E_F$ is plotted onto abscissa axis; zero is the Fermi energy, above which excited states are located (after [26]).

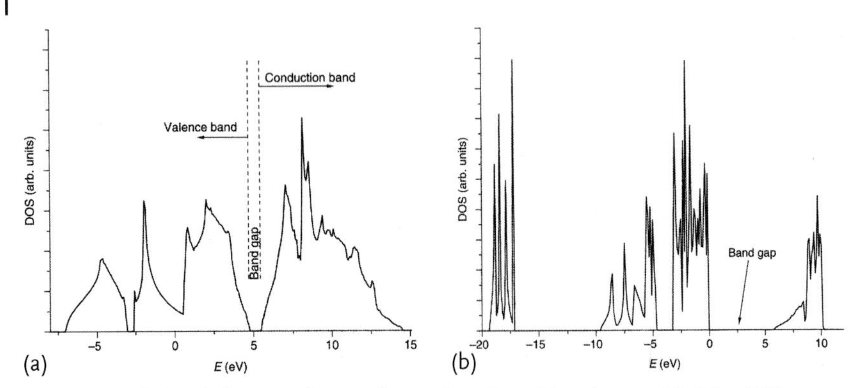


**Figure 9.2** Calculated electronic density of states for silicon (a) and quartz (b) (after [26]).

The densities of states for silicon and quartz are presented in Figure 9.2. The region of energy in Figure 9.2a can be divided into two separated regions, the valence band and the conduction band. These two regions are separated by a region that contains no electron states at all. This is a forbidden band or band gap. Silicon is semiconductor because its band gap is relatively narrow, 1.1 eV. As can be seen from Figure 9.2b a band gap of quartz equals to 6 eV. Quartz is an insulator because electrons cannot surmount the gap and go over the conduction band. Unlike silicon, the valence band of quartz includes several separate energy bands with distinct energy gaps between them.

Figure 9.3 illustrates that theoretical calculations of the copper band structure are in a good agreement with experiment. Although these results were obtained as early as 1960s, more recent calculations produce similar results. The lower parabola bounding the wide energy band corresponds to s electrons. The bands shown in Figure 9.3 consist of five narrow d bands that correspond essentially to a filled d shell.

## 9.4 Properties of Intermetallic Compounds

Compounds of aluminum with transition elements attracts wide attention due to the need of high-performance structural materials in the aerospace, aircraft and automobile industries. The most promising candidates are compounds of aluminum with scandium, titanium, vanadium, and chromium. These transition metals have the outer electron shells $3d^1 4s^2$, $3d^2 4s^2$, $3d^3 4s^2$, and $3d^5 4s^1$, respectively. Aluminum has the $3s^2 3p^1$ outer electron shell. The interatomic bonding in structures of trialuminides results from the overlap d-orbitals of transition metals with the hybridized sp orbital of aluminum.

The $Al_3(Sc, Ti, V, Cr)$ series of trialuminides crystallize in the cubic $L1_2$ (Figure 10.2) and tetragonal $DO_{22}$ structure. The tetragonal elementary cell of $DO_{22}$

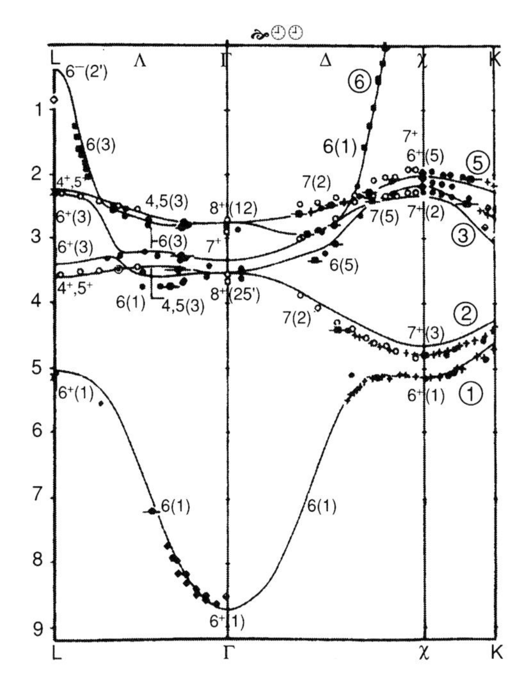

**Figure 9.3** Comparison of the calculated energy bands of copper (curves) with experimental data (points) measured by angle-resolved photoemission (after [1]).

structure can be considered as consisting of two $L1_2$ cubes stacked along $Oz$ direction with a [1/2 1/2 0] shift between these cubes.

The intermetallic compounds $Al_3$(Sc, Ti, V, Cr) have been investigated using the first-principle calculations in both the $L1_2$ and $DO_{22}$ structures [27]. The strong directional metallic-covalent bond exists in aluminum-rich compounds. In order to achieve good mechanical properties, a certain degree of covalency is desired, since it increases bonding strength and consequently leads to higher mechanical strength and hardness. However, it also leads to undesired high brittleness. The series $Al_3$(Sc, Ti, V, Cr) allows one to study the gradual formation of covalent bonds with increasing d-band filling leading to substantial changes in the mechanical properties.

Each atom of the transition metal in $L1_2$ structure is surrounded by 12 equivalent aluminum-nearest neighbors forming a regular cuboctahedron. Such a configuration leads to the formation of $sd^5$ hybrids promoting the metal–aluminum bonding. The main bonding character comes from the saturation of dominant hybrid orbitals located on the transition metal atoms. The bonding is formed by $d^3$ hybrids for $L1_2$ structure and $d^4$ hybrids for the $DO_{22}$ structure. In the series $Al_3Sc$–$Al_3Ti$–$Al_3V$–$Al_3Cr$ the bonding configuration becomes gradually saturated with electrons. The formation of strong directional interatomic bondings leads to an increase in computed elastic constants. The bulk modulus for the cubic structures increases in the same series as 87–107–118–126 GPa. Charge is concentrated in the tetrahe-

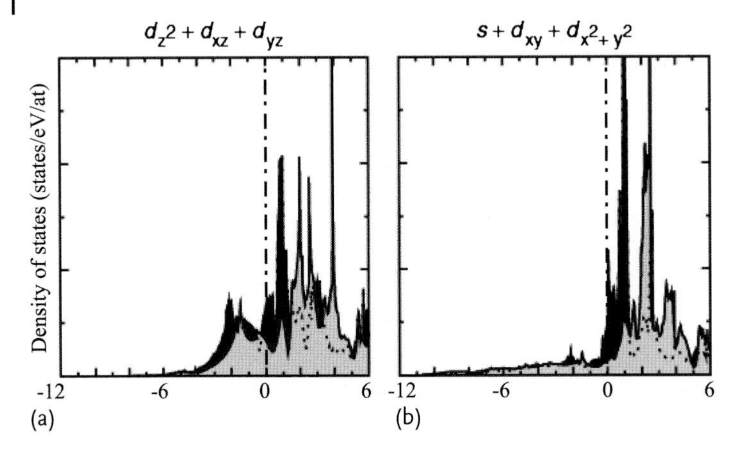


**Figure 9.4** Partial density of states of $Al_3Sc$ (gray) and $Al_3Ti$ (black) in the cubic $L1_2$ structure. The $d^3$ hybrid orbitals are concentrated on the transition metal (a), the $sd^3$ hybrid orbitals are concentrated on aluminum atom (b). The atomic orbitals that are a source of hybridization are shown above the graphs (after [27]).

dral holes along the ⟨111⟩ directions and leads to the formation of covalent bonds between close-packed {111} planes which are strongest for the $Al_3V$ compound.

Figure 9.4 illustrates the hybridization of electrons by formation of trialuminides. The area between *X*-axis and the curves in Figure 9.4a (LDOS for Sc and Ti atoms) is larger than the analogous areas (LDOS for Al atom) in Figure 9.4b. The $d^3$ hybrids are stored up at atoms of transition metals scandium and titanium. The $sd^2$ hybrids are stored up at aluminum atom.

The authors of [27] underline: "Our present study of the elastic anisotropy represent only a first step towards an understanding of the origin of the brittleness of most TM trialuminides." I would like to add that the ultimate judgment may only be done by comparing with the experiment.

## 9.5
## Structure, Electron Bands, and Superconductivity of $MgB_2$

Superconductivity is a phenomenon of exactly zero electrical resistance occurring in certain materials when cooled below some critical temperatures $T_c$. Superconductivity attracts attention of physicists and engineers because of prospects of its applications in industry. The matter is of research applications, defense and space, instruments for science and medicine, machine tools, materials processing, transportation, and energy.

In a normal conductor, an electric current may be visualized as a fluid of electrons moving through a lattice ready-built of heavy ions. The electrons constantly collide with the ions in the lattice, and during each collision some part of the energy carried by the current is absorbed by the lattice. The energy of electrical current is converted into heat, which is essentially the vibrational kinetic energy of the crystal lattice ions. As a result, the energy carried by the current is constantly being dissipated. This is the phenomenon of electrical resistance.

The situation is different in a superconductor. In a conventional superconductor, the electric current cannot be resolved into individual electrons. Instead, it consists of bound pairs of electrons known as Cooper pairs. The Cooper pairs are named for physicist Leon N. Cooper who, with John Bardeen and John Robert Schrieffer, formulated the first successful model explaining superconductivity in conventional superconductors. A key conceptual element in this theory is the pairing of electrons close to the Fermi level into pairs through interaction with the crystal lattice. This pairing results from a slight attraction between the electrons related to lattice vibrations; the coupling to the lattice is a phonon interaction.

The electron–phonon interaction is the capital ingredient in superconductivity. Phonons mediate the coupling of two electrons at a very low temperature, although the electrons have minus charge repelling each other. The Cooper pair and phonon move across the crystal lattice without a resistance.

The electron pairs have a slightly lower energy and leave an energy gap above them of the order of $\Delta E = -0.001$ eV, which inhibits the kind of collision interactions which lead to ordinary resistivity. This pairing is caused by an attractive force between electrons arisen from the exchange of phonons. The energy spectrum of this Cooper pair fluid possesses an energy gap, meaning there is a minimum amount of energy $\Delta E$ that must be supplied in order to excite the fluid. Therefore, if a gain in energy due to the electron coupling is larger than the thermal energy of the lattice, given by $k_B T$, where $k_B$ is the Boltzmann constant and $T$ is the temperature ($|\Delta E| > k_B T$), the fluid will not be scattered by the lattice. The Cooper pair fluid is thus a superfluid, that is, it can flow without energy dissipation.

The Cooper pairs form and break up constantly, but the overall effect perpetuates itself down the line, enabling electrons to zip through the superconductor.

A relatively high superconducting transition temperature $T_c = 39$ K of magnesium diboride $MgB_2$ was discovered in 2001. Investigation of this composition is an excellent example of combination of theory and experiment to understand the mechanism of superconductivity. The electron structures of atoms are

$$^{5}\mathrm{B}\colon 1s^2 2s^2 2p^1\ ,\ ^{12}\mathrm{Mg}\colon 1s^2 2s^2 2p^6 3s^2\ ;\quad \text{compare}\quad ^{6}\mathrm{C}\colon 1s^2 2s^2 2p^2\ .$$

$MgB_2$ is a compound with the metal bond that has a layer structure (Figure 9.5a). It is useful to consider $MgB_2$ in the light of similarities with and differences from its cousin hexagonal graphite. The structure of two materials can be understood in terms of honeycomb graphite structure shown in Figure 14.14. The simple hexagonal form of graphite consists of these planes stacked with hexagons over one another in the three-dimensional simple hexagonal structure. This is also the structure of $MgB_2$ which is illustrated in Figure 9.5a. The boron atoms form graphite-like planes in the hexagonal lattice and the magnesium atoms occupy sites in the centers of the hexagon between the layers. Since each atom of magnesium provides two valence electrons, the total electron valence count per cell is the same for graphite and $MgB_2$. Thus we can expect the band structures to be closely related and the bands near the Fermi level to be similar.

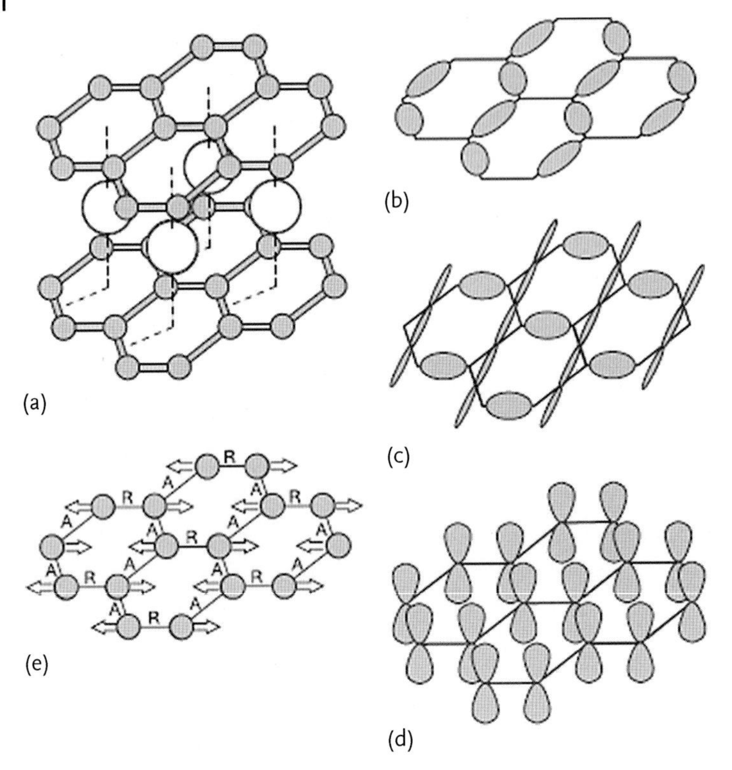


**Figure 9.5** Crystal structure of magnesium diboride, electronic states and a vibrational mode of boron atoms. (a) Crystal structure of $MgB_2$, hexagonal honeycomb layers of boron atoms alternate with layers of magnesium atoms, centered on the hexagons. (b), (c) Strong $\sigma$-bonds lie in the plane. Not all $\sigma$-bonds in the boron layers of $MgB_2$ are occupied. (d) Weak $\pi$-bonds extend above and below the boron plane. (e) A vibrational mode of boron atom that couples strongly to $\sigma$-bonding electronic states near the Fermi level. As boron atoms move in the arrow direction, shortened bonds, marked with A become attractive to electrons, whereas elongated bonds, marked with R, become repulsive. The $\sigma$-bonding states (b), (c) couple strongly to vibrational mode because they are mainly located in either the attractive or repulsive bonding of the mode. The $\pi$-bonding states (d) do not couple strongly to this mode (after [28]).

In the hexagonal planes of graphite, each carbon atom, which has four valence electrons, is bonded to three others, occupying all available planar bonding states forming the $\sigma$-bondings; its remaining fourth electron moves in orbitals above and below the plane, forming $\pi$-bondings.

Magnesium diboride $MgB_2$, like graphite, has strong $\sigma$-bondings in the planes and weak $\pi$-bondings between them, but since boron atoms have fewer electrons than the hybridized carbon atoms, not all the $\sigma$-bondings in the boron planes are occupied. And because not all the $\sigma$-bondings are filled, lattice vibration in the boron planes has a much stronger effect on electronic states near Fermi level resulting in the formation of strong electron pairs confined to the planes.

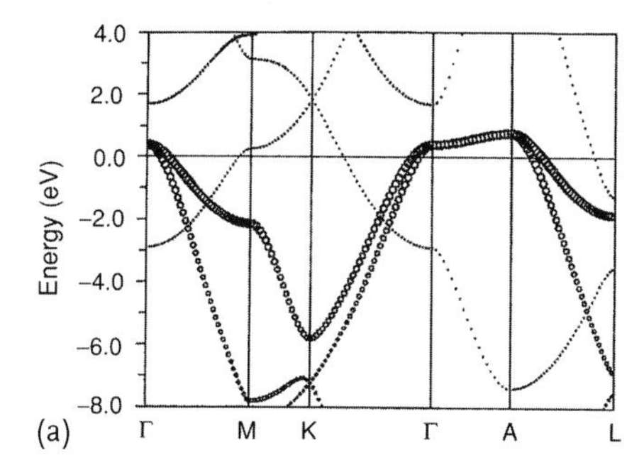


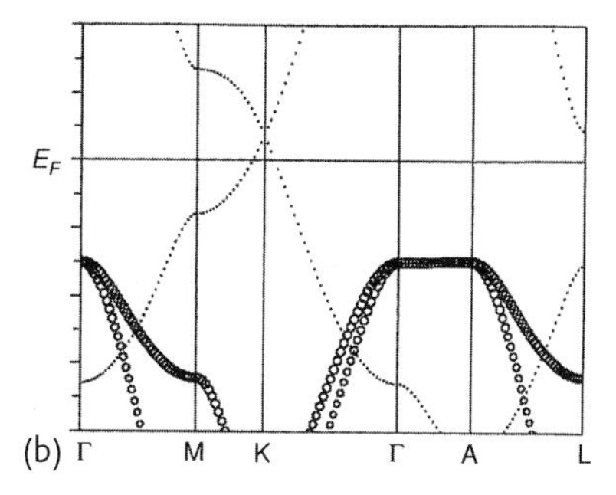


**Figure 9.6** Band structure along main hexagonal symmetry lines for $MgB_2$ (a) and hexagonal graphite (b). Zero of energy is the Fermi level. The planar $\sigma$-bonding states, highlighted with larger symbols, are higher in energy in $MgB_2$ so that they are partially unoccupied (after [29]).

Partially occupied $\sigma$-bondings driving superconductivity in a layered structure is one of the new concepts that appeared from the theoretical studies; generally speaking, nature does not like unoccupied $\sigma$-bonding states. Not all the boron electrons are needed in strong pair formation to achieve high $T_c$. In addition to the strongly bonded $\sigma$-pairs, the boron electrons involved in $\pi$-bondings form much weaker pairs.

An important feature is that the electronic states dominated by orbitals in the boron plane couple strongly to specific phonon modes, making pair formations favorable. This explains the high transition temperature. The analysis of the authors [28] suggests comparable or higher transition temperatures may result in layered materials based on boron, carbon, and nitrogen with partially filled planar orbitals.

The bands have been calculated by many groups of researchers with similar results. Figure 9.6 shows the comparison of the band structures of $MgB_2$ and graphite. The linearized augmented plane waves method was used for calculations. The symmetry point notations are those for the simple hexagonal Brillouin zone, presented in Figure 4.15.

For each, two sets of bands are identifiable: the highlighted $\sigma(sp_xp_y)$ states and $\pi(p_z)$ states. The $\sigma$ bonding states are completely filled in graphite and provide the strong covalent bonding. They are unfilled in $MgB_2$ and hence metallic, with a concentration of 0.067 holes/boron atom. There are correspondingly more electron carriers in $\pi$ bands. The relative shift of the $\sigma$ and $\pi$ bands between graphite and $MgB_2$ is identified by $\approx$ 3.5 eV.

In graphite, the $\sigma$-bonding states are shifted well below the Fermi energy, whereas in $MgB_2$ the bonding is weaker so that the $\sigma$-bands cross the Fermi energy. In addition, there is greater dispersion perpendicular to the layers (that is, $\Gamma \rightarrow$ A), especially for the $\pi$-bands, which are rather three-dimensional in nature. The authors of [29] found that the boron bonding stretching modes dominated in the electron–phonon coupling.

The shading of points indicates the degree of $\sigma$-bonding character of the states, that is, the strong in-plane bonding states. The graphite bands are only slightly

**Table 9.4** Work of separation of grains $W_{sep}$ and energy of segregation $E_{seg}$ in copper. ML is the monolayer of the segregant (after [30]).

| Parameter | Pure Cu | Cu + 0.5ML Bi | Cu + 1ML Bi | Cu + 0.5ML Pb | Cu + 1ML Pb |
|---|---|---|---|---|---|
| $E_{seg}$ (eV) | – | −1.32 | −1.58 | −1.40 | −1.47 |
| $W_{sep}$ (J m$^{-2}$) | 3.10 | 2.07 | 1.33 | 2.21 | 1.59 |

modified from those of a single plane of graphite, which has the Fermi energy exactly at the K points where the $\pi$-bands touch to give a zero gap and a Fermi surface that is a set of points in two dimensions.

## 9.6
## Embrittlement of Metals by Trace Impurities

It has been known for over 100 years that adding small amounts of certain impurities to polycrystal copper can change the metal from being ductile to a material that will fracture in a brittle way (that is, without plastic deformation before the fracture). This occurs, for example, when bismuth is present in copper at concentrations below 100 ppm. Similar effects were observed with lead and mercury impurities. Qualitatively, when the impurities cause brittle fracture, the fracture tends to occur at grain boundaries, so something about the impurities changes the properties of grain boundaries in a dramatic way. That this can happen at very low concentrations of bismuth is not completely implausible because bismuth is almost completely insoluble in bulk copper. This means that it is very favorable for bismuth atoms to segregate to grain boundaries rather than to exist inside grains, meaning that the local concentration of bismuth at grain boundaries can be much higher than the net concentration in the material as a whole.

Embrittlement by the segregation of impurity elements to grain boundaries leads to failure by fast fracture. A question that has been debated over hundred years is: How can minute traces of bismuth cause this ductile metal to fail in a brittle manner? Three hypotheses for this embrittlement exist.

Two of them assign an electronic effect to either a strengthening or weakening of the interatomic bonding, the third one postulates a simple atomic-size effect. Distinguishing between these proposed mechanisms would be very difficult using direct experiments.

The authors of [30] have performed the first-principles calculations that allowed them to support a size effect explicitly and to reject the electronic hypotheses.

The work of fracture may be characterized by a critical energy-release rate for cleavage $G_c^{cleav}$. A crack propagates in the absence of a plastic strain if the equality $G_c^{cleav} = 2\gamma_s$ holds, where $\gamma_s$ is the surface energy. If a nucleation and motion of dislocations occur at the tip of the crack the stress concentration relives. A critical

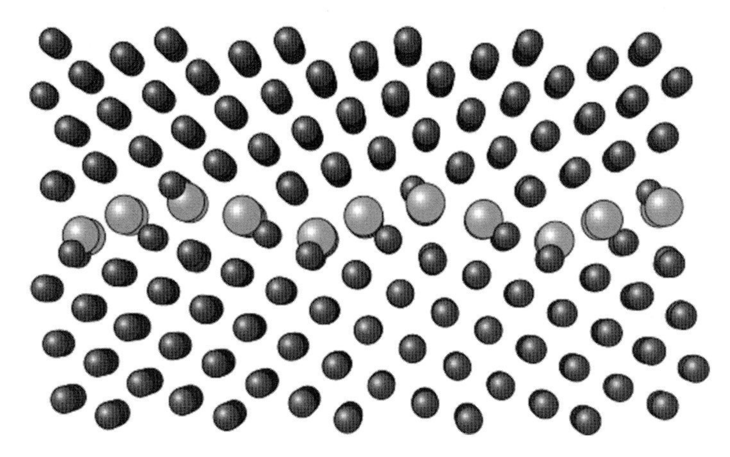

**Figure 9.7** Structure of a one-monolayer bismuth-segregated grain boundary in copper. Small spheres are copper atoms, big spheres are bismuth atoms (after [30]).

energy-release rate $G_c^{disl} = 2\gamma_s$ may be defined at the elastic crack-tip opening. A ductile to brittle transaction occurs on condition $G_c^{cleav} < G_c^{disl}$.

The local density approximation to density functional theory allowed one to find out that for a typical grain boundary in pure copper $G_c^{cleav} = 3.10\,J\,m^{-2}$. Upon segregation of one monolayer of bismuth to the same grain boundary $G_c^{cleav}$ decreases to $1.33\,J\,m^{-2}$. As a result occurs the ductile-to-brittle transition.

The segregant must satisfy two conditions for the grain boundary embrittlement in copper: it must be virtually insoluble, and its atoms have much larger radius than that of copper atoms. The segregant then segregates to the boundary, where it literally pushes apart the copper atoms across the interface and weakens the interatomic bonding. Lead and bismuth have been found to be practically indistinguishable in making copper brittle. This is contrary to the expectations of an electronic picture of embrittlement.

Table 9.4 illustrates the effect of impurities on the intercrystalline strength and the segregation energy of copper. The $E_{seg}$ value depends on the concentration of a segregant on the grain boundaries. According to calculations, the maximum boundary concentration is 22 atom $nm^{-2}$, equivalent to 1.2 monolayer on a copper(111) plane. The resultant loss of the intercrystalline strength is evident from $W_{sep}$ values (Table 9.4) and is a consequence of an excess of volume at the grain boundaries.

Figure 9.7 shows a monolayer of the bismuth atoms segregated at grain boundaries in copper.

Crucially, the density functional theory allows one directly to examine the structure and composition of the grain boundaries.

# 10
# *Ab initio* Simulation of the $Ni_3Al$-based Solid Solutions

*Ab initio* or first-principle methods are used to derive macroscopic observable features under the controlled condition of a "computational experiment" and rooted in the quantum description of interacting nuclei and electrons. Density functional theory has become the method of choice for most applications, due to its combination of reasonable scaling with system size and good accuracy in reproducing most properties of solids.

In this chapter we consider the interatomic bonding in an technologically important intermetallic compound. For our study, we use the experimental methods as well as the computer simulation technique. I think that this consideration is an example of examination of the interatomic bonding on four levels: electronic, atomic, microscopic, and macroscopic.

Simulation of intermetallic phases has been performed together with Dr. Oleg Rubel. The author would like to express the deep gratitude to Dr. O. Rubel for his help concerning the computer simulation.

## 10.1
## Phases in Superalloys

Development of engineering industry leads to the increase in requirements for the high-temperature strength of materials. The continual need for better fuel efficiency has resulted in faster-spinning, hotter-running gas turbine engines. This has created a need for alloys that can withstand higher temperatures and stresses in the hot zones of modern gas turbines. The development has led, during the past decades, to a steady increase in the turbine entry temperatures (5 K per year averaged over the past 20 years) and this trend is expected to continue.

Such alloys – superalloys – have been developed. The largest applications of superalloys are in aircraft and industrial gas turbines, rocket engines, space vehicles, submarines, nuclear reactors, and landing apparatus. The nickel-based alloys possess many attractive properties for structural applications, such as high-temperature creep and excellent oxidation resistance. The superalloys are used in a temperature interval approximately from 1023 to 1273 K.

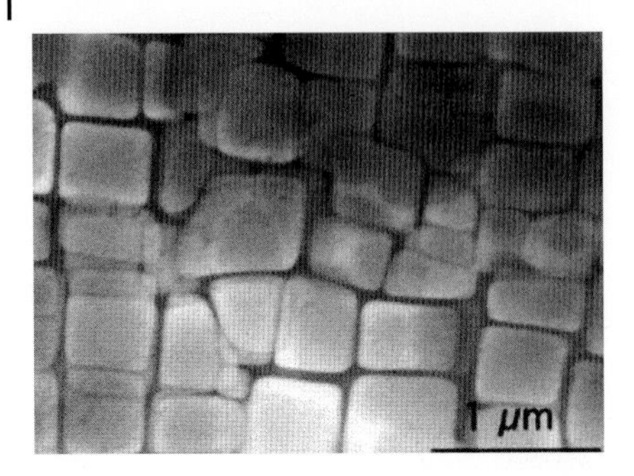


**Figure 10.1** The electron microphotograph showing the microstructure of a nickel-based superalloy containing (at.%) 17.1 Cr + 9.7 Al + 1.7 Ti + 2.3 W + 6.3 Co. The cuboidal particles of $\gamma'$ phase in $\gamma$ matrix are readily seen.

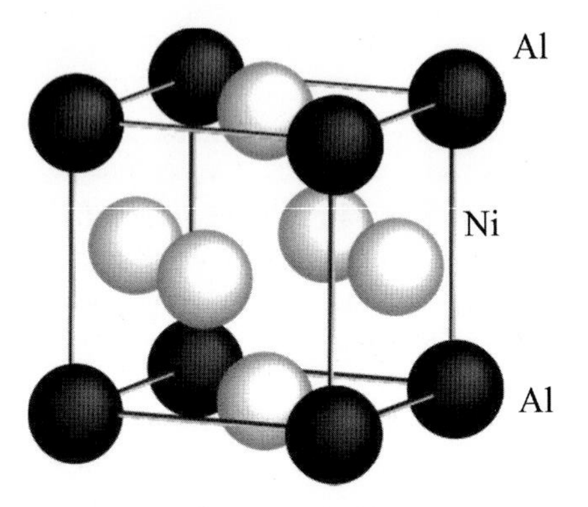


**Figure 10.2** An elementary cell of the $Ni_3Al$ intermetallic compound in $L1_2$ crystal structure. Positions of nickel atoms are represented by white spheres, positions of aluminum atoms are represented by black spheres.

The structure of the majority of nickel-based superalloys consists of a matrix, that is of the $\gamma$-phase and particles of the hardening $\gamma'$-phase (Figure 10.1).

The $\gamma$-phase is a solid solution with a face-centered crystal lattice and randomly distributed different species of atoms. By contrast, the $\gamma'$-phase has an ordered crystalline lattice of $L1_2$ type (Figure 10.2). In pure intermetallic compound $Ni_3Al$ the atoms of aluminum are placed at the vertices of the cubic cell and form the sublattice A. Atoms of nickel are located at the centers of the faces and form the sublattice B. The $\gamma'$-phase has remarkable properties, in particular, an anomalous dependence of strength on temperature. The $\gamma'$-phase first hardens, up to about 1073 K, and then softens. The interatomic bondings Ni–Al are covalent.

## 10.2 Mean-Square Amplitudes of Atomic Vibrations in $\gamma'$-based Phases

In alloy production, the $\gamma'$-phase is a phase of a variable composition. The industrial alloys never contain the phase of $B_3A$-type in the stoichiometric relation of two chemical elements. It solutes other elements, which one adds to the superalloy composition in order to ensure desired properties.

Sublattices B and A of the $\gamma'$-phase can dissolve a considerable amount of other elements. Many of the industrial nickel-based superalloys also contain, in addition to chromium, aluminum, and titanium, molybdenum, tungsten, niobium, tantalum, and cobalt. The addition of rhenium and ruthenium are under investigation for superalloys, which must withstand high stresses at temperatures from 1273 to 1373 K.

The strength of interatomic bonding in $Ni_3Al$-based phases depends appreciably upon their composition.

In our research group, we studied the mean-square amplitudes of the atomic vibrations in $Ni_3Al$-based solid solutions. The $\gamma'$-phases were extracted electrolytically from aged industrial superalloys. An X-ray technique for the measurement of mean-square amplitudes separately for each of the two sublattices of a $\gamma'$-phase has been developed [31]. We determined the chemical compositions of the $\gamma'$-phases by chemical analysis. The measurements of intensity of X-ray irradiation diffracted by specimens were carried out at room temperature and at temperatures from 723 to 1023 K. An experimental installation for X-ray structural investigations of solids at high temperatures has been designed. We measured the intensities $I$ of the (100) and (200) reflections and compared ratios $I_{100}/I_{200}$ with calculated values. The ratios have been calculated for the distribution of all kinds of elements between the B and A sublattices. The ratio under study depends on the average factors of the X-ray scattering in the sublattices, $\overline{f_B}$ and $\overline{f_A}$, respectively. The final formulas of the phases have been established under the condition of a coincidence of the experimental and calculated values of this ratio.

The obtained data are presented in Table 10.1. The first column of the Table is the composition of alloys, the second column is the composition of strengthening phases. Nickel is the rest in composition of superalloys, lines from 2 to 4.

It is seen from the second column of Table 10.1 that titanium atoms occupy places in sublattice A replacing aluminum atoms. Iron and cobalt atoms are located preferentially in sublattice B replacing nickel atoms. Tungsten, molybdenum, and chromium atoms partition between the two sublattices, but tungsten and molybdenum mainly replace the aluminum atoms.

In pure $Ni_3Al$ (the first line in Table 10.1) the mean-square amplitudes of the aluminum atoms (sites A) appreciably exceed the amplitudes of the nickel atoms (sites B) at all tested temperatures, $\overline{u_A^2} > \overline{u_B^2}$. The replacement of a part of the aluminum atoms with other atoms decreases the amplitudes of the heat vibrations in the aluminum sublattice, so that now $\overline{u_A^2} < \overline{u_B^2}$.

**Table 10.1** Mean-square amplitudes of atom vibrations in $Ni_3Al$-based ($B_3A$) phases.

| Composition of alloy (wt%) | Phase | $\overline{\Delta u_B^2}$ (pm²) 873 | 1023 K | $\overline{\Delta u_A^2}$ (pm²) 873 | 1023 K |
|---|---|---|---|---|---|
| 86.71Ni + 13.29Al | $Ni_{2.98}$ $Al_{1.00}$ | 470 | 660 | 580 | 810 |
| 14.0Cr + 1.7Al + 2.7Ti +3.0Mo + 2.0Nb | $Ni_{2.83}$ $Cr_{0.07}$ $Fe_{0.04}$ $Mo_{0.06}$ ($Al_{0.51}$ $Ti_{0.31}$ $Cr_{0.03}$ $Nb_{0.07}$ $Mo_{0.08}$) | 380 | 550 | 230 | 330 |
| 19.8Cr + 2.1Al + 1.4Ti +9.1W + 4.5Mo | $Ni_{2.83}$ $Cr_{0.08}$ $Fe_{0.04}$ $W_{0.03}$ $Mo_{0.02}$ ($Al_{0.43}$ $Ti_{0.28}$ $Cr_{0.10}$ $W_{0.08}$ $Mo_{0.03}$) | 340 | 450 | 60 | 80 |
| 9.5Cr + 4.5Al + 5.3W +9.8Mo + 5.1Co | $Ni_{2.76}$ $Co_{0.09}$ $Fe_{0.01}$ $Cr_{0.06}$ $Mo_{0.02}$ ($Al_{0.79}$ $W_{0.08}$ $Mo_{0.11}$ $Cr_{0.03}$) | 340 | 580 | 40 | 50 |

Thus, the alloying of the intermetallic compound $Ni_3Al$ results in the decrease of mean-square amplitudes of the atom vibration. This is a direct experimental evidence of increase in strength of interatomic bonding.

The effect of adding molybdenum and tungsten to the aluminum sublattice is the largest. These two elements change the properties of the intermetallic compound appreciably. One can see from Table 10.1 that the values $\overline{u_A^2}$ at 1023 K are 10–16 times less for phases containing both these elements than for pure $Ni_3Al$.

## 10.3 Simulation of the Intermetallic Phases

The properties of the $\gamma'$-phase depend upon its composition. In turn, the mechanical properties of the $\gamma'$-phase have an significant effect on the high-temperature strength of superalloys. It is of interest to calculate the properties of the strengthening phases proceeding from the first principles.

We have performed these calculations for $Ni_3Al$-based alloyed phases of different composition. Our goal was to find parameters of unit cell, bulk modulus, and the elastic constants.

Figure 10.3 presents a model structure of the basic intermetallic phase under study. The $2 \times 2 \times 2$ supercell was used for *ab initio* simulation based on density functional theory, concept of pseudopotential, and the Kohn–Sham equations. The supercell consists of 8 unit cells of $Ni_3Al$ phase. The $3 \times 3 \times 3$ supercell is shown in Figure 10.4. The computation time is affected by the external dimensions of the model structure.

The input data for calculations only include a specified composition and a previous value of the parameter of the crystal lattice. We entered the coordinates of every of atoms (Figures 10.3, 10.4) and a conjectural value of the unit cell parameter. Besides that, we specified a step of the strain, with which the supercell volume will be changed. This step equaled to 1%, from −4 to +4%.

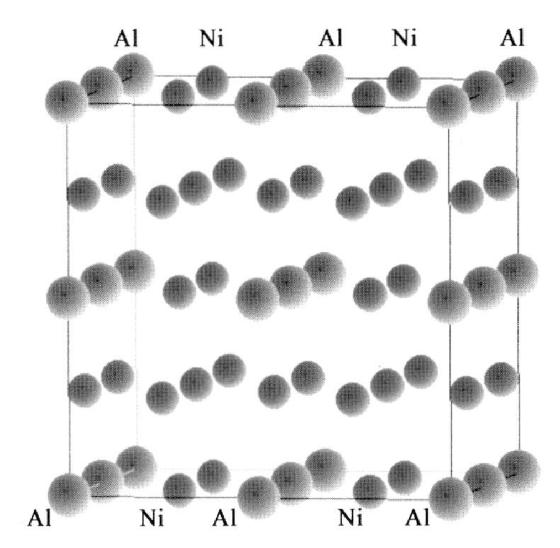


**Figure 10.3** Structure of the 2 × 2 × 2 supercell of $Ni_3Al$ intermetallic compound for *ab initio* calculations. Positions of nickel atoms are represented by small spheres, position of aluminum atoms are represented by large spheres (compare with Figure 10.2).

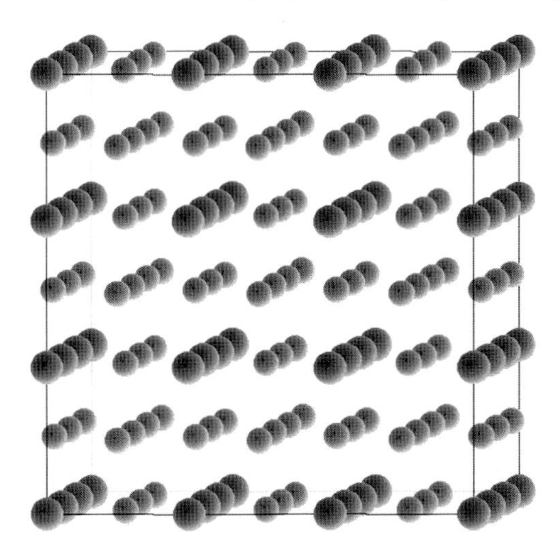

**Figure 10.4** Structure of the 3 × 3 × 3 supercell of $Ni_3Al$ intermetallic compound for *ab initio* calculations. Positions of nickel atoms are represented by small spheres, position of aluminum atoms are represented by large spheres.

The equilibrium structure must have minimum of energy as compared to others.

The dependence energy of the phase on volume was calculated using density functional theory and the linear augmented plane wave method (see Section 8.6). We made use of the Wien2k program for calculating the crystal properties [19].

Calculations have been performed and the atoms were displaced step-by-step until a minimum of energy of the structure has been reached. At first, we obtained an asymmetrical curve "energy-volume" and a corresponding minimum value of

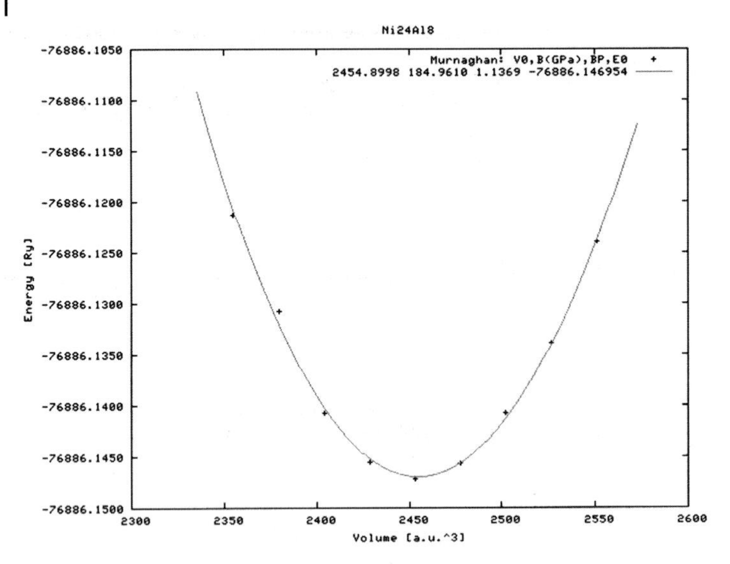


**Figure 10.5** Energy of the $Ni_3Al$ intermetallic compound as a function of the volume. The data calculated from the first principles. A minimum value of energy of −76 886.146 954 Ry is reached at an equilibrium volume of 2454.8998 $(a.u.)^3$.

energy and volume. The corrected value of the unit cell parameter was specified for the next series of calculations and all the computational processes were repeated until the curve became symmetrical. Thus, the calculations allowed one to find a structure that possessed a minimum of energy.

Figure 10.5 presents the computational result. The dependence of the energy on the volume for $Ni_3Al$ intermetallic compound is a parabola. One can see that the minimum of energy corresponds to structure of a volume of 2454.2998 $a_0^3$. Thus, the parameter of unit cell is $a = 0.3569$ nm. Experimental value of $a$ for $Ni_3Al$ phase equals to 0.3572 nm. The fit of calculated and experimental data 0.08% is surprisingly good.

Bulk modulus $B$ is related to the curvature of the $E(V)$ dependence, $B = V(d^2E/dV^2)$ (Section 7.1). Bulk modulus for pure $Ni_3Al$ is found from the curve of Figure 10.5 to be equal to $B = 184.9610$ GPa.[1)]

Alloying of the intermetallic compound $Ni_3Al$ results in a replacement of a part of Ni or Al atoms by atoms of other chemical elements. Metallic elements form the substitutional solid solutions in the intermetallic compound. We have investigated the solid solutions of Mo, Nb, W, Cr, Re, Ru, Ti, and Co. All these elements are practically used for alloying of nickel-based superalloys.

A $2 \times 2 \times 2$ supercell is shown in Figure 10.6, in which a part of aluminum atoms (37.5 at.%) is replaced by rhenium and ruthenium atoms. All the nickel atoms retain their positions. The next one, Figure 10.7, illustrates the influence of this alloying on the total energy of the intermetallic compound.

1) Our result correlates well with that of authors [27], $B = 182$ GPa.

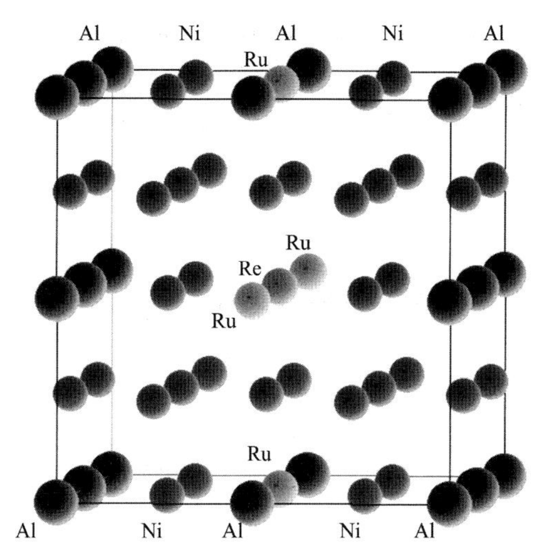


**Figure 10.6** Structure of the 2 × 2 × 2 supercell of the $Ni_3(Al_{0.625}Re_{0.125}Ru_{0.250})$ intermetallic compound. Ni atoms retain their positions in B-sublattice. Rhenium and ruthenium atoms replace a part of aluminum atoms in A-sublattice of the compound.

One can see by comparing the curves of Figures 10.5 and 10.7 that the alloyed structure becomes more rigid. The curve of energy increases from the minimum of function more rapidly in the case of intermetallic phase, in which a part of the aluminum atoms has been replaced. As a consequence of this alloying of the intermetallic phase, the bulk modulus increases from 185.0 to 209.5 MPa. The parameter of unit cell increases from 0.3569 to 0.3585 nm.

The supercell of a solid solution of tungsten, molybdenum, rhenium, and ruthenium in $Ni_3Al$ intermetallic phase is shown in Figure 10.8.

A cubic crystal possesses two shear moduli, $C_{11} - C_{12}$ and $C_{44}$. The method of the elastic moduli calculation is described in Appendix D.

Figures 10.9 and 10.10 present the graphs that were used for computation of elastic constants. We obtained linear dependences of total energy of the structures on the volume strain squared. It is seen that the angle of the line inclination is greater for the alloyed intermetallic phases than for pure $Ni_3Al$. The cause of this is greater values of elastic constants for solid solutions in $Ni_3Al$ than for pure compound $Ni_3Al$. It is possible to find all elastic constants of the phases using graphs like ones shown in the Figures 10.9 and 10.10 as the bulk modulus values of the phases are known from the curves "total energy-volume" (see Appendix D).

Table 10.2 presents the elastic properties of phases under study. Alloying of the intermetallic phase results in an improvement of properties. Elastic characteristics of the phases, which were calculated from first principles, increase.

The greatest value of bulk modulus, $B = 222.1$ GPa, has been found for a complex-alloyed phase $Ni_{2.5}Co_{0.5}(Al_{0.5}Re_{0.125}Ru_{0.125}Ta_{0.125}W_{0.125})$. Thus, bulk modulus increases by 19.5% as compared with pure $Ni_3Al$. Elastic properties of this phase are presented in the fifth line of Table 10.2. The greatest elastic moduli were

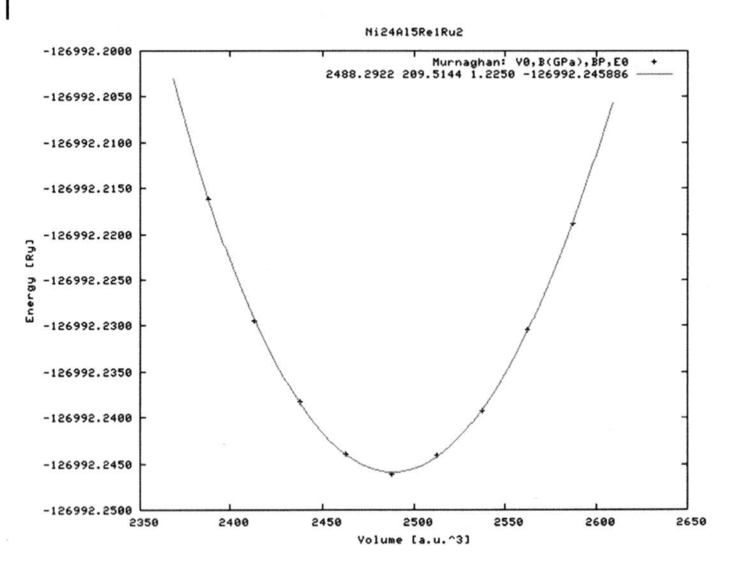


**Figure 10.7** Energy of the $Ni_3(Al_{0.625}Re_{0.125}Ru_{0.250})$ intermetallic compound as a function of volume. A minimum value of energy of −126 192.245 886 Ry is reached at the equilibrium volume of 2488.2922 $(a.u.)^3$.

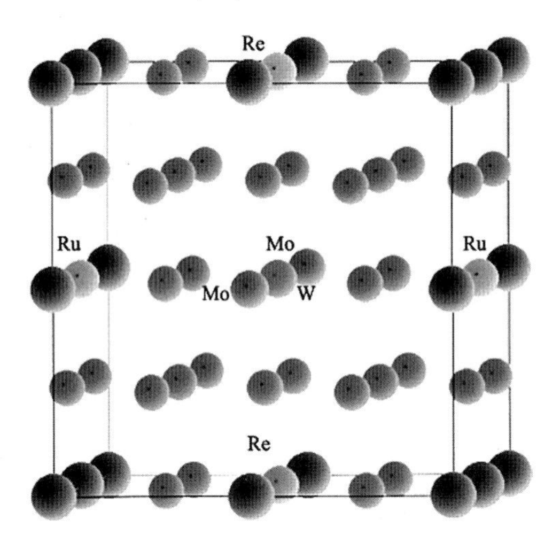


**Figure 10.8** Structure of the $Ni_3(Al_{0.5}W_{0.125}Mo_{0.125}Re_{0.125}Ru_{0.125})$ solid solution. The $2 \times 2 \times 2$ supercell. W, Mo, Re and Ru atoms replace Al atoms in the A-sublattice. Ni atoms retain their positions in the B-sublattice.

determined for this phase to be equal to 316.4, 175.0 and 160.3 GPa for $C_{11}$, $C_{12}$, and $C_{44}$, respectively. For pure $Ni_3Al$ phase the corresponding values are 243.1, 155.9, and 131.9 GPa.

Bulk modulus increases considerably for the phases $Ni_3(Al_{0.625}W_{0.125}Mo_{0.250})$, $Ni_3(Al_{0.625}W_{0.125}Mo_{0.125}Re_{0.125})$ and $Ni_{2.5}Cr_{0.5}(Al_{0.75}Re_{0.125}Ru_{0.125})$. The values of constants $C_{11}$, $C_{12}$, $C_{44}$ also increase. It is important that the anisotropy factor de-

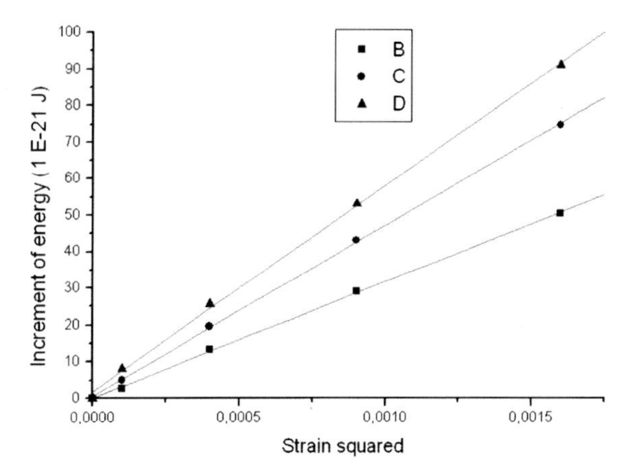


**Figure 10.9** Calculated energy as a function of the square of the orthorhombic strain used to determine the value of $C_{11} - C_{12}$ in structures $Ni_3Al$ (curve B), $Ni_3(Al_{0.625}Re_{0.125}Ru_{0.250})$ (curve C), and $Ni_3(Al_{0.625}W_{0.125}Mo_{0.250})$ (curve D). (See Appendix D for details).

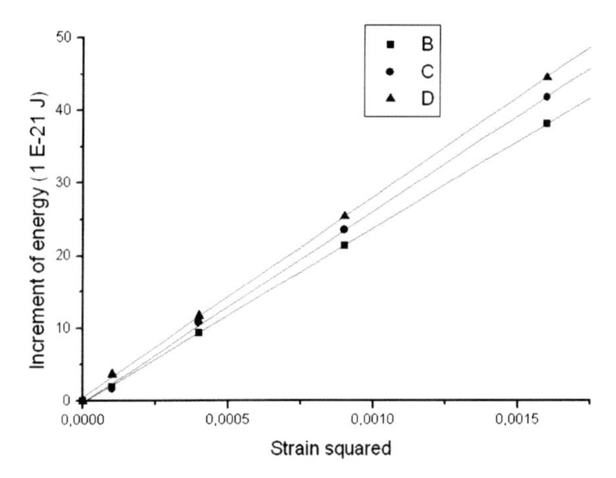


**Figure 10.10** Calculated energy as a function of the square of the monoclinic strain used to determine the value of $C_{44}$ in structures $Ni_3Al$ (curve B), $Ni_3(Al_{0.625}Re_{0.125}Ru_{0.250})$ (curve C), and $Ni_3(Al_{0.625}W_{0.125}Mo_{0.250})$ (curve D). (See Appendix D for details).

creases from 3.03 to 2.26. This fact justifies a growth of a part of metallic bond in the structures. The bonding of the metallic type is known not to be directed. On the contrary, the covalent bond is a directed bond. Metals are plastic, as a rule, whereas covalent compounds are brittle.

The least values of anisotropy factor of 1.60 and 1.74 are characteristic for phases containing chromium atoms that replace nickel atoms, $Ni_{2.5}Cr_{0.5}(Al_{0.75}Re_{0.125}Ru_{0.125})$ and $Ni_{2.5}Cr_{0.5}(Al_{0.75}W_{0.125}Mo_{0.125})$, respectively.

Compare the effect of elements 4d and 5d series on bulk modulus of the phase under study (Table 10.3). W, Ta, Re, and Os replacing Al atoms increase bulk mod-

**Table 10.2** The parameters of unit cells, bulk moduli, and elastic constants of $Ni_3Al$-based solid solutions. The data were calculated from first principles. Values of $B$, $C_{11}$, $C_{12}$, and $C_{44}$ in GPa.

| Phase | $a$ (nm) | $B$ | $C_{11}$ | $C_{12}$ | $C_{44}$ | $A$ |
|---|---|---|---|---|---|---|
| $Ni_3Al$ | 0.3569 | 185.0 | 243.1 | 155.9 | 131.9 | 3.03 |
| $Ni_3(Al_{0.625}Re_{0.125}Ru_{0.250})$ | 0.3585 | 209.5 | 294.0 | 167.3 | 143.0 | 2.26 |
| $Ni_3(Al_{0.625}W_{0.125}Mo_{0.250})$ | 0.3594 | 209.2 | 310.1 | 158.7 | 148.4 | 1.96 |
| $Ni_3(Al_{0.625}W_{0.125}Mo_{0.125}Re_{0.125})$ | 0.3592 | 210.8 | 309.6 | 161.4 | 143.3 | 1.93 |
| $Ni_{2.5}Co_{0.5}(Al_{0.5}Re_{0.125}Ru_{0.125}Ta_{0.125}W_{0.125})$ | 0.3598 | 222.1 | 316.4 | 175.0 | 160.3 | 2.27 |
| $Ni_{2.5}Cr_{0.5}(Al_{0.75}Re_{0.125}Ru_{0.125})$ | 0.3599 | 206.6 | 275.2 | 172.2 | 82.5 | 1.60 |
| $Ni_{2.5}Cr_{0.5}(Al_{0.75}W_{0.125}Mo_{0.125})$ | 0.3610 | 203.9 | 273.2 | 169.2 | 90.8 | 1.74 |
| $Ni_{2.75}Cr_{0.25}(Al_{0.875}Mo_{0.125})$ | 0.3589 | 188.9 | – | – | – | – |
| $Ni_{2.75}Cr_{0.25}(Al_{0.875}Nb_{0.125})$ | 0.3598 | 188.4 | – | – | – | – |
| $Ni_{2.75}Cr_{0.25}(Al_{0.875}W_{0.125})$ | 0.3591 | 188.3 | – | – | – | – |

**Table 10.3** The effect of elements of 4d and 5d series on the unit cell parameter and bulk modulus of $Ni_3Al$-based solid solutions. The data were calculated from first principles.

| Phase | Dissolved elements | $a$ (nm) | $B$ (GPa) |
|---|---|---|---|
| $Ni_{2.5}Co_{0.5}(Al_{0.5}Nb_{0.125}Zr_{0.125}Mo_{0.125}Tc_{0.125})$ | of 4d series | 0.3616 | 203.5 |
| $Ni_{2.5}Co_{0.5}(Al_{0.5}W_{0.125}Ta_{0.125}Re_{0.125}Os_{0.125})$ | of 5d series | 0.3599 | 225.2 |

ulus from 185 to 225 GPa. However, 4d elements Nb, Zr, Mo, Tc only increase the bulk modulus to 204 GPa.

Consequently, the purposeful variation of chemical composition enables us to improve mechanical properties of the compound under consideration.

## 10.4 Electron Density

We have computed the electron density in intermetallic phases of interest also proceeding from first principles.

Figure 10.11 presents the isolines of density of valence electrons in the (001) plane for the pure $Ni_3Al$-phase and for a solid solution in this phase.

It is clear that the replacement of the aluminum atoms by rhenium and ruthenium atoms and the nickel atoms by chromium atoms has a significant effect on the distribution of electrons. The isolines of the outer electronic shells are isolated for $Ni_3Al$; they each belong to its own atom, Figure 10.11a. For the $(Ni_{2.75}Cr_{0.25})(Al_{0.625}Re_{0.125}Ru_{0.250})$ solid solution the same isolines are shearing by the neighboring atoms. It is well seen in Figure 10.11b.

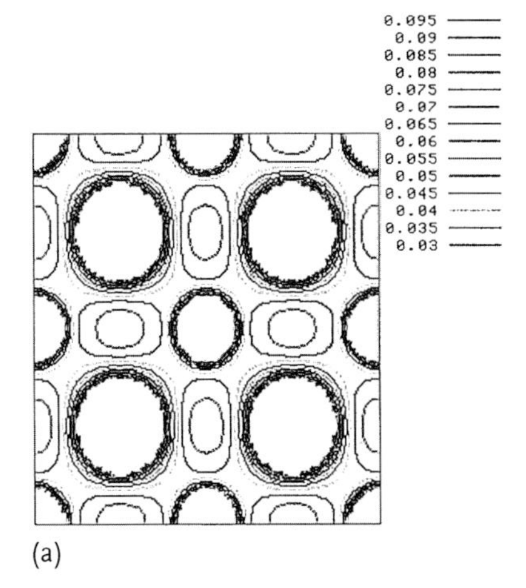


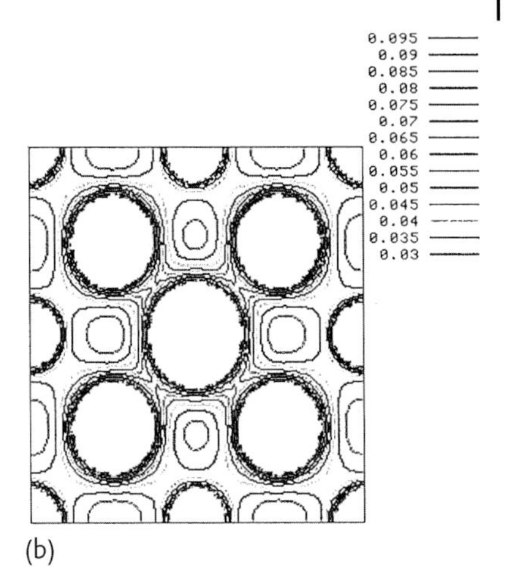


**Figure 10.11** The electron density in the (001) plane of intermetallic compounds; *Ox* [100] axis is directed to the right, *Oy* [010] axis is directed upwards. Numbers denote the concentration of electrons in electrons/(a.u.)$^3$ units; (a) pure $Ni_3Al$, isolines of electron density of the neighboring atoms are separated from each other; (b) the $(Ni_{2.75}Cr_{0.25})(Al_{0.625}Re_{0.125}Ru_{0.250})$ solid solution, isolines of electron density of 0.035, 0.040 and 0.050 electrons/(a.u.)$^3$ are shearing by the neighboring atoms.

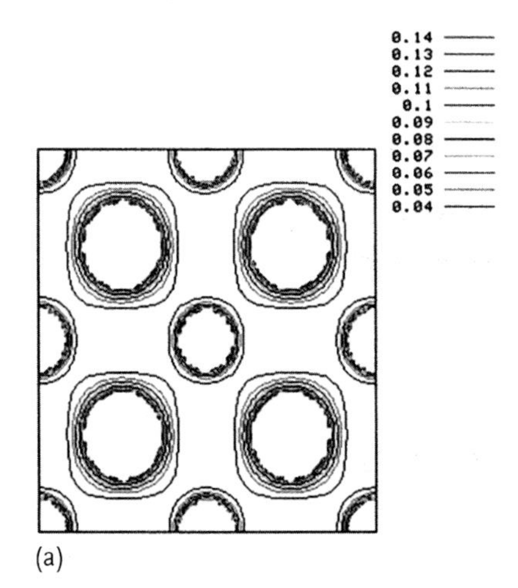


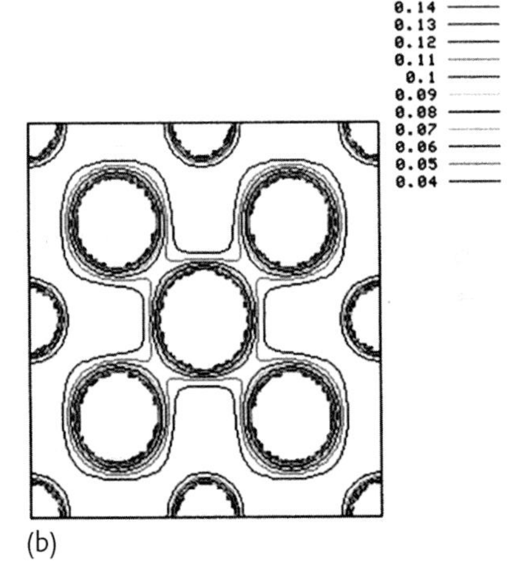


**Figure 10.12** The same as in Figure 10.11. (a) The $Ni_3Al$ intermetallic compound; (b) the $Ni_3(Al_{0.625}W_{0.125}Mo_{0.125}Re_{0.125}Ru_{0.125})$ solid solution. In the case of the solid solution the isolines of valence electrons of 0.04 and 0.05 electrons/(a.u.)$^3$ are shearing by the neighboring atoms.

Figure 10.12 illustrates the effect of replacing Al atoms in the B-sublattice of intermetallic compound $Ni_3Al$ by rhenium, ruthenium, tantalum, and tungsten atoms on distribution of the valence electrons. Comparing Figure 10.12a and b, one can see that the valence electrons resonate between atoms in the solid solution. The degree of a delocalization of the valence electrons increases.

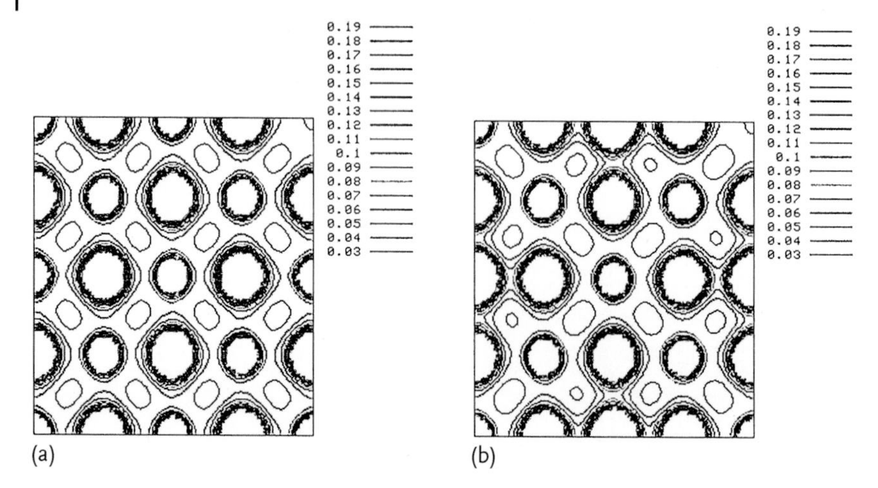


**Figure 10.13** The electron density in the (001) plane for the $Ni_3Al$ intermetallic compound (a) and for the $(Ni_{2.75}Cr_{0.25})(Al_{0.5}W_{0.125}Mo_{0.125})$ phase (b). *Ox* axis is directed along [110], *Oy* axis along [$\bar{1}$10]. The shearing of electrons by neighboring atoms for solid solution is evident.

The effect of replacing aluminum atoms in intermetallic compound $Ni_3Al$ by chromium, tungsten, molybdenum, rhenium, and ruthenium atoms on distribution of valence electrons is shown in Figure 10.13. One can see from this Figure that some isolines in solid solution are shearing by the neighboring atoms. Thus, the role of the delocalized electrons increases in the solid solutions. This means that a fraction of metallic bond in structure grows. This is the cause of the material strengthening.

Thus, aluminum atoms in the $Ni_3Al$ structure are bonded with other atoms weaker than nickel atoms. The mean-square amplitude of the former is considerably greater than the amplitude of the last. Anisotropy factor of $Ni_3Al$-phase equals to 3. This is typical for compounds with covalent bond.

The replacement of aluminum atoms in the aluminum sublattice by tungsten, molybdenum, rhenium, ruthenium atoms results in an increase in bulk modulus and elastic constants. The strength of interatomic bonding increases. The values of $\overline{u_A^2}$ amplitudes become 10–16 times less for phases containing these elements than for pure $Ni_3Al$.

Formation of solid solutions and replacement of the aluminum atoms decreases anisotropic factor because of occurrence of an additional quantity of collectivized electrons. A part of metallic bond increases. All the alloying elements are transition metals, that is, they have an unfilled d shell. These elements add electrons in the d band of the $Ni_3Al$ intermetallic compound.

# 11
# The Tight-Binding Model and Embedded-Atom Potentials

Tight-binding potentials are approaches that lie, in a sense, between *ab initio* local density functional theory and empirical isotropic interatomic pair potentials. The tight-binding method is the simplest possible way to incorporating quantum mechanical principles into the calculation of the potential. In contrast to the free-electron model it is based on using non- or weakly overlapping atomic eigenfunctions. Consequently, tight-binding solutions are based on formulating eigenfunctions as linear combinations of atomic orbitals or basis sets of plane waves. In contrast to the most of the empirical and semiempirical potentials, tight-binding potentials account for the directional nature of the bonding (particularly under the influence of the p- and d-atomic orbitals), for the consideration of the bonding and antibonding states, and for energy changes as a result of atomic displacements.

The tight-binding model is an approach to the electronic band structure from the atomic borderline case. It describes the electronic states starting from the limit of an isolated atom. It is assumed that the Fourier transform of the Bloch function can be approximated by the linear combination of atomic orbitals (LCAO). Thus, the band structure of solids is investigated starting from the Hamiltonian of an isolated atom centered at each lattice site of the crystal lattice.

This simple model gives good quantitative results for energetic bands derived from strongly localized atomic orbitals, where eigenfunctions practically vanish at half the distance to the neighboring atom.

## 11.1
## The Tight-Binding Approximation

The approach regards a solid as a collection of weakly interacting neutral atoms. The approximation is most useful for describing the energy bands that arise from the partially-filled d shells of transition metal atoms and for describing the electronic structure of insulators.

"In the tight-binding model we assume the opposite limit to that used for the nearly-free-electron approach, that is the potential is so large that the electrons spend most of their lives bound to ionic cores, only occasionally summoning the quantum-mechanical wherewithal to jump from atom to atom [4]." This forms a

*Interatomic Bonding in Solids: Fundamentals,Simulation,andApplications*, First Edition. Valim Levitin.

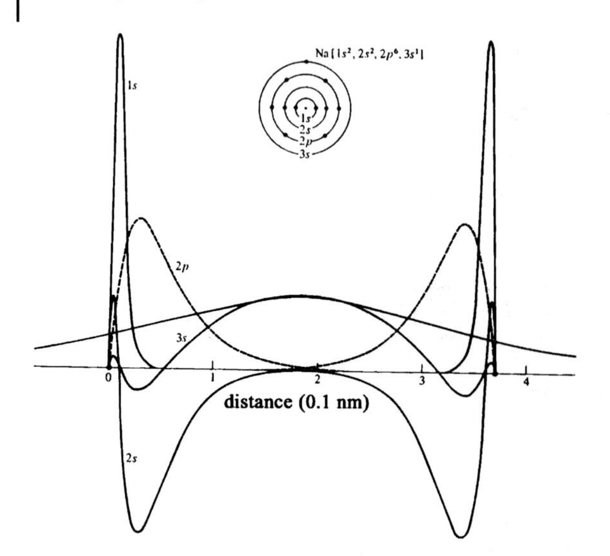


**Figure 11.1** Calculated eigenfunctions for the levels of atomic sodium plotted about two neighboring nuclei separated by a distance of 0.37 nm in bulk sodium. The solid curves are $r\psi(r)$ for the 1s, 2s, 3s levels. The dashed curve is $r$ times the radial eigenfunction for the 2p levels. Note that 1s orbitals have essentially no overlap, the 2s and 2p curves overlap only a little and 3s curves overlap extensively. Data of D.R. Hartree and W. Hartree (after [7]).

central tenet of the tight-binding approximation – the nearest neighbor – is defined as a two center approximation.

Figure 11.1 illustrates the calculated electron eigenfunctions for 1s, 2s, 2p, and 3s levels of sodium atom. It is seen from Figure 11.1 that only overlap of 3s electrons is substantial. The 3s electrons are valence electrons, and the rest are localized electrons.

In the tight-binding model, it is assumed that the full periodic Hamiltonian $H$ of the crystal may be approximated by the Hamiltonian $H_{\text{at}}$ of an isolated atom located at each lattice site. A correction to the atomic potential $\Delta V$ is required to obtain the full Hamiltonian of the system. This correction includes potential generated by all the others ions in the lattice except for the one already considered:

$$H = H_{\text{at}} + \Delta V(\boldsymbol{r}) \,. \tag{11.1}$$

The $\Delta V(\boldsymbol{r})$ value is substantial in the interatomic area where eigenfunction $\psi_n(\boldsymbol{r})$ is small and on the contrary potential correction vanishes in the lattice site. $\psi_n(\boldsymbol{r})$ is quite a good approximation to a stationary state of the crystal for all $\boldsymbol{R}$ in the Bravais lattice. So will also be for $\psi_n(\boldsymbol{r} - \boldsymbol{R})$, where $\boldsymbol{R}$ is the lattice translation vector. $n$ is the energetic band level and collectively represents a full set of atomic quantum numbers.

The eigenfunctions $\psi_n(\boldsymbol{r})$ are defined by the Schrödinger equation,

$$H_{\text{at}}\psi_n(\boldsymbol{r}) = E_n\psi_n(\boldsymbol{r}) \,, \tag{11.2}$$

where $H_{at}$ is the Hamiltonian of a single atom.

For periodic lattice of $N$ atoms the eigenfunction is a linear combination of the atomic orbitals (LCAO). It may be expressed as [3]

$$\psi_n(\boldsymbol{r}) = \frac{1}{\sqrt{N}} \sum_{\boldsymbol{R}} \exp(i\boldsymbol{k} \cdot \boldsymbol{R}) \psi_{at}(\boldsymbol{r} - \boldsymbol{R}) , \tag{11.3}$$

where $\psi_{at}(\boldsymbol{r})$ is not necessarily an exact atomic stationary-state eigenfunction, but one may expect that the function $\psi_{at}(\boldsymbol{r})$ in a solid is quite close to the atomic eigenfunction. It is one to be determined by further calculation. Based on this expectation, one seeks a $\psi^{(n)}(\boldsymbol{r})$ that can be expanded in a relatively small number of localized atomic eigenfunctions:

$$\psi^{(n)}(\boldsymbol{r}) \equiv \psi_n(\boldsymbol{r}) = \sum_{i=1}^{N} b_i^{(n)} \psi_i^{(n)}(\boldsymbol{r}) , \tag{11.4}$$

where $N$ is the number of atomic bases functions, $(n)$ is a level of the electron, $\psi_i^{(n)}(\boldsymbol{r})$ is the $i$th basic eigenfunction, and coefficients $b_i^{(n)}$ are to be determined.

The application of the tight-binding method to bulk systems is most easily introduced by first considering a lattice of atoms with overlapping s orbitals $\psi_s$ and corresponding free atomic energy levels $E_s$. For s band we have a single equation giving an explicit expression for the energy of the s band arising from this s level.

For an atomic p level which is triply degenerate we get a set of three homogeneous equations, whose eigenvalues would give the $E(\boldsymbol{k})$ for the three bands. Tight-binding approximation for p electrons may be obtained by writing $\psi_{\boldsymbol{k}}(\boldsymbol{r})$ as a linear combination of the atomic $p_x$, $p_y$, and $p_z$ orbitals (see Figure 3.8). That is,

$$\psi_n^{(p)}(\boldsymbol{r}) = \frac{1}{\sqrt{N}} \sum_{\alpha=p_x,p_y,p_z} b_\alpha \sum_{\boldsymbol{R}} \exp(i\boldsymbol{k} \cdot \boldsymbol{R}) \psi_\alpha(\boldsymbol{r} - \boldsymbol{R}) . \tag{11.5}$$

A linear combination of $\psi_n(\boldsymbol{r} - \boldsymbol{R})$ is used to produce a Bloch function $\psi_n(\boldsymbol{r})$, which must obey the Bloch theorem

$$\psi_n(\boldsymbol{r} + \boldsymbol{R}) = e^{i\boldsymbol{k}\cdot\boldsymbol{R}} \psi_n(\boldsymbol{r}) . \tag{11.6}$$

To get a d band from atomic d levels we should have to solve a $5 \times 5$ secular problem.

For more interesting bands, the conduction bands, the results of tight-binding are usually in rather poor agreement with the experiment. Tight binding could be systematically improved by including additional levels, so that the accuracy of the calculated bands increases, at the expense of the simplicity and transparency of the model.

## 11.2
## The Procedure of Calculations

The single-particle Schrödinger equation is expressed as

$$-\frac{\hbar^2}{2m}\nabla^2\psi_n(\boldsymbol{r}) + V(\boldsymbol{r})\psi_n(\boldsymbol{r}) = E_n\psi_n(\boldsymbol{r}) \,. \tag{11.7}$$

To determine the total energy, one writes the potential as $V(\boldsymbol{r}) + \Delta V(\boldsymbol{r})$, where $\Delta V(\boldsymbol{r})$ contains all corrections to the atomic potential required to produce the periodic potential of the crystal.

Without going into details, the sequence of operations can be described according [7] as follows:

The Schrödinger equation (11.7) $\leftrightarrow$ substitution $H = H_{\text{at}} + \Delta V(\boldsymbol{r}) \leftrightarrow$ equations (11.3), (11.4) $\rightarrow$ multiplying the Schrödinger equation by the complex conjugate eigenfunction $\psi^*_{\text{at}}(\boldsymbol{r})$ and integration over all $\boldsymbol{r}$ $\rightarrow$ calculation of the $b_i^{(n)}$ coefficients $\rightarrow$ calculation of energy $E_n$.

The independent electron approximation takes into account the electron–electron interactions indirectly. It assumes that this important effect can be allowed for a sufficiently clever choice of the periodic potential $V(\boldsymbol{r})$ appearing in the one electron Schrödinger equation. Thus $V(\boldsymbol{r})$ contains not only the periodic potential due to the ions alone, but also periodic effects due to interactions of the electron with all other electrons.

The latter interaction depends on the configuration of the other electrons; that is, it depends on their eigenfunctions, which are also determined by a Schrödinger equation of the form (11.7). Thus, we obtain a "closed disk." To find out the potential appearing in (11.7) one must first know all the solutions to (11.7). However, one must know the potential in order to do this.

The procedure starts with a shrewd guess, a value of $V_0(\boldsymbol{r})$ for $V(\boldsymbol{r})$, calculate from (11.7) the eigenfunctions for the occupied electronic levels, and from these recompute $V(\boldsymbol{r})$. If the new potential, $V_1(\boldsymbol{r})$ is the same as (or very close to) $V_0(\boldsymbol{r})$ one says that self-consistency has been achieved and takes $V = V_1$ for the actual potential. If $V_1$ differs from $V_0$, one repeats the procedure starting with $V_1$, taking $V_2$ as the actual potential if it is very close to $V_1$, and otherwise continuing on to calculation of $V_3$. The hope is that this procedure will eventually converge, yielding a self-consistent potential that reproduces itself.

## 11.3
## Applications of the Tight-Binding Method

Figure 11.2 shows the electron band structure of vanadium. In these calculations, the crystal potential has been approximated by a superposition by atomic potentials. We can compare the energies of two atomic configurations with five valence electrons. A potential which results from an atomic configuration containing an additional 4s electron and one less 3d electron (Figure 11.2a) leads to a narrowing of

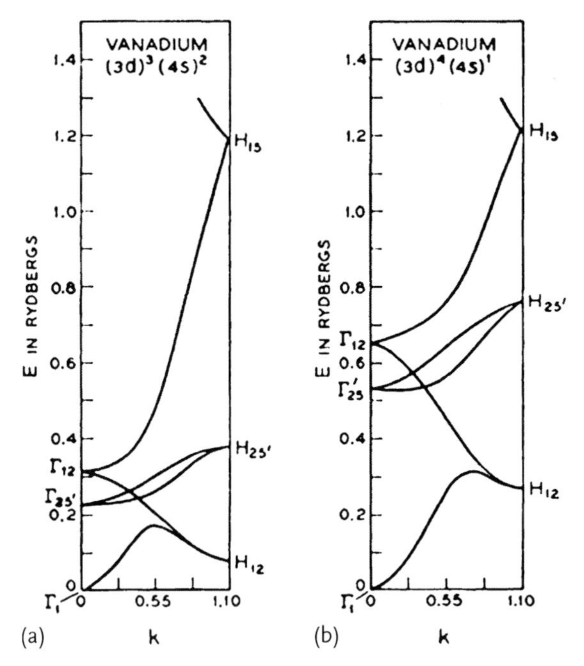


**Figure 11.2** Energy bands for vanadium along the line $\Delta(\Gamma \rightarrow H)$ of the first Brillouin zone obtained from two different potentials; (a) bands resulted from an atomic configuration of $3d^34s^2$; (b) bands that involved a $3d^44s^1$ atomic configuration (after [32]).

the 3d band and a decrease in the energy separation between the top of the 3d band and the bottom of 4s band. It is evident that for the $^{23}$V: (Ar)$3d^34s^2$ structure all the obtained curves correspond to lesser values of the energy than $^{23}$V: (Ar)$3d^44s^1$. Consequently, the structure with an unfilled d shell and a filled next s shell is a stable structure (Figure 11.2b). This phenomenon is typical for transition metals.

The authors of [33] presented the data on the accuracy of the tight-binding method. They evaluated the structural energy differences, elastic constants, vacancy formation energies, and surface energies, comparing to first-principle calculations and experimental data. In most cases there is a good agreement between experimental and tight-binding results.

Vacancy formation energies are most easily determined by the supercell total-energy method. One atom in the supercell is removed and neighboring atoms are allowed to relax around this vacancy while preserving the symmetry of the lattice. The great advantage of the tight-binding method over first-principle calculations is that we can do the calculation for a very large supercell, eliminating the possibility of vacancy–vacancy interactions. It was found that a supercell containing 108 atoms was sufficient to eliminate the interaction. The results are presented in Table 11.1. The calculated vacancy formation energies of copper, niobium, silver, tantalum, and iridium are in a satisfactory agreement with the experiment. The tight-binding parameterizations of rhodium and tungsten give the incorrect vacancy formation energies, 1–2 eV above the experimental value.

**Table 11.1** Comparison of the energy of vacancy formation $E_v$ and the surface energy $\gamma$ calculated from tight-binding theory (TB) with an experiment. The first-principle data obtained from the local density approximation (LDA) are shown, where available. The experimental column shows a range of energies if several experiments have been tabulated. Otherwise the estimated error in the experiment is given. Data of [33].

| Metal | $E_v$ (eV) | | | $\gamma$ for (111) plane (J m$^{-2}$) | | |
|---|---|---|---|---|---|---|
| | TB | LDA | Exper. | TB | LDA | Exper. |
| Cu | 1.18 | 1.41; 1.29 | $1.28 \pm 1.42$ | 1.73 | 1.94 | 1.77 |
| Nb | 2.82 | | $2.65 \pm 0.03$ | 2.44 | | 2.30 |
| Mo | 2.46 | | $3.0 - 3.6$ | 2.84 | | 2.90; 6.00 |
| Rh | 3.35 | 2.26 | 1.71 | 2.46 | 2.54; 2.53 | 2.60 |
| Pd | 2.45 | 1.57 | $1.85 \pm 0.25$ | 1.57 | 1.64 | 2.00 |
| Ag | 1.24 | 1.20; 1.06 | $1.11 - 1.31$ | 1.14 | 1.21 | 1.32 |
| Ta | 2.95 | | $2.9 \pm 0.4$ | 3.14 | | 2.78 |
| W | 6.43 | | $4.6 \pm 0.8$ | 6.75 | | 2.99 |
| Ir | 2.17 | | 1.97 | 2.59 | | 3.00 |
| Pt | 1.79 | | $1.35 \pm 0.09$ | 2.51 | | 2.49 |
| Au | 1.12 | | $0.89 \pm 0.04$ | 1.48 | | 1.54 |

The tight-binding model can be used to calculate surface energies by the supercell technique. A slab of metal is formed by cleaving the crystal along the desired plane, creating two identical free surfaces. The distance between the two surfaces is increased, creating a set of slabs that repeat periodically in the direction perpendicular to the surfaces. The slabs must be thick enough so that the two surfaces on the same slab cannot interact with each other. Numerical error in the TB surface energies was found to be about 0.1 J m$^{-2}$.

The results for the fcc metals are good. Results for bcc metals are satisfactory except for tungsten.

Satisfactory results have been obtained as a result of tight-binding approximation for elastic constants of metals (Table 11.2)

## 11.4 Environment-Dependent Tight-Binding Potential Models

The first-principle methods that have been described in Chapters 8–10 are the most exact because the only approximations are the treatment of exchange and correlation within density functional theory and preliminary assumptions about a structure.

The common feature of tight-binding theory is the use of a minimal local basis of local orbitals. This usually means at most just one s-orbital, three p-orbitals, and five d-orbitals per atom (Figure 3.8). Some authors believe that the tight-binding

**Table 11.2** Elastic constants (GPa) for cubic elements. Comparison of results of calculations according to the tight-binding theory and experimental data (after [33]).

| Metal | $C_{11}$ | | $C_{12}$ | | $C_{44}$ | |
|---|---|---|---|---|---|---|
| | calculation | experiment | calculation | experiment | calculation | experiment |
| V | 224 | 228 | 106 | 119 | 92 | 43 |
| Cr | 432 | 346 | 88 | 66 | 250 | 100 |
| Cu | 161 | 168 | 108 | 121 | 55 | 75 |
| Nb | 204 | 246 | 137 | 139 | 34 | 29 |
| Mo | 453 | 450 | 147 | 173 | 120 | 125 |
| Rh | 491 | 413 | 171 | 194 | 260 | 184 |
| Pd | 233 | 227 | 163 | 176 | 63 | 72 |
| Ag | 133 | 124 | 86 | 93 | 42 | 46 |
| Ta | 275 | 261 | 140 | 157 | 78 | 82 |
| W | 529 | 523 | 170 | 203 | 198 | 160 |
| Ir | 694 | 590 | 260 | 249 | 348 | 262 |
| Pt | 380 | 347 | 257 | 251 | 71 | 76 |
| Au | 184 | 189 | 154 | 159 | 43 | 42 |

approximation has advantages of rapid computation, a large number of atoms in model crystals under study and a single physical insight.

The tight-binding model has existed for many years as a convenient and transparent model for the description of electronic structures in molecules and solids. In the early literature one adopted the two-center approximation almost invariably. Tight binding was regarded as a simple empirical scheme for the construction of Hamiltonians by placing atomic-like orbitals at atomic sites and allowing electrons to hop between atoms. One terms the energies of hopping electrons as "hopping energies" and methods of its calculation as "hopping integrals."

"It was later realized that the tight binding approximation may be directly deduced as a rigorous approximation to the density functional theory," is the opinion of the author of [34]. The other possibilities in principle are three center terms and an overlap of orbitals on distant lattice sites. One computes the hopping integral types of $V_{ss\sigma}$, $V_{ss\pi}$, $V_{pp\sigma}$, $V_{pp\pi}$, and a pairwise repulsive potential between atoms.

A large part of tight-binding modeling is concerned with how to approximate the Kohn–Sham equations, which authors consider in a matrix form. The matrix elements involve a kinetic energy and potential energy parts. The first one depends only on the vector $R_{ij}$ and the particular orbitals, not on the environment of the atom in the material. The kinetic energies are calculated as integrals that are called two-center integrals. Figure 11.3 illustrates the integrals within the tight-binding approach.

More difficult is computation of the potential energy because the potential $V_{\text{eff}}$ depends of the local environment. One uses a simplification, that is, to write $V_{\text{eff}}$ as a sum of atom-centered contributions.

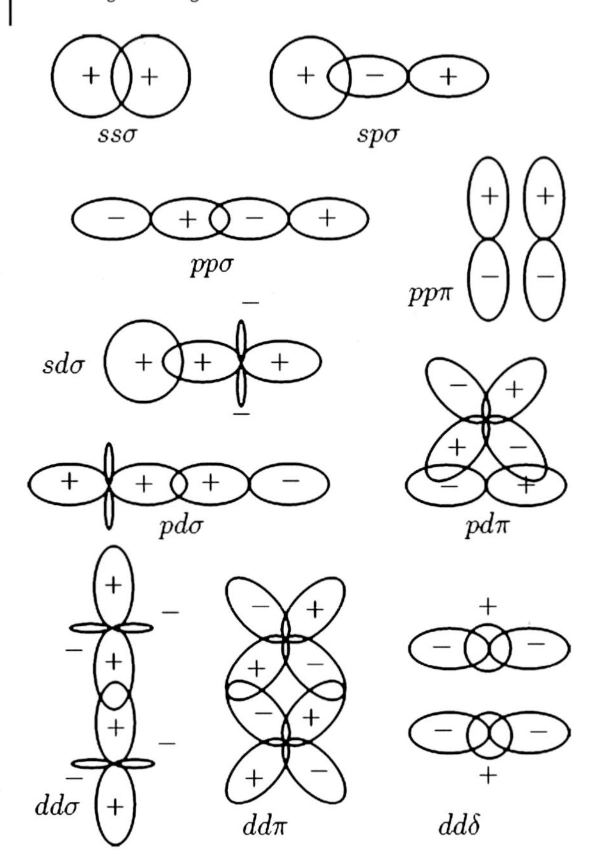


**Figure 11.3** Bonding integrals, that is, atomic orbitals of various s, p, d types joined along a bonding. Radial symmetry is assumed leading to the designation of the bonding as $\sigma$, $\pi$ or $\delta$. Data of Majewsky and Vogl (after [34]).

Another approach goes beyond traditional two-center approximations and allows the tight-binding parameters and the repulsive potential to be dependent on the bonding environment [36, 37]. In this approach, the environment dependence of the hopping parameters is modeled through incorporating two new scaling functions into the traditional two-center integrals. The first one is a screening function, which mimics the electronic screening effects in solids such that the interaction strength between two atoms in the solid becomes weaker if there are intervening atoms located between them. This approach allows one to distinguish between first- and farther-neighbor interactions within the same interaction potential without having to specify separate interactions for first and second neighbors. The second function scales the distance between two atoms according to their effective coordination numbers. Longer effective bonding lengths are assumed for higher-coordinated atoms.

The expression for the binding energy of a system with $M$ atoms and $N$ valence electrons in tight-binding molecular dynamics (TBMD) is given by

$$E_{\text{binding}} = E_{\text{bs}} + E_{\text{rep}} - E_0 \,. \tag{11.8}$$

The first term on the right-hand side of (11.8) is the band structure energy which is equal to the sum of the one-electron eigenvalues $E_i$ of the occupied states given by a tight-binding Hamiltonian $H_{\text{TB}}$,

$$E_{\text{bs}} = \sum_i f_i E_i \,, \tag{11.9}$$

where $f_i$ is the Fermi–Dirac function[1] and $\sum_i f_i = N$.

The second term on the right-hand side of (11.8) is a repulsive energy usually expressed as a sum of short-ranged pairwise interactions

$$E_{\text{rep}} = \frac{1}{2} \sum_{i,j,i \neq j} \phi(r_{i,j}) \tag{11.11}$$

or a functional of sum of pairwise interactions

$$E_{\text{rep}} = \sum_i F_i \left[ \sum_j \phi(r_{i,j}) \right] , \tag{11.12}$$

where $F$ is a function which, for example, can be a fourth order polynomial.

The term $E_0$ in (11.8) is a constant which represents the sum of the energies of the individual atoms.

The two-center approximation could be justified only when a system is governed by strong covalent interactions. For systems where metallic bond effects are significant, contributions beyond pairwise interactions should not be neglected. The environment-dependent tight-binding models have been developed in order to go beyond the initial theory [37]. Table 11.3 shows a good agreement between calculated and experimental data for diamond and silicon. Silicon is a strong covalent bonded material best described by the tight-binding scheme.

Figure 11.4 illustrates the cohesive energy for different crystalline structures of silicon. It is evident that data of a tight-binding approach fit the results obtained from the first-principles calculations well.

1) The Fermi–Dirac function of the electron energy $E$ distribution is given by

$$f(E) = \frac{1}{\exp((E-\mu)/(k_{\text{B}} T)) + 1} \,, \tag{11.9}$$

where $\mu$ is the chemical potential, $k_{\text{B}}$ is the Boltzman constant, and $T$ is temperature.

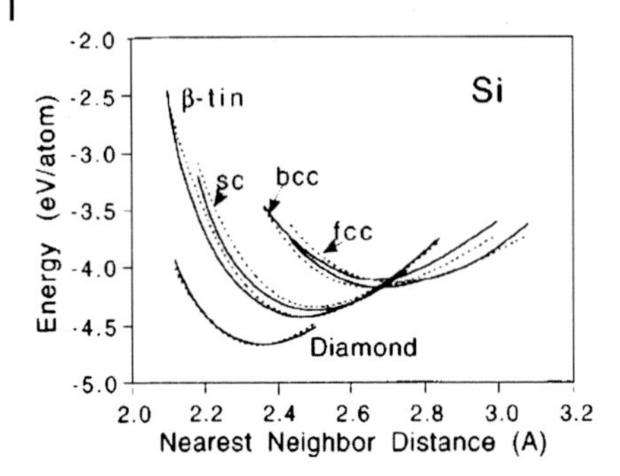


**Figure 11.4** The cohesive energies per atom as a function of nearest-neighbor distance for silicon in different crystalline structures. The solid curves are the tight-binding approximations and the dashed curves are the local density approximations (after [35]).

**Table 11.3** Parameters of unit cell and elastic constants for diamond and silicon. Comparison of results of calculations according to the tight-binding theory and experimental data.

| Element | *a* (nm) | | *B* (GPa) | | $C_{11}-C_{12}$ (GPa) | | $C_{44}$ (GPa) | |
|---|---|---|---|---|---|---|---|---|
| | calc. | exper. | calc. | exper. | calc. | exper. | calc. | exper. |
| C | 0.3585 | 0.3567 | 4.19 | 4.42 | 9.25 | 9.53 | 5.55 | 5.76 |
| Si | 0.5450 | 0.5430 | 0.900 | 0.978 | 0.993 | 1.012 | 0.716 | 0.796 |

## 11.5 Embedded-Atom Potentials

The embedded-atom method (EAM) overcomes the limitations of the pair potential technique. It is considered to be practical enough for calculations of defects, impurities, fractures, and surfaces in metals. In this model, an impurity (that is, a quasi-atom) is assumed to experience a locally uniform or only slightly nonuniform, environment. The energy of the quasi-atom can be expressed as

$$E_{\text{quas}} = E_Z[\rho_{\text{host}}(\boldsymbol{R})] \,, \tag{11.13}$$

where $\rho_{\text{host}}(\boldsymbol{R})$ is the electron density of the host without impurity at the site $\boldsymbol{R}$, where the impurity is to be placed. $E_Z$ is the quasi-atom energy of the impurity with atomic number $Z$. Relaxation of the host crystal lattice can be treated by calculating the lattice energy from pair potentials. Within the embedded-atom approach all atoms are viewed as being embedded in the host lattice containing of all other atoms.

When an impurity is introduced, the total potential is a sum of the host and the impurity potentials. Because the host potential is uniquely determined by the unperturbed host electron density, and the impurity potential is determined by the position and charge of the impurity nucleus, the energy of the host with impurity $E_{\text{h,imp}}$ is a functional of the host electron density and also a function of the impurity type and position. That is,

$$E_{\mathrm{h,imp}} = F_{Z,R}[\rho_{\mathrm{h}}(r)] , \tag{11.14}$$

where $Z$ and $R$ are the type and position of the impurity, $\rho_{\mathrm{h}}(r)$ is the unperturbed host electron density. The new feature of the embedded-atom method is that the energy of an impurity is determined by the electron density before the impurity is added.

Total energy of a system of atoms in the embedded-atom method is given by [40]

$$E = \frac{1}{2} \sum_{i,j,i \neq j} \Phi_{i,j}(\boldsymbol{R}_{ij}) + \sum_{i} F(\overline{\rho}_i) , \tag{11.15}$$

where $\Phi_{i,j}(\boldsymbol{R}_{ij})$ is a short-range pair potential, $\boldsymbol{R}_{ij}$ is distance between atoms $i$ and $j$, $F(\overline{\rho}_i)$ is the embedding energy that is dependent on the electron density of the host at position $\boldsymbol{R}_i$ but without atom $i$. The sums are over the atoms $i$ and $j$. The electron density in the site $i$ is

$$\overline{\rho}_i = \sum_{j(\neq i)} \rho_{\mathrm{at}}(R_{ij}) . \tag{11.16}$$

In general, functions $\Phi$ and $F$ are not known. To determine these functions one may[2)] use as the input values such experimental data as lattice parameters, elastic constants, energy of the vacancy formation, sublimation energy, and the stacking fault energy. Thermal expansion factors, heat of solution, phonon frequencies, surface energy, and energy of the vacancy diffusion can also be included in quantities used for determination of the functions $\Phi$ and $F$.

In contrast to pair potentials, the embedded-atom method incorporates a term that accounts for the many-body interaction between atoms in an approximate manner. The introduction of this term enables a semiquantitative, and in some cases even quantitative, description of metallic systems.

The authors of [39] determined the functions semiempirically that enter in (11.15) and applied them to the following problem: surface energy in nickel and palladium, binding of the hydrogen atoms to vacancies in these metals, structure of surface layers, and effect of hydrogen on fracture.

Twelve parameters were required for every specific element in order to develop the function within the modified embedded-atom method [40]. The listed values include the sublimation energy of elements, nearest-neighbor distance, embedding energy, factors for atomic densities, and so on. The modified embedded-atom method was successfully applied to calculating various bonding characteristics.

The functional form (11.15) was originally derived as a generalization of the effective medium theory and the second moment approximation to tight-binding theory. Later, however, it lost its close ties with original physical meaning and came to be treated as a semiempirical expression with adjustable parameters.

2) See, for instance, [38, 39].

A complete description of an $n$-component system requires $n(n+1)/2$ pair interactions functions $\rho_{at}(R_{ij})$ and $n$ embedding functions $F(\overline{\rho_i})$ $(i = 1, 2, \ldots, n)$.

An elemental metal is described by three functions $\Phi(R), \rho(R)$ and $F(\overline{\rho})$, while a binary system A–B requires seven functions $\Phi_{AA}, \Phi_{AB}, \Phi_{BB}, \rho_A(R), \rho_B(R)$, $F_A(\overline{\rho}), F_B(\overline{\rho})$. Over the past two decades the embedded-atom potentials have been constructed for many metals and a number of binary systems. EAM functions are usually defined by analytical expressions. The pair-interaction and electron-density functions are normally forced to vanish together with several higher derivatives at a cutoff radius $R_c$. Typically, $R_c$ covers 3–5 coordination shells.

For elemental metals different authors use polynomials, exponents, Morse, Lennard-Jones, or Gaussian functions or their combinations. Authors believe that in the absence of strong physical leads, any reasonable function can be acceptable as long as it works [41]. It is important, however, to keep the functions as simple and as smooth (as possible).

Author of [41] writes: "The increasing of fitting parameters should be done with great caution. The observed improvement in accuracy of fit can be inclusive as the potential may perform poorly for properties not included in the fit. Many sophisticated potentials contain hidden flaws that only reveal themselves under certain simulation conditions. As a rough rule of thumb, potentials whose $\Phi(R)$ and $\rho(R)$ together contain over 15 fitting parameters may lack reliability in applications. At the same time, using too few (say, $< 10$) parameters may not take full advantage of the capability of EAM. Since the speed of atomistic simulations does not depend on the complexity of potential functions or the number of fitting parameters, it makes sense to put efforts in optimizing them for the best accuracy and reliability."

The author of this book would like to emphasize a fact of fundamental importance. The matter concerns so-called fitting parameters. One validates a hypothesis or theory comparing theoretical and experimental data. A researcher chooses a set of the theory parameters in order to provide the best possible fit of the data. Sometimes these parameters are coefficients that have no physical meaning. Moreover, one uses up to 20 undefined fitting parameters. The point is that a varying of numerous parameters enables one to obtain a satisfactory fit between experimental and calculated results. However, one should not draw any conclusions about the correctness of the suggested hypothesis from this fit.

## 11.6 The Embedding Function

There are two ways of constructing the embedding function $F(\overline{\rho})$. One way is to describe it by an analytical function (or cubic spline) with adjustable parameters. Another way is to postulate an equation of state of the ground-state structure.

Most authors use the universal binding curve,

$$E(a) = E_0(1 + ax)e^{-ax} , \tag{11.17}$$

where $E(a)$ is the crystal energy per atom as a function of the lattice parameter $a$, $x = a/a_0 - 1$, $a_0$ is the equilibrium value of $a$.

The pair potential $U(R)$, atomic charge density $\rho_{at}(\boldsymbol{r})$, and embedding function $F(\rho)$ are usually fitted to reproduce the known equilibrium atomic volume, elastic moduli, and ground-state structure of the defect-free lattice.

An embedding curve with a minimum is often constructed by adding an attractive square root-dependent term to the repulsive linear term of the closed shell atoms, namely [3],

$$F(\rho_i) = C\rho_i - D\rho_i^{\frac{1}{2}} , \tag{11.18}$$

where $C$ and $D$ are positive constants. The nonpairwise behavior is most easily demonstrated by the coordination number dependence of the binding energy. The binding energy per atom of a lattice with coordination number $\kappa$ may be written in the form

$$U(\kappa) = A\kappa - B\kappa^{\frac{1}{2}} , \tag{11.19}$$

where

$$A = \frac{1}{2}U(r_0) + a\rho_{at}(r_0) \tag{11.20}$$

and

$$B = b\rho_{at}^{\frac{1}{2}}(r_0) \tag{11.21}$$

with $r_0$ being the nearest-neighbor distance.

The binding energy is affected by the coordination number. It is seen from Figure 11.5 for the linear chain ($\kappa = 2$), graphite ($\kappa = 3$), diamond ($\kappa = 4$), simple

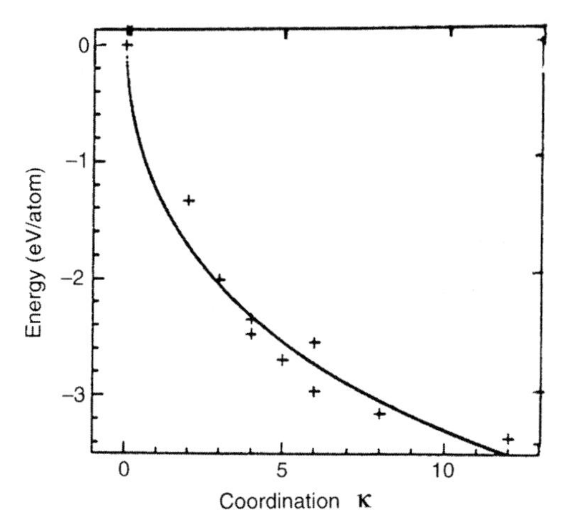


**Figure 11.5** The binding energy per atom $U$ as a function of the coordination number $\kappa$. The crosses denote the local density approximation, whereas the curve is a least-squared fit of the form of (11.19). Data of Heine, Robertson and Payne (after [3]).

**Table 11.4** Results calculated by the embedded-atom method compared with the experimental data for some metals. $E_{coh}$ is the cohesive energy, $B$ is the bulk modulus, $E_v$ is the energy of the vacancy formation, $\gamma_{(111)}$ is the surface energy of the (111) plane, $N_{fit}$ is a number of fitting parameters (or input parameters) which were used by the authors for calculations.

| Material | $E_{coh}$ (eV/atom) | | $B$ (GPa) | | $E_v$ (eV) | | $\gamma_{(111)}$ (mJ/m$^2$) | | $N_{fit}$ | Ref. |
|---|---|---|---|---|---|---|---|---|---|---|
| | EAM | Exper. | EAM | Exper. | EAM | Exper. | EAM | Exper. | | |
| Al | −3.35 | −3.36 | 85 | 79 | 1.20 | 0.75 | 805 | 823–830 | 11 | [42] |
| Ag | −2.85 | −2.85 | 140 | 140 | 1.103 | 1.1 | 862 | 1140 | 30 | [43] |
| Ni | −4.45 | −4.45 | 181.0 | 181.0 | 1.57 | 1.60 | 1759 | 2280 | 30 | [44] |
| Al | −3.36 | −3.36 | 79.0 | 79.0 | 0.71 | 0.68 | 601 | 980 | 13 | [44] |
| Cu | −3.54 | −3.54 | 138.3 | 138.3 | 1.27 | 1.272 | 1239 | 1790 | 28 | [45] |

cubic ($\kappa = 6$), vacancy lattice ($\kappa = 8$), and face centered cubic ($\kappa = 12$). For example, the binding energy of aluminum that predicted local density approximation coincides with the curve of (11.19) well.

The square root-dependence of the attractive contribution to the binding energy is a consequence of the unsaturated nature of the metallic bond. If all the bondings were saturated then we would expect the binding energy to be directly proportional to the number of bondings present as electrons in each bonding would be localized between their parent atoms. However, in a sp bonded metal there are not enough valence electrons to form saturated bondings with all close-packed neighbors so that the electrons resonate between all bondings in a delocalized fashion.

Increasing the local coordination around a given atom reduces the strength of neighboring bondings as the electrons are spread more evenly between them. This can be seen from (11.19) because the attractive binding energy per bonding decreases as $1/\kappa^2$ as the coordination number $\kappa$ increases.

Table 11.4 presents the parameters that were calculated for some metals by the embedded-atom method. The parameters are compared with the experimental data. Data in the sixth column $N_{fit}$ require some remarks. In order to ensure favorable results, the authors made use of input from 11 to 30 and fitting parameters and thus diverted their models rather far from the physical reality (see the authors' remark at the end of Section 11.5).

## 11.7 Interatomic Pair Potentials

In models employing simple pair potentials (Morse, Lennard-Jones, Buckingham) only the direct interaction between two atoms is considered. These potentials are radially symmetric and ignore the directional property of the interatomic bond. They make the best use for molecules. One may estimate the total energy of a solid by the use of pair potentials though they involve no further cohesive term.

The energy of an atomic system expressed trough a classical pair potential may be written

$$E_{\text{tot}} = \frac{1}{2} \sum_{i=1, i \neq j}^{N} \sum_{j=1, j \neq i}^{N} \psi_{ij}(\boldsymbol{r}_{ij}) \,. \tag{11.22}$$

The second type of pair potential describes energy changes due to configurational variations at a constant average atom density, rather than the total energy of the system. It can be described by a more general equation,

$$E_{\text{tot}} = \frac{1}{2} \sum_{i=1, i \neq j}^{N} \sum_{j=1, j \neq i}^{N} \psi_{ij}(\boldsymbol{r}_{ij}) + U(\rho) \,, \tag{11.23}$$

where $U(\rho)$ is the cohesive contribution to the total energy and $\rho$ is the average density of the material. This model takes into consideration a pseudopotential representing the ion cores in simple sp bonded metals, such as lithium, sodium, potassium, magnesium, and aluminum. The total energy is then assumed to be consisting of a large density-dependent (but structure-independent) term $U(\rho)$, and structure-dependent term $\psi_{ij}(\boldsymbol{r}_{ij})$. One should note that the pseudopotential approach differs from the various many-body potentials such as the embedded-atom method. The latter method considers the local rather than the average electronic density of the material.

Many properties of materials, such as plastic deformation, fracture, diffusion, and phase transformation, require statistical averaging over many atomic events. Computer modeling of such processes is facilitated by usage of semiempirical interatomic potentials allowing fast calculations of the total energy and classical interatomic forces.

Due to their computational efficiency, the interatomic potentials enable molecular dynamics simulation for systems of millions of atoms. "State-of-the-art potentials capture the most essential features of interatomic bond, reaching the golden compromise between computational speeds and accuracy of modeling [41]."

Molecular dynamics, Monte-Carlo, and other simulation methods require multiple evaluations of Newtonian forces $\boldsymbol{F}_i$ acting on individual atoms $i$ or (as in the case of Monte-Carlo simulations) the total energy of the system, $E_{\text{tot}}$. Atomistic potentials, also referred to as force fields, parameterize the configuration space of a system and represent its total energy as a relatively simple function of a configuration point. The interatomic forces are then obtained as coordinate derivatives of $E_{\text{tot}}$, $\boldsymbol{F}_i = -\partial E_{\text{tot}}/\partial \boldsymbol{r}_i$, $\boldsymbol{r}_i$ being the radius-vector of an atom $i$.

The calculation of $E_{\text{tot}}$ and $\boldsymbol{F}_i$ is a simple and fast numerical procedure that does not involve quantum mechanical calculations, although the latter are often used when generating potentials. Potential functions contain fitting parameters, which are adjusted to give desired properties of the material known from experiment or first-principles calculations. Once the fitting procedure is complete, the parameters are not subject to any further changes and the potential thus defined is used in all subsequent simulations of the material. The underlying assumption is that a

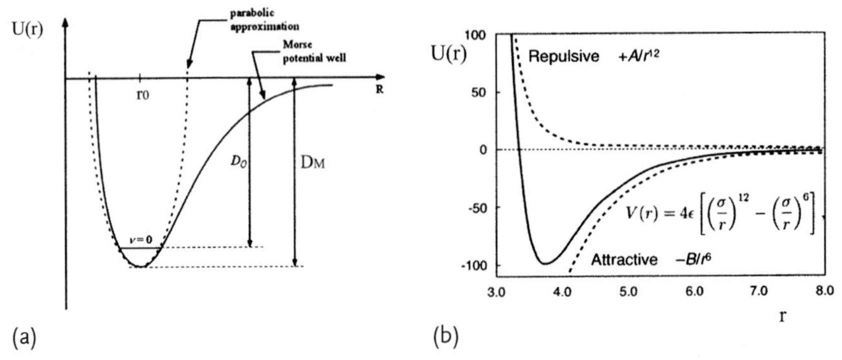


**Figure 11.6** Potentials of Morse (a) and Lennard-Jones (b).

potential providing accurate energies and forces at configuration points used in the fit will also give reasonable results for configurations between and beyond them. This property of potentials, often refereed to "transferability" is probably the most adequate measure of their quality.

Early atomistic simulations employed pair potentials, usually of the Morse or Lennard-Jones type (Figure 11.6). Although such potentials have been and still are a useful model for fundamental studies of generic properties of materials, the agreement between simulation results and experiment can only be quantitative at best. While such potentials can be physically justified for inert elements and perhaps some ionic solids, they do not capture the nature of interatomic bonding even in simple metals, not to mention transition metals or covalent solids.

If $U(r)$ is the energy of an interaction of two atoms a distance $r$ apart, then, in order $U(r)$ to represent the interatomic potential of two atoms in a stable crystal, it must satisfy the following conditions:

- the force $-\partial U/\partial r$ must be attractive at large $r$ and repulsive at small $r$;
- $U(r)$ must have a minimum at some point $r = r_0$, $(\partial U(r)/\partial r)_{r=r_0} = 0$;
- the magnitude of $U(|r - r_0|)$ must increase more rapidly with $|r - r_0|$ than $r^3$;
- all elastic constants are positive;
- $C_{11} - C_{12} > 0$.

Four classical interatomic pair potentials are widely used to describe solid states. Various crystal properties can be expressed in terms of the Morse potential $U_M$. The pair interatomic potential of Morse is expressed as

$$U_M = D_M\{\exp[-2\alpha(r - r_0)] - 2\exp[-\alpha(r - r_0)]\} , \tag{11.24}$$

where $D_M$ and $\alpha$ are constants with dimensions of energy and reciprocal distance, respectively, and $r_0$ is an equilibrium distance of approach of two atoms. Since $U_M(r_0) = -D_M$, $D_M$ is the dissociation energy per atom. In order to obtain a potential energy of a crystal whose atoms are at rest, it is necessary to sum (11.24) over the entire crystal. This is most easily done by choosing one atom in the crystal lattice as an origin, calculating its interaction with all the other atoms in the crystal,

**Table 11.5** Morse potential constants for the pairwise atomic interaction in cubic metals (after [46]).

| Metal | $D_M$ (eV) | $\alpha$ (nm$^{-1}$) | $r_0$ (nm) |
|---|---|---|---|
| Na | 0.06334 | 5.8993 | 0.5336 |
| K | 0.05424 | 4.9767 | 0.6359 |
| Rb | 0.04644 | 4.2981 | 0.7207 |
| Ca | 0.1623 | 8.0535 | 0.4569 |
| Sr | 0.1513 | 7.3776 | 0.4988 |
| Ba | 0.1416 | 6.5698 | 0.5373 |
| Al | 0.2703 | 11.646 | 0.3253 |
| Cr | 0.4414 | 15.721 | 0.2754 |
| Fe | 0.4174 | 13.885 | 0.2845 |
| Ni | 0.4205 | 14.199 | 0.2780 |
| Cu | 0.3429 | 13.588 | 0.2866 |
| Ag | 0.3323 | 13.680 | 0.3115 |
| Mo | 0.8032 | 15.079 | 0.2976 |
| W | 0.9906 | 14.116 | 0.3032 |

and then multiplying by $N/2$, where $N$ is the total number of atoms in the crystal. Thus, the total energy is given by

$$U_{\text{M,total}} = \frac{1}{2} N D_{\text{M}} \sum_j \{\exp[-2\alpha(r_j - r_0)] - 2\exp[-\alpha(r_j - r_0)]\} , \qquad (11.25)$$

where $r_j$ is the distance from the origin to the $j$th atom. Calculation of the constants $D_{\text{M}}$, $\alpha$ and $r_0$ has been performed [46] using experimental values for the energy of sublimation, the compressibility, and the lattice constant. The results are listed in Table 11.5.

The Lennard-Jones interatomic potential is defined as

$$U_{\text{LJ}} = D_{\text{LJ}} \left[ \left(\frac{r_0}{r}\right)^{12} - 2\left(\frac{r_0}{r}\right)^6 \right] . \qquad (11.26)$$

The $r^{-12}$ term, which is the repulsive term, describes repulsion at short ranges due to overlapping electron orbitals and the $r^{-6}$ term, which is the attractive long-range term, describes attraction at long ranges.

The Rydberg potential is given by

$$U_{\text{Ryd}} = -D_{\text{Ryd}} \left[ 1 + a\left(\frac{r - r_0}{r_0}\right) \right] \exp\left[ -a\left(\frac{r - r_0}{r_0}\right) \right] . \qquad (11.27)$$

The Rydberg potential has been applied by some authors to body-centered and face-centered cubic metals. The parameters of the potential were calculated using the experimental values for the energy of vaporization, the lattice constant, and the compressibility. Results have been obtained for the elastic constants, for the equation-of-state curves, and for the volume dependence of the cohesive energy. They were compared with those obtained using the Morse potential, and with the experimental data. For the elastic constants, the results by the two potentials are not very different, but a significant improvement has been obtained in several cases for the equation of state and for the volume dependence of the cohesive energy.

The Buckingham potential can be expressed as

$$U_{\mathrm{B}} = A \exp(-B r) - \frac{C}{r^6} , \tag{11.28}$$

where $A$, $B$, and $C$ are constants. The Buckingham potential was proposed, as a simplification of the Lennard-Jones potential, in a theoretical study of the equation of state for gaseous helium, neon and argon.

# 12
# Lattice Vibration: The Force Coefficients

Vibrational spectra provide a wealth of information about interatomic bonding in solids. These spectra are measured experimentally.

The forces, which act on both electrons and nuclei by a vibrational motion, are of the same order of magnitude due to their electric charge. The changes, which occur in their momenta as a result of these forces, must also be the same. One might therefore, assume that the actual momenta of the electrons and nuclei are of similar magnitude. In this case, since the nuclei are so much more massive than the electrons, accordingly, they must have much smaller velocities. Thus it is plausible that on the typical time-scale of the nuclear motion, the electrons will very rapidly relax remaining in their ground state. This separation of electronic and nuclear motion is known as is the adiabatic, or the Born–Oppenheimer approximation (Section 8.1).

The fundamental quantities of interest for interatomic bonding are the energy $\Phi$, forces acting on the atom $F_i$, and the force constants $\Phi_{\alpha\beta}$.

Potential energy of an oscillating crystal lattice is a function of positions of atoms (ions),

$$\Phi = \Phi(\boldsymbol{r}_i) . \tag{12.1}$$

A force acting on an atom is given by

$$F_i = -\frac{d\Phi}{du_i} , \tag{12.2}$$

where $u_i$ is the displacement of the $i$th atom relative to other atoms. The force constants can be expressed as

$$\Phi_{\alpha\beta} = \frac{\partial^2 \Phi}{\partial u_\alpha \partial u_\beta} , \tag{12.3}$$

where $u_\alpha$ and $u_\beta$ are the atom displacements in $\alpha$ and $\beta$ directions.

Long waves propagate in solids as if the medium is a continuum. In this case one obtains the simple relation between these quantities,

$$v = \lambda \cdot \nu = \frac{2\pi}{k} \cdot \frac{\omega}{2\pi} = \frac{\omega}{k} , \tag{12.4}$$

*Interatomic Bonding in Solids: Fundamentals,Simulation,andApplications,* First Edition. Valim Levitin.

where $v$ is the wave velocity, $k$ is the magnitude of wave vector, $\omega$ is the angle frequency of oscillation. At long wavelengths (low values of $k$) dependence of the frequency on $1/\lambda$ is linear and the velocity of the wave is independent on the wavelength; this is the case of sound waves. At a short wavelength the wave propagation is affected by the discreteness of crystal lattice, and such a relationship (12.4) does not hold. Instead, an equation for the group velocity takes place,

$$v_g = \frac{d\omega}{dk} \,. \tag{12.5}$$

The group velocity is the velocity of the envelope of the wave (see Section 2.2 and Appendix B).

## 12.1 Dispersion Curves and the Born–von Karman Constants

Measurements of dispersion curves provide information about the interatomic forces in solids. In fact, a dispersion curve is a function of the vibration frequency $\nu$ on the wavelength $\lambda$. The methods of neutron spectroscopy based on the phenomenon of diffraction of heat neutrons by crystals enables one to graph the dispersion curves of solids. The most accurate measurements of these curves are obtained by the inelastic neutron scattering using triple axis spectrometers.

It is generally accepted to consider atom vibrations in crystals in terms of vibrational modes or also as the propagation of phonons as quanta of the normal oscillation. Keeping in mind the wave particle duality, the wave motion can be presented as a particle-like motion as well. A particle in question is called a phonon. Thus, the phonon is a quantum mechanical description of vibrational motion, in which a lattice uniformly oscillates at the same frequency. One can consider the plane traveling wave as the phonon flux. Energy of the phonon is given by

$$E_{\text{phonon}} = h\nu = \frac{h}{2\pi} \cdot 2\pi\nu = \hbar\omega \,. \tag{12.6}$$

Figure 12.1 presents the dispersion curve of sodium in $[\zeta\zeta\zeta]$, that is in [111] direction.

A particular presentation of experimental data is used. In the theory of lattice vibration it is generally accepted to denote the phonon wave vector by $\boldsymbol{q}$. The reduced wave vector $\zeta$ is plotted along abscissa. This dimensionless quantity equals to the ratio of the magnitude of the phonon wave vector to its maximal magnitude in the given crystallographic direction. By definition,

$$\zeta = \frac{q}{q_m} \,, \tag{12.7}$$

where

$$q_m = \frac{2\pi}{2b} \,. \tag{12.8}$$

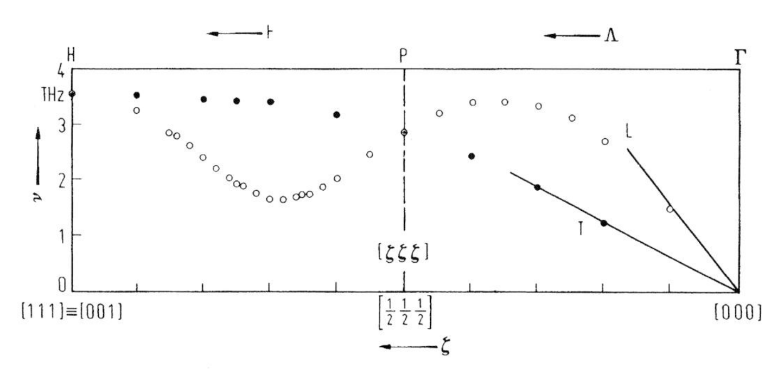


**Figure 12.1** Dispersion curves for sodium that were measured by the neutron diffraction technique at 90 K [9]. The ratio $\zeta = q/q_m = 2b/\lambda$ is plotted along the *x*-axis; *b* is the lattice spacing; $\zeta$ is changed from 0 on the right to 1 on the left with a step of 0.1. Frequency of vibration $\nu$ is plotted along the *y*-axis. $\Gamma$, *P*, and *H* are the points in the Brillouin zone (see Figure 4.13). L is a longitudinal branch, T is a transverse branch of oscillations. The same frequency of vibration is characteristic at the point $[\frac{1}{2}\frac{1}{2}\frac{1}{2}]$ of bcc lattice for longitudinal waves as well as for transverse ones; the curves L and T cross each other at the center of unit cell.

$b$ is the smallest distance between the projections of the atoms onto the direction of the vibration. A wavelength shorter than $2b$ can not exist in the crystal lattice. Consequently,

$$\zeta = \frac{q}{q_m} = \frac{2b}{\lambda} . \tag{12.9}$$

The wavelength decreases from the right to the left in Figure 12.1. One can see that a linear dependence between $\nu$ and $1/\lambda$ holds if $0 < \zeta < 0.4$ for the T-branch and $0 < \zeta < 0.2$ for the L-branch (sound waves). However, as soon as the wavelength decreases so that it becomes comparable with the interatomic distance, the curves deviate from straight lines. The T-curve tends to a value of 3.6 THz, The L-curve passes inflection points. Note that the slope of a tangent $d\nu/d\zeta$ is positive near the small values of $\zeta$, however, further it becomes zero and even negative!

The dependence of wave velocity on wavelength is called the dispersion of velocity. This phenomenon is related to displacements of oscillating atoms in antiphase.

Researchers extensively use the Born–von Karman theory in order to estimate the strength of interatomic bonding [47]. This traditional method is to make a formal expansion of energy of a vibrating crystal in a series and to treat the expansion coefficients as force constants. The main assumptions of this theory are as follows.

The given atom interacts pairwise with some number of neighbor atoms. The restoring force affecting the given atom depends linearly on the relative displacement of the other atom. The forces affecting on the given atom is the sum of forces exerted by displaced atoms. In the expansion of the potential energy of the atom interaction into a series of powers of displacements, one neglects terms of the third degree and higher. Actually, the Born–von Karman model does not consider any physical nature of the interatomic forces between atoms in solids.

An important question to the Born–von Karman model is how many neighbor atoms should be considered as acting on the given atom chosen as the origin. The major limitation of this approach is the large increase in the number of fitting parameters with increasing in interaction range. “The coupling parameters determined by such a fit have not necessary any direct physical meaning. They only provide a merely phenomenological description of the dispersion [9].”

The potential energy of a crystal is a function of mutual positions of all atoms. In a general case a three-dimensional crystal consists of unit cells with a basis. The cell contains $n$ atoms. The position of a cell number $l$ in the space is determined by the vector $\boldsymbol{r}(l)$. The position of a chosen atom number $k$ in the cell is determined by the vector $\boldsymbol{r}(k)$. It is convenient to introduce the notation $\boldsymbol{r}(lk)$ where the first letter in parenthesis determines the position of the cell, and the second letter determines the location of the atom in this cell. Thus,

$$\boldsymbol{r}(lk) = \boldsymbol{r}(l) + \boldsymbol{r}(k) . \tag{12.10}$$

The potential energy of the crystal lattice is the sum of all pairwise potential energies,

$$\Phi = \sum_{lk<l'k'} \Phi[\boldsymbol{r}(l'k') - \boldsymbol{r}(lk)] . \tag{12.11}$$

One denotes the directions of the atom displacements by $\alpha, \beta$. $\alpha, \beta = 1, 2, 3$, along $Ox$, $Oy$, $Oz$ axes, respectively. The displacement of the atom $(lk)$ in the direction $\alpha$ is denoted by $\boldsymbol{u}_\alpha(lk)$. Similarly, if another atom $(l'k')$ deviates along $\beta$ its displacement is denoted by as $\boldsymbol{u}_\beta(l'k')$.

Typical lattice vibrations involve small atomic excursions of the order of 10 pm or smaller, thus we may expand the expression for potential energy $\Phi$ into the Taylor series about the equilibrium position of the ions. We restrict ourselves by the powers of displacements not exceeding the second one. This is the so-called harmonic approximation. We obtain

$$\begin{aligned} \Phi = \Phi_0 &+ \sum_{l,k,\alpha} \left[ \frac{\partial \Phi}{\partial u_\alpha(lk)} \right]_0 u_\alpha(lk) \\ &+ \frac{1}{2} \sum_{l,k,\alpha,l',k',\beta} \left[ \frac{\partial^2 \Phi}{\partial u_\alpha(lk) \partial u_\beta(l'k')} \right]_0 u_\alpha(lk) u_\beta(l'k') . \end{aligned} \tag{12.12}$$

The quantity $\Phi_0$ on the right side of the equation does not depend on the displacement, it is the balanced potential energy of the crystal. The second term in (12.12) is zero since it is the first derivative of a potential being evaluated at the equilibrium position.

Simplifying notations we arrive at

$$\Phi = \frac{1}{2} \sum_{l,k,\alpha,l',k',\beta} \Phi_{\alpha\beta}(lk; l'k') u_\alpha(lk) u_\beta(l'k') . \tag{12.13}$$

**Table 12.1** The Born–von Karman force constants for sodium at temperature 90 K (after [9]).

| Atom | $\alpha\beta$ | $\Phi_{\alpha\beta}$ (N/m) | Atom | $\alpha\beta$ | $\Phi_{\alpha\beta}$ (N/m) |
|---|---|---|---|---|---|
| $\frac{1}{2}\frac{1}{2}\frac{1}{2}$ | $xx$ | 1.178 | $\frac{3}{2}\frac{1}{2}\frac{1}{2}$ | $xx$ | 0.052 |
| | $xy$ | 1.132 | | $yy$ | −0.07 |
| 100 | $xx$ | 0.472 | $\frac{3}{2}\frac{1}{2}\frac{1}{2}$ | $yz$ | 0.003 |
| | $yy$ | 0.104 | | $xz$ | 0.014 |
| 110 | $xx$ | −0.038 | 111 | $xx$ | 0.017 |
| | $zz$ | −0.0004 | | $xy$ | 0.033 |
| | $xy$ | −0.065 | | – | – |

The force acting onto the atom $lk$ is equal to a minus gradient of the potential energy of its interaction with all other atoms. Differentiating expression (12.13) with respect to the displacement $u_\alpha(lk)$, in accordance with Newton's Second Law, we obtain the following equation of motion:

$$m\ddot{u}_\alpha(lk) = -\frac{1}{2}\sum_{l',k',\beta} \Phi_{\alpha\beta}(lk;l'k')u_\beta(l'k') \,. \tag{12.14}$$

The solution to the equation of motion is given by

$$u_\alpha(lk) = u_{\alpha 0}e^{[i(\boldsymbol{k}\,\boldsymbol{r}(lk)-\omega t)]} \,, \tag{12.15}$$

where $u_{\alpha 0}$ is the amplitude. Equation (12.15) describes the wave displacement of the atoms, see (2.12).

The coefficients $\Phi_{\alpha\beta}$ in (12.14), which represent the second derivatives of the potential energy with respect to the atomic displacements determined at the equilibrium points, are called atomic force constants. By definition, they have an explicit physical meaning. The coefficient $\Phi_{\alpha\beta}(lk;l'k')$ is equal to the minus force which acts on the atom $(lk)$ in the direction $\alpha$, when the other atom $(l'k')$ deviates per unit distance in the direction $\beta$. The Born–von Karman model implies that all other atoms stay at their equilibrium positions.

One uses a large number of force constants in order to analyze the experimental dispersion curves, assuming the atomic two-body interactions. We have noticed that this approach imposes no restrictions on the nature of such interactions.

Table 12.1 illustrates the force constants for sodium that has the body-centered crystal lattice. We see that a force of 1.178 N/m acts on the atom at origin [000] in the direction $Ox$ when an atom at the center of unit cell displaces along $Ox$ per unit of length. The force acting along $Oy$ equals to 1.132 N/m. Only a displacement of the nearest-neighbor atoms $[\frac{1}{2}\frac{1}{2}\frac{1}{2}]$ and [100] have a significant effect on the atom at origin [000]. In sodium, the fitted force constants decrease in magnitude rapidly with spacing. The fifth neighbor constants are only about 2% of the nearest-neighbor ones. Thus, within the Born–von Karman theory one may draw

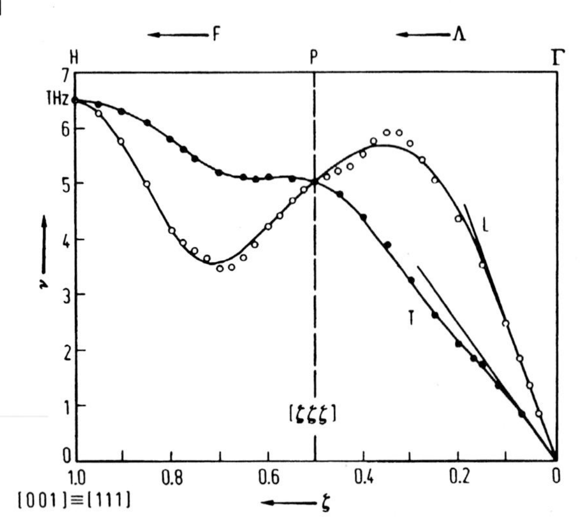


**Figure 12.2** Dispersion curves of Nb in [ζζζ] direction. L and T denote longitudinal and transverse branches of the lattice vibration, respectively. The open circles and filled circles are experimental points. The solid lines represent data calculated according to the Born–von Karman model with eight neighbors.

a conclusion that atoms in crystal lattice of the simple metal sodium interact only with nearest neighbors.

The transition metal niobium attracts a lot of attention of researchers because of its relatively high superconducting transition temperature. It turned out that niobium shows a number of pronounced anomalies in the phonon dispersion, which are also typical for vanadium and tantalum. The anomalies in [00ζ], [0ζζ] directions appear as a crossover of the longitudinal and transverse branches at $\zeta = 0.7$ or $\zeta = 0.3$ as well as additional maxima and minima.

Dispersion curves for niobium in [ζζζ] direction at 296 K are shown in Figure 12.2. Shapes of L- and T-curves look like that of sodium (Figure 12.1). The fit between experimental and calculated data seems to be satisfactory. However, niobium, which has the same body-centered crystal lattice, requires very long-ranging force constant models. The values of the constants depend critically on the range allowed in the fitting procedure and no physical significance can be attributed to them.

The force constants for niobium are listed in Table 12.2. It is natural that the values $\Phi_{\alpha\beta}$ for niobium are of a higher order of magnitude than for sodium. The existence of long-range forces is typical for niobium. Whereas atom $[\frac{3}{2}\frac{1}{2}\frac{1}{2}]$ is placed at the fourth coordination sphere, the force constant equals to 3.61 N/m (25.5% of the first coefficient).

## 12.2
## Fourier Transformation of Dispersion Curves: Interplanar Force Constants

Foreman and Lomer were the first to report that if the vibrations propagate in the cubic crystal lattice along the high symmetrical directions ⟨111⟩, ⟨110⟩, ⟨100⟩, the mathematical description of the process can be reduced to a linear chain of atoms [48]. The problem of vibrations of parallel atomic planes is reduced to the problem of interacting oscillating points of the equal mass. Every oscillatory mode can be treated as related to a separate linear chain.

The dispersion curves provide many useful data that can be extracted by the Fourier transformation of these curves. A basic design formula (see derivation of the formula in Appendix D) can be expressed as

$$\omega^2(\zeta) = \frac{2}{m}\left[\sum_{p=1}^{N} F_{\mathrm{p}} - \sum_{p=1}^{N} F_{\mathrm{p}}\cos(\pi p\zeta)\right] , \tag{12.16}$$

where $\omega(\zeta)$ is the angular frequency of vibration, coefficients $F_{\mathrm{p}}$ are the interplanar force constants, $\zeta = q/q_{\mathrm{m}} = 2b/\lambda$, (see (12.7)–(12.9)), $m$ is the mass of the atom, $N$ is the number of atoms that interacts with the atom at origin.

Let us consider a periodic function $f(x)$ having a period $L$. Thus, $f(x) = f(x \pm L)$ for every $-L \leq x \leq L$. Any even periodic function can be expanded in a Fourier series,

$$f(x) = \frac{a_0}{2} + \sum_{n=1}^{\infty} a_n \cos\left(\pi n\frac{x}{L}\right) , \tag{12.17}$$

where the Fourier coefficients are given by

$$a_n = \frac{1}{L}\int_{-L}^{L} f(x)\cos\left(\pi n\frac{x}{L}\right) dx . \tag{12.18}$$

**Table 12.2** The Born–von Karman force constants for niobium.

| Atom | $\alpha\beta$ | $\Phi_{\alpha\beta}$ (N/m) | Atom | $\alpha\beta$ | $\Phi_{\alpha\beta}$ (N/m) |
|---|---|---|---|---|---|
| $\frac{1}{2}\frac{1}{2}\frac{1}{2}$ | $xx$ | 14.14 | $\frac{3}{2}\frac{1}{2}\frac{1}{2}$ | $xx$ | 3.61 |
| | xy | 8.84 | | $yy$ | −0.75 |
| 100 | $xx$ | 14.16 | $\frac{3}{2}\frac{1}{2}\frac{1}{2}$ | $yz$ | −0.95 |
| | $yy$ | −3.64 | | $xy$ | 1.26 |
| 110 | $xx$ | 2.27 | 111 | $xx$ | −1.16 |
| | $zz$ | −6.38 | | $xy$ | −1.33 |
| | $xy$ | 0.79 | | – | – |

**Table 12.3** The interplanar force constants (N m$^{-1}$) for metals with a bcc crystal lattice. Oscillation mode is L[111]. The values of $F_p$ were calculated by the author in accordance with (12.21) proceeding from the experimental dependences $\nu - \zeta$ [9].

| Metal | $F_1$ | $F_2$ | $F_3$ | $F_4$ | $F_5$ | $F_6$ | $F_7$ |
|---|---|---|---|---|---|---|---|
| Na | 0.94 | 0.82 | 3.94 | −0.098 | 0.13 | 0.11 | −0.05 |
| K | −0.11 | 1.13 | 2.02 | 0.41 | −0.53 | 0.57 | −0.48 |
| Fe | 23.67 | 16.81 | 46.38 | 3.83 | 1.20 | −0.15 | −0.73 |
| Nb | 17.50 | 8.93 | 41.14 | −4.27 | 5.03 | −3.94 | −1.03 |
| Mo | 25.44 | 41.47 | 33.94 | 16.18 | 0.44 | 1.75 | 2.25 |

Comparing (12.16) and (12.17) we obtain

$$\frac{a_0}{2} = \frac{2}{m} \sum_{p=1}^{N} F_p \tag{12.19}$$

and

$$\sum_{n=1}^{\infty} a_n \cos\left(\frac{\pi n x}{L}\right) = -\frac{2}{m} \sum_{p=1}^{N} F_p \cos(\pi p \zeta) . \tag{12.20}$$

Changing the variables in (12.18), $f(x) \to \omega^2(\zeta)$, $-L \to -1$, $L \to 1$, $x/L \to \zeta$, $dx \to d\zeta$, $1/L \to 1$, $\pi n x/L \to \pi p \zeta$, $a_n \to F_p$ we arrive at a formula suitable for computational purposes,

$$F_p = -\frac{m}{2} \int_{-1}^{+1} \omega^2(\zeta) \cos(\pi p \zeta) , \tag{12.21}$$

where the quantities $F_1, F_2, \ldots, F_N$ are the interplanar force constants. It is logical to interpret $F_p$ as a force per atom acting between $p$ neighboring planes normal to the $\langle 111 \rangle$, $\langle 110 \rangle$, $\langle 100 \rangle$ symmetrical directions [48]. For example, $F_3$ is the force per atom in the first plane when the third plane is displaced at the unit distance along the normal to this plane.

Table 12.3 shows the values of the calculated interplanar force constants for five metals with the body-centered cubic crystal lattice. For simple metals sodium and potassium the first interplanar coefficients are of the order of several N m$^{-1}$. These coefficients of niobium and molybdenum make tens of N m$^{-1}$. The $F_3$ coefficient is maximal.

The arrangement of atoms is repeated in every fourth $(\bar{1}\,\bar{1}\,1)$ plane as illustrated in Figure 12.3. The atoms in the zeroth and the third planes (labeled by circles or crosses in this Figure) are situated as close to each other as possible in the direction $[\bar{1}\,\bar{1}\,1]$ of the bcc crystal lattice. The coefficient $F_3$ correspondingly has the maximum value for all metals mentioned in Table 12.3 except molybdenum. It is clear that the atom planes interact with each other stronger when the atoms are situated closer.

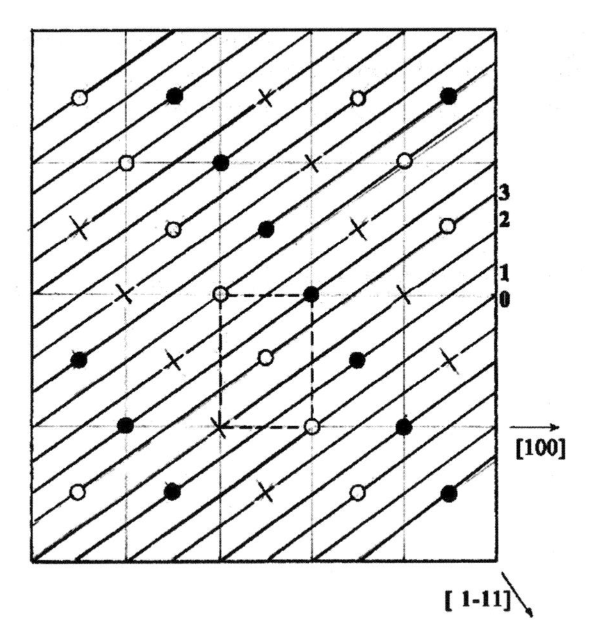


**Figure 12.3** Atoms in the (011) plane of bcc crystal lattice. The solid lines are intersections of the set of the parallel planes $(\bar{1}\bar{1}1)$ with the (011) plane that is perpendicular to them. Circles, shaded circles and crosses represent atoms in planes 0123..., respectively. Dotted lines outside the diagonal plane.

## 12.3 Group Velocity of the Lattice Waves

Differentiating (12.16) we obtain an expression for the group velocity of the lattice wave

$$v_i = \frac{d\omega_i}{dq} = \frac{b}{m}\left[\frac{\sum_{p=1}^{N} p\, F_p \sin(\pi p\, \zeta_i)}{2\pi \nu_i}\right] . \qquad (12.22)$$

The equation (12.22) allows us to calculate the dependence of group velocity of lattice waves on the wave vector (on the wavelength). It is possible if the dispersion curve, that is the dependence frequency $\nu$ on reduced wave vector $\zeta$, has been measured experimentally.

Potassium is a typical simple metal. The model of the electronic gas (Chapter 5, Table 5.2) describes the cohesive energy of the bulk potassium well. The completely delocalized $s$ electrons ensure the interatomic bond in potassium.

Dispersion curves of longitudinal and transverse oscillations along direction a type of $[\zeta\,\zeta\,\zeta]$, that is L[111] and T[111] for potassium are presented in Figure 12.4. This direction is chosen because the distance between neighboring atoms is minimal along it in the body-centered cubic crystal lattice. Similar curves are typical for other alkali elements.

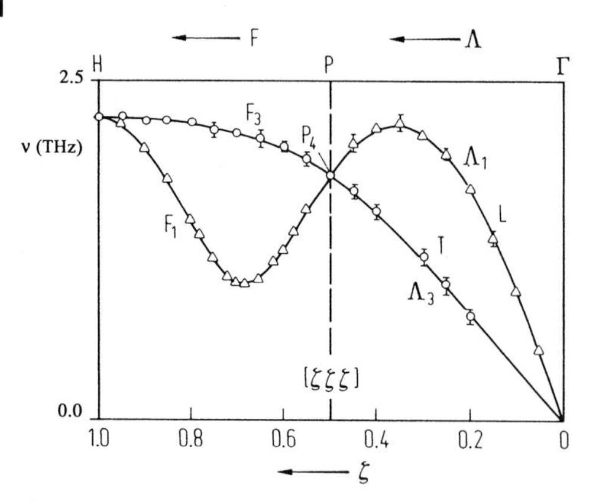


**Figure 12.4** Dispersion curves of longitudinal L[ζ ζ ζ] and transverse T[ζ ζ ζ] branches in potassium at 9 K. $\Gamma$, $P$ and $H$ are points in the Brillouin zone, $\Lambda$ and $F$ are directions [ζ ζ ζ], see Figure 4.13 and Table 4.1. Circles and triangles are experimental points obtained by the neutron diffraction method. The solid curves represent the fifth neighbor Born–von Karman fit.

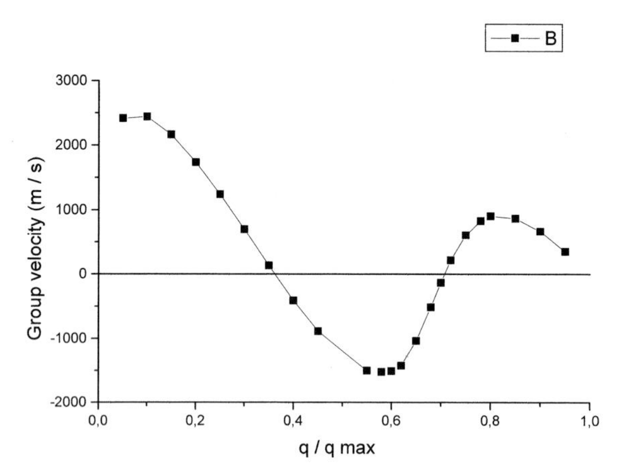


**Figure 12.5** The group velocity as a function of the wave vector for the longitudinal L[ζ ζ ζ] branch in potassium.

Figure 12.5 illustrates the dependence of the group velocity on $\zeta$ for potassium calculated according to (12.22). This curve has a typical shape for simple metals. Near point $\zeta = 0$ the dependence $\nu(\zeta)$ (Figure 12.4) is linear. This area corresponds to long waves $\lambda$. The product $\lambda\nu$ equals to the velocity of sound wave $v_{\text{sound}}$ in bulk potassium. Consequently, the angle of inclination of curve near zero $2b \cdot d\nu/d\zeta$ determines the velocity of sound wave. It is seen from Figure 12.5 that velocity of the longitudinal sound wave in potassium equals to 2400 m/s.

As value of $\zeta$ increases, the wavelength $\lambda$ decreases and becomes comparable with the interplanar spacings in crystal lattice. Backward wave (reflected mode) interacts with the forward wave and the velocity of the forward wave decreases (at

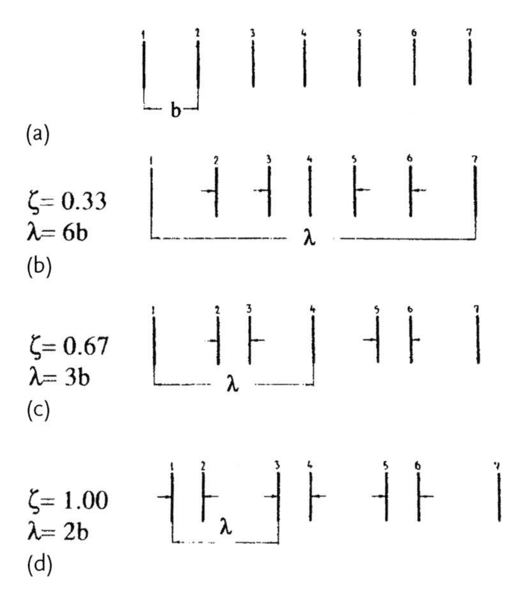


**Figure 12.6** The scheme of displacements of parallel atomic planes {111} during longitudinal vibrations in the crystal lattice. The panel (a) illustrates the undisturbed lattice; the panels (b)–(d) correspond to $\zeta = 0.33, 0.67, 1.00$, respectively, and illustrate the formation of standing waves. The arrows show the displacements of planes.

$0.15 < \zeta < 0.33$, Figure 12.5). At $\zeta = 0.33$ the group velocity equals to zero. This means that the forward wave and the backward wave are in antiphases. A standing wave is installed in the crystal lattice. In other words, the velocity of the phonon $L[\frac{1}{3}\frac{1}{3}\frac{1}{3}]$ equals to zero. As $\lambda$ continues to decrease the velocity becomes negative. Further the velocity increases and vanishes again at $\zeta = 0.67$ (phonon $L[\frac{2}{3}\frac{2}{3}\frac{2}{3}]$). This means that the corresponding planes are oscillating in antiphase again.

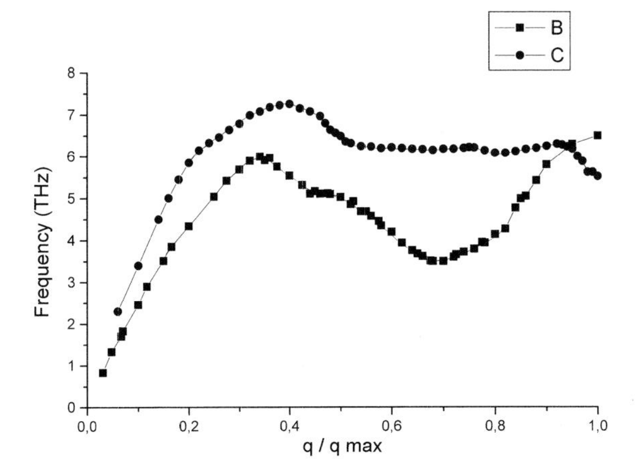


**Figure 12.7** The phonon dispersion curves for the longitudinal $L[\zeta\zeta\zeta]$ branch measured by the inelastic neutron technique: B, niobium; C, molybdenum.

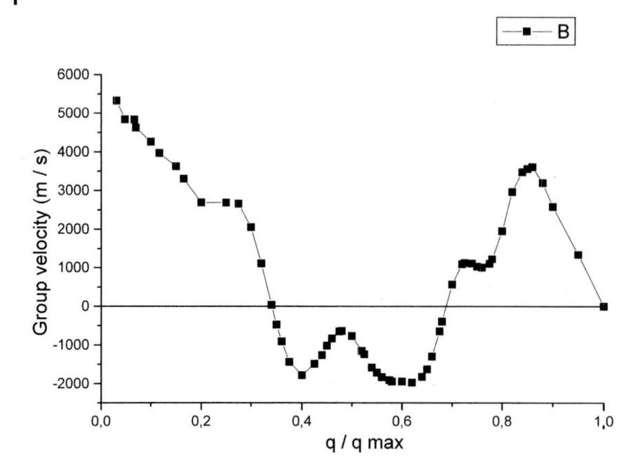


**Figure 12.8** The group velocity of the oscillation mode L[$\zeta\zeta\zeta$] as a function of the $q/q_m = \zeta$ ratio for niobium. The velocity was calculated from the experimental curve of $\nu(\zeta)$ for each $\zeta$ value according to (12.21) and (12.22). $N = 20$.

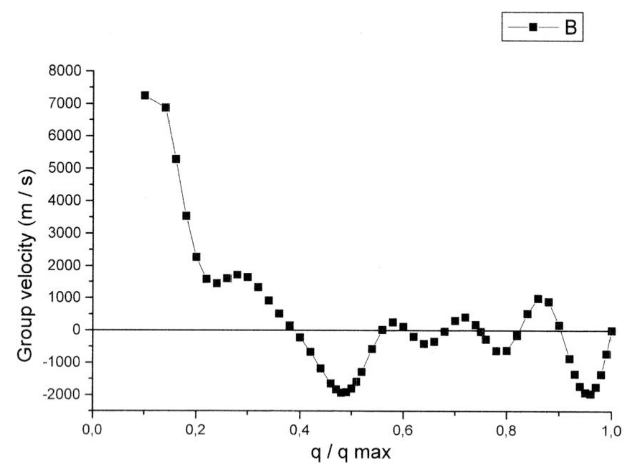


**Figure 12.9** The same as in Figure 12.8 for molybdenum. $N = 14$.

Figure 12.6 illustrates the vibration of crystallographic planes {111}. The normal mode $\zeta = 0.67$ of lattice vibration corresponds to the phonon L[$\frac{2}{3}\frac{2}{3}\frac{2}{3}$]. It can be seen that planes labeled 1 and 4 are at rest whereas the planes 2 and 3 are displacing in opposite directions.

In the normal mode $\zeta = 0.33$ planes 1 and 4 are at rest, but planes 2 and 3 move in the same direction. We see in Figure 12.5 that the group velocity equals to zero in both cases.

The lattice vibration in transition metals niobium and molybdenum turns out to have more complicated behavior. Figure 12.7 shows the dispersion curves for the longitudinal [111] branch measured by the inelastic neutron scattering technique for niobium and molybdenum. For niobium the branch exhibits a dip near $\zeta =$ 0.67, whereas the same branch for molybdenum is flat near $\zeta = 0.67$ but shows a dip near $\zeta = 1$.

The dependence of group velocity on $\zeta$ is shown in Figures 12.8 and 12.9 for niobium and molybdenum, respectively. The group velocity vanishes for niobium at $\zeta = 0.33$ and $\zeta = 0.67$ as well as for potassium. However, the dependence $v - \zeta$ for these transition metals consists of several maxima and minima. This is most probably related to involvement in a simultaneous displacement of a large number of crystal planes.

## 12.4 Vibration Frequencies and the Total Energy

An advanced approach to lattice dynamics involves the precise determination of the crystalline total energy as a function of the lattice displacement associated with a particular phonon. This method is commonly referred to as the frozen-phonon method. It utilizes first-principles band-structure techniques to obtain the total energy for each frozen-in position of the lattice. The phonon frequency can then be obtained from the resultant "potential-energy" curve.

The frozen-phonon method is an efficient method for investigating the interactions in solids. It is a completely adequate first-principle method for obtaining the lattice dynamics data of transition metals. Transition metals are an excellent example where electronic structure calculations of "frozen phonon" energies can provide much information on the interatomic bonding and the states near the Fermi energy that couple strongly to the phonons [49]. The phonon energies for many transition metals have been shown to be well described by calculations at wave vectors $\boldsymbol{q}$ along high-symmetry directions. For example, there is an interesting anomaly in the longitudinal frequency for $L(\frac{2}{3}, \frac{2}{3}, \frac{2}{3})$ in the bcc structure crystals zirconium, niobium, and molybdenum [50].

The motivation for studying the $L(2/3, 2/3, 2/3)$ is the marked difference in the phonon spectra of these metals at these wave vectors (compare frequencies for niobium and molybdenum at $\zeta = 0.67$ in Figure 12.7).

For a phonon mode of arbitrary $\zeta = q/q_{\mathrm{m}}$ (except the Brillouin zone boundary) the total energy per atom is given by

$$E = \frac{1}{4} m \omega_q^2 u_{0q}^2 , \tag{12.23}$$

where $m$ is the atomic mass, $\omega_q$ is the phonon frequency, $u_{0q}$ is the amplitude of the distorted wave. For a zone boundary phonon,

$$E = \frac{1}{2} m \omega_q^2 u_{0q}^2 . \tag{12.24}$$

One can calculate the total energy of a crystal as a function of lattice displacement based on first principles. The phonon frequency can then be obtained from the curvature of the curve energy-displacement for small displacements, (12.23) and (12.24).

In Figure 12.10 three neighboring atomic planes normal to the $[\bar{1}\bar{1}1]$ direction in the bcc structure are shown. The diagonal of the cube equals to $a\sqrt{3}$, where $a$ is

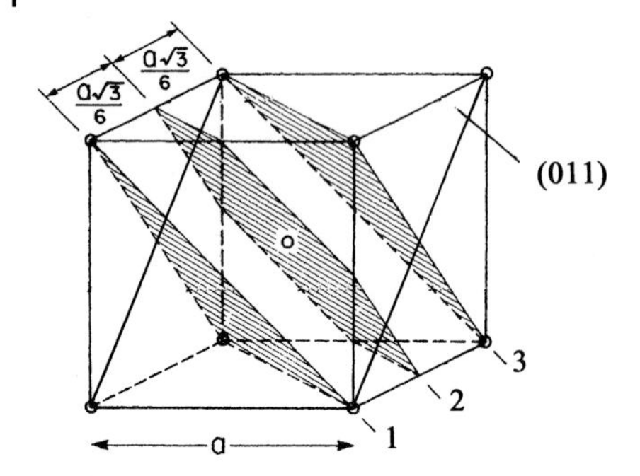


**Figure 12.10** The parallel $(\bar{1}\bar{1}1)$ planes in bcc crystal lattice. The (011) plane is perpendicular to a set of planes $(\bar{1}\bar{1}1)$ (see Figure 12.3). (After [50] by permission of American Physical Society).

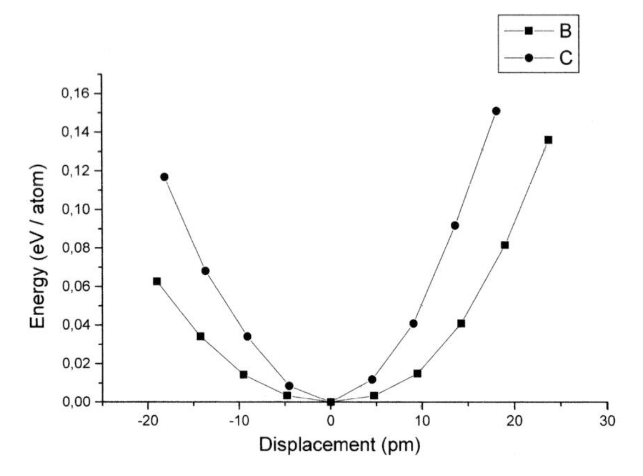


**Figure 12.11** Energy as a function of atomic displacements for niobium (squares and curve B) and molybdenum (circles and curve C). The points are the calculated data for $L(\frac{2}{3}, \frac{2}{3}, \frac{2}{3})$.

the crystal lattice parameter. It can be seen that minimal distance $b$ between the $(\bar{1}\bar{1}1)$ parallel planes equals to $a\sqrt{3}/6$. The $b$ values equal to 94.97 and 90.64 pm for Nb and Mo, respectively.

The phonon $L(\frac{2}{3}\frac{2}{3}\frac{2}{3})$ distortion corresponds to leaving every of 1 and 4 plane stationary and moving the remaining planes 2 and 3 toward each other or apart from each other (Figure 12.6c).

Figure 12.11 presents the calculated dependence of energy (per atom) of crystal lattice on atomic displacements. It can be seen that molybdenum has faster lattice than niobium. One can note also an anharmonicity of the dependence. Energy increases faster when atoms come closer to each other than when they move off. Figure 12.12 shows the forces per atom as a function of the atom displacement.

The level of interatomic forces for molybdenum is higher than that of niobium. If the displacement equals to 10 pm, for instance, the restoring force equals to 0.0045 and 0.0096 eV/(pm atom) for niobium and molybdenum, respectively (that is, 0.721 and 1.538 nN/atom).

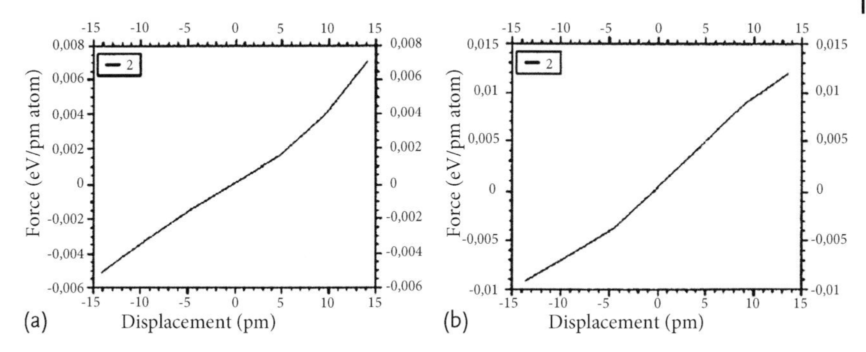


**Figure 12.12** The force acting on an atom of the metal as a function of the atom displacements: (a) niobium, (b) molybdenum.

Independently of any model one would expect a relative decrease in the phonon frequency of L[111] branch in the vicinity of the $(\frac{2}{3}, \frac{2}{3}, \frac{2}{3})$ position for any monoatomic bcc crystal, since for this vibration mode, the nearest-neighbor displacement between atoms in the [111] direction is preserved (this is illustrated in Figure 12.3, Figure 12.6c and Figure 12.14).

Thus, the corresponding restoring force vanishes for potassium, and the atomic displacements for this mode can be viewed as a shearing motion between chains of atoms along [111] direction. The absence of the dip for molybdenum (also in isoelectronic chromium and tungsten) is probably due to some special aspects of the electronic structure. An increase in the $L(\frac{2}{3}, \frac{2}{3}, \frac{2}{3})$ phonon frequency as one goes from niobium and zirconium to molybdenum is accompanied by an increase of the bonding d-like charge density along the nearest-neighbor [111] direction in the bcc lattice. The authors of [50] have determined using the force analysis that the increased frequency of this phonon for molybdenum over niobium is caused by the development of directional bonding from the additional occupied d states. Compare the outer electron shells:

$$^{41}\text{Nb: [Kr]}4\text{d}^4 5\text{s}^1\,, \quad ^{42}\text{Mo: [Kr]}4\text{d}^5 5\text{s}^1\,.$$

This gives rise to bond-bending forces which restores the equilibrium position much like in covalently bonded semiconductors. This is opposite of the behavior of simple metals where free-electron screening merely acts to reduce the ionic restoring force.

Figure 12.13 shows the calculated energy versus the displacement of this phonon for Mo, Nb, and Zr. The curvature agrees well with measured phonon frequencies and corresponds to the sharp dip in the phonon dispersion curves, which is a precursor to the phase transition that actually occurs in zirconium.

Let us look again at Figure 12.7. The frequencies that correspond to $\zeta = 0.67$ are 3.5 and 6.5 THz for niobium and molybdenum, respectively. Remember the simple formula for the spring pendulum, $\omega = \sqrt{k_{\text{el}}/m}$, where $\omega$ is the angle frequency, $k_{\text{el}}$ is the coefficient of elasticity and $m$ is the load mass. Vibration frequency increases when $k_{\text{el}}$ increases.

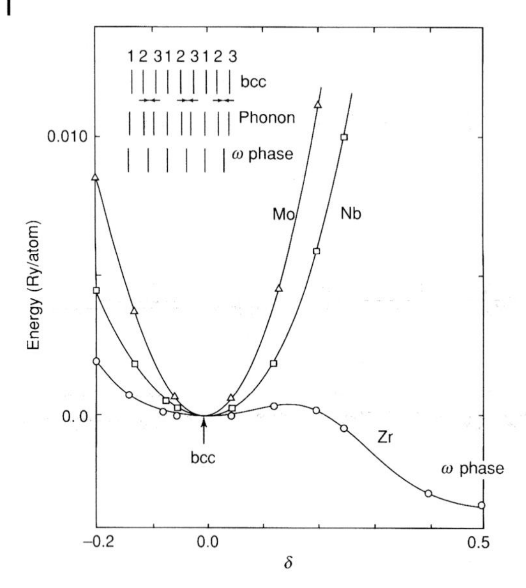


**Figure 12.13** Calculated total energy as a function of the displacement corresponding to the longitudinal vibration mode $L(\frac{2}{3}, \frac{2}{3}, \frac{2}{3})$. Body-centered cubic structure has a minimum energy for niobium and molybdenum. The minimum energy structure for zirconium at low temperature is called the "$\omega$ phase." Inset shows the displacements for the (111) planes of the bcc crystal. Planes 2 and 3 coincide for the $\omega$ phase (after [50] by permission of American Physical Society).

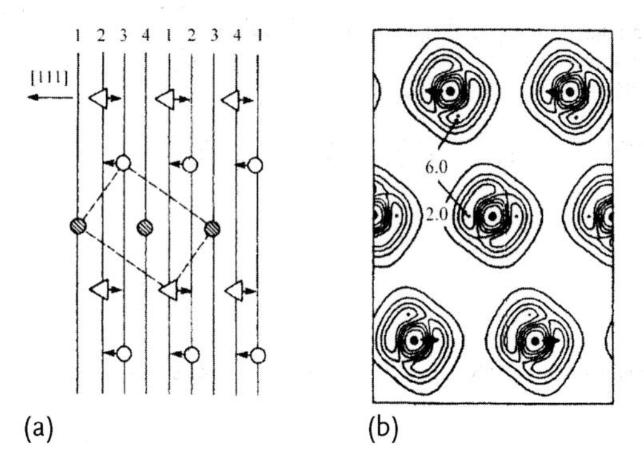


**Figure 12.14** Atomic displacements corresponding to $L[\frac{2}{3}\frac{2}{3}\frac{2}{3}]$ phonon in the (110) plane of bcc crystal lattice. Arrows indicate the directions of atomic displacements associated with the distortion caused by this phonon. (a) Scheme of vibrations: shaded circles, triangles, and circles denote atoms in planes 1, 2, and 3, respectively. (b) Contour plots of the valence electron density for molybdenum; atoms in planes 2 and 3 have been displaced by a small amount $\delta = 0.05$ (after [50] by permission of American Physical Society).

Figure 12.14a shows a side-view of the planes perpendicular to the [111] direction in the body-centered cubic crystal. The displacements of the atomic planes is similar to Figure 12.6c. Atoms in planes 1 and 4 are at rest, atoms in planes 2 and 3 displace in the opposite direction. From the figure it can be seen that the interatomic distances for neighboring atoms along the [111] direction remain fixed for the $L[\frac{2}{3}\frac{2}{3}\frac{2}{3}]$ phonon. Thus, one may consider this phonon as a shear motion of chains running along the [111] axis. This was the idea of Foreman and Lomer [48], see Section 11.2.

Contour plots of the electronic valence density for molybdenum are shown in Figure 12.14b. With small displacements the charge density for molybdenum shows strong d lobes giving a clear density maxima between neighboring atoms.

Increased frequency of the $L(\frac{2}{3}, \frac{2}{3}, \frac{2}{3})$ phonon for molybdenum compared to niobium is caused by the development of directional bonding from the additional occupied d states. This gives rise to bonding forces, which help to restore the equilibrium position of molybdenum atoms.

We see in Figure 12.9 that a relatively great number of planes in molybdenum are involved in the oscillation process in the 0.5–1.0 interval of $\zeta$. This confirms the existence of the strengthened interatomic d bonding between molybdenum atoms.

# 13
# Transition Metals

The transition metals are not describable by the conventional near-free electron model of the metallic bond since the valence d electrons remain relatively tightly bound to their parent atoms forming unsaturated bondings with their neighbors. These d bondings are responsible for the structural and cohesive properties of transition metals.

Figure 13.1 illustrates the dependence of the melting temperature of transition metals on number of electrons in the d shells. The transition metals have open d shells. Maximal number of electrons in d shell is known to be equal to 10. One can easily see the general "parabolic" trend of the melting temperature across a transition-metal row. A formation of a partially filled d band promotes an increase in the melting temperature. The largest melting temperatures have tantalum, tungsten, rhenium, and osmium, which contain 3, 4, 5, and 6 electrons in the 5d shell, respectively. Niobium, molybdenum, technetium, and ruthenium with 4, 5, 6, and 7 electrons in the 4d shell, respectively, are the high-melting elements compared with yttrium ($4d^1 5s^2$), zirconium ($4d^2 5s^2$), and rhodium ($4d^8 5s^1$).

It is readily seen from Figure 13.2 that energy of vacancy formation has the largest values for molybdenum, niobium, and rhodium with 5, 4, and 8 electrons in their 4d shell, respectively.

The regularity of this behavior, also found for cohesive energy, clearly suggests that the strength of interatomic bonding in transition metals must be related to the formation of the d band. The three periods of transition metal series also display the same pattern in bcc, fcc, hcp structures with the filling of the d shell. The effect reveals itself weaker for 3d shell. This can be related to magnetic properties of iron, cobalt and nickel.[1)]

1) The exchange interaction between valence electrons in these metals results in aligning their spins, so the spins of these electrons tend to line up. The exchange force is stronger than the opposite electrostatic force.

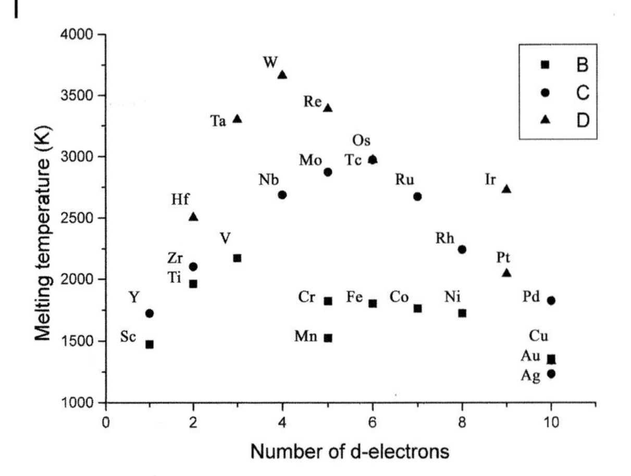


**Figure 13.1** Relationship between melting temperature of the transition elements and the number of electrons in the d shells. The symbols B show the elements with the 3d shells; the symbols C show the elements with the 4d shells, and the symbols D represent the elements with the 5d shells.

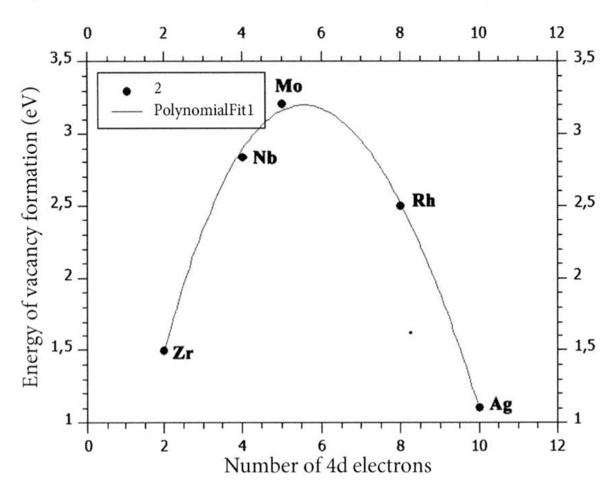


**Figure 13.2** Energy of vacancy formation $E_V$ of the 4d transition metals as a dependence on the number of d electrons. The experimental $E_V$ data from [11].

## 13.1 Cohesive Energy

The modern electron theory provides a firm basis upon which it is possible to build qualitative and quantitative models of solids. The density functional theory and the pseudopotential methods are applicable techniques for the investigation of transition metals.

Several effects occur as atoms condense to form a solid:

- the discrete energy levels of the atom broaden into bands;
- a shift in energy results;

- the hybridization of electrons takes place;
- the electrons change their occupation;
- finally, the motion of electrons in the solid becomes correlated.

The local density functional theory enabled authors of [51] to treat these effects successfully. The only input to their calculations was the atomic number. The authors used the theory of Kohn and Sham for all the electrons (core, valence, s, p, and d) focusing on exchange and correlation.

By means of elaborate numerical calculations, the enumerated above effects have been treated essentially without approximations. The systematic application of the theory to the third and the fourth row metals exhibits the important trends both within and between the rows.

Figure 13.3 shows the equilibrium nuclear separation, cohesive energy, and bulk modulus for 3d and 4d rows of transition metals. The atomic number increases in step of one from 19 to 31 in the left-hand column and from 37 to 49 in the right-hand column. Figure 13.3a shows the expected "parabolic" dependence of internuclear separation on the d shell filling. The computed values are correct within a few percent. The structures of metals with shells of $3d^1 4s^2$ (scandium), $3d^2 4s^2$ (titanium), $4d^2 3s^2$ (zirconium), $4d^{10} 5s^1$ (silver) are more "loose."

In Figure 13.3b the trend with atomic number is once again reproduced showing, in particular, a rapid increase in cohesion in both transition series as the bonding d states are occupied. The error in prediction of the cohesive energy is rather large for chromium, manganese, and ferromagnetic metals: iron, cobalt, and nickel.

Figure 13.3c compares the calculated bulk modulus with measured values. With the exception of four 3d transition elements that possess the strong magnetic effects the agreement is truly remarkable.

Once again we ascertain that the interatomic bonding is maximal at half-filling of the d shell leading to the maximum density, cohesive energy, and bulk modulus of transition metals.

A renormalized-atom method has been worked out by authors of [52] in order to investigate in detail and calculate the cohesive energy of the 3d and 4d transition metals. This method is structured in such a way that it is possible to separate the energy of formation of the metal starting from free atoms into a number of terms and examine the importance of each.

In the renormalization scheme one utilizes the free-atom s and d wave functions, truncates them at the radius of the Wigner–Seitz sphere and normalizes them within this sphere, thereby preserving charge neutrality. In this way the atoms are prepared approximately in the form in which they actually enter the solid metal therefore placing them together.

The Wigner–Seitz radius $r_{WS}$ is defined by

$$\frac{4}{3} r_{WS}^3 = \frac{V}{N} \tag{13.1}$$

where $V/N$ is the volume per atom.

The predicted cohesive energy of a transition metal is decomposed into five physical contributions as follows:

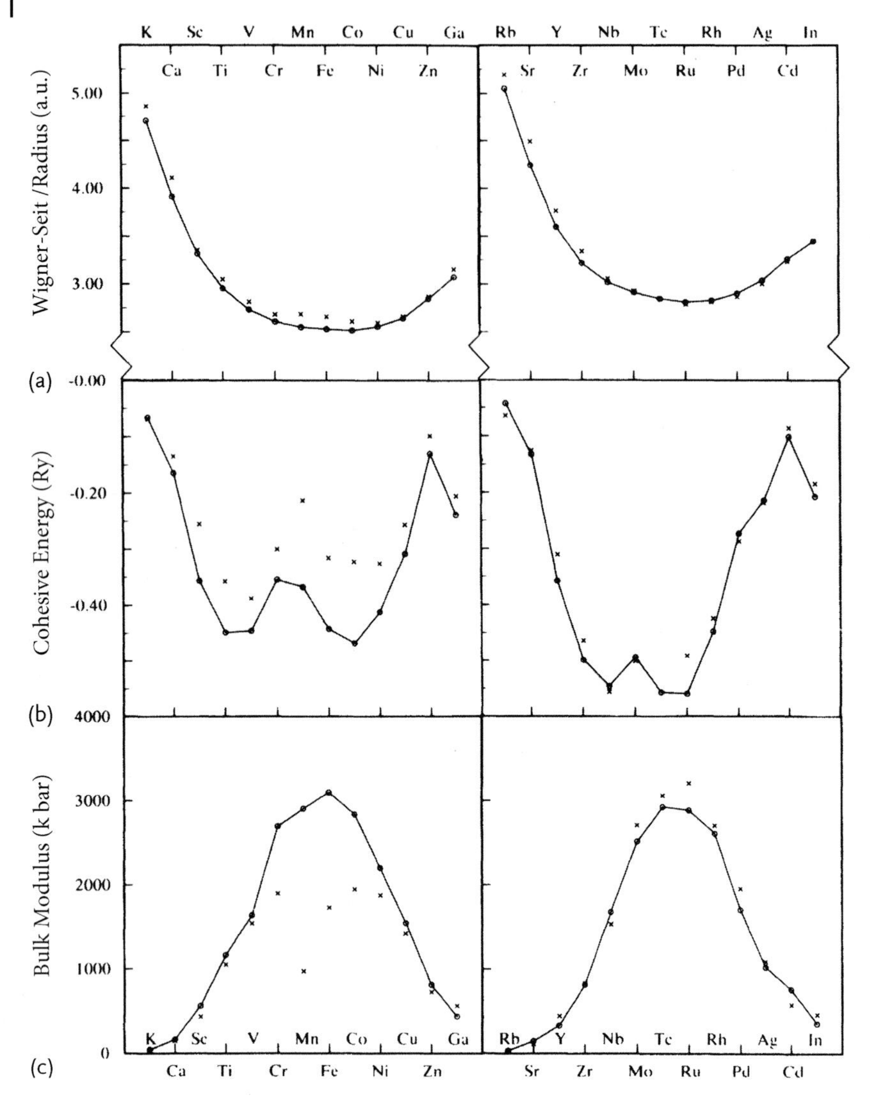


**Figure 13.3** Dependence of cohesive properties of transition metals on atomic number. Calculated data are indicated by circles and curves; measured values are indicated by crosses. (a) shows equilibrium nuclear separation; (b) presents cohesive energy in Ry/atom; (c) shows bulk modulus in kbar (1 kbar = 0.1 MPa). Data of authors [51].

- the atomic preparation energy required to excite a free atom from its ground state $d^n s^2$ electronic shell to average energy of the $d^{n-1}s^1$ shell, that is, to the configuration appropriate to the solid;
- the difference in total Hartree–Fock energy between the free $d^{n-1}s^1$ atom and the renormalized atom;
- the difference between the average energy of the free-electron band containing one electron and that of the renormalized atom s level;

- the change in one-electron energy per unit cell which results from the broadening of the renormalized atom d level in the d band;
- the change in one-electron energy per unit cell due to the hybridization of the conduction band and d band.

This method of calculation allows the cohesive energy to be decomposed into a number of contributions whose relative importance can be investigated as a function of valence and as a function of density. The separation of the calculated cohesive energy into its components leads to a qualitative understanding between attractive and repulsive forces which determines the equilibrium density.

Figure 13.4 displays the density of states of titanium at each of the five stages in the development of the full crystalline density of states. The average energy of the $3d^3s^1$ configuration is 0.141 Ry atom$^{-1}$ (1.92 eV atom$^{-1}$) above the $3d^24s^2$ state of the free atom. When the atom is renormalized, approximately 1.06 electrons are forced inside the Wigner–Seitz radius, decreasing the one-electron binding energy of the s and d orbitals. Because of the substantial cancellation of the double counted electron–electron repulsion terms, the net energy cost of renormalization is only 0.064 Ry atom$^{-1}$ (0.87 eV atom$^{-1}$).

In Figure 13.4d, the s level has been allowed to broaden into a free-electron band. The difference in one-electron energy between this band and the renormalized atom s level is $-0.033$ Ry atom$^{-1}$ ($-0.45$ eV atom$^{-1}$). The largest contribution to the cohesion of a transition metal with a particularly filled d band such as titanium is due to the broadening of the renormalized-atom d level into the d band. For titanium, the calculated contribution of d band broadening is $-0.375$ Ry atom$^{-1}$ ($-5.10$ eV atom$^{-1}$). In Figure 13.4f, it is shown that the hybridization of the s and d bands has occurred. Its effect is to push both s and d states away from the center of the d band. Since only the states whose energy has been decreased are filled, the net effect is a bonding contribution of $-0.161$ Ry atom$^{-1}$ ($-2.19$ eV atom$^{-1}$).

Summation of the listed contributions gives the cohesive energy of titanium $-0.364$ Ry atom$^{-1}$ ($-4.95$ eV atom$^{-1}$). The authors give an experimental value of $-0.37$ Ry atom$^{-1}$ ($-5.03$ eV atom$^{-1}$).

No evidence of d–d repulsion is found for the transition or noble metals. Instead the "spring" which holds the atoms apart is the result of the increasing kinetic energy of the conduction electrons as the density is increased. The d–d interaction is uniformly attractive and produces the minimum in the Wigner–Seitz radius near the center of the transition period.

Figure 13.5 illustrates in detail the calculated cohesive energy and its components for the 3d and 4d transition metals. The experimental value of metal density and fcc structure were assumed by calculations. For each element the contributions from atomic preparation, renormalization, conduction-band formation, and d band broadening plus s–d hybridization are indicated from left to right. The final calculated cohesive energy is represented by the filled block, while the experimental value is marked by the open block.

The atomic preparation energy is the energy required to take the observed ground state of the free atom and promote it into a singlet state with one valence

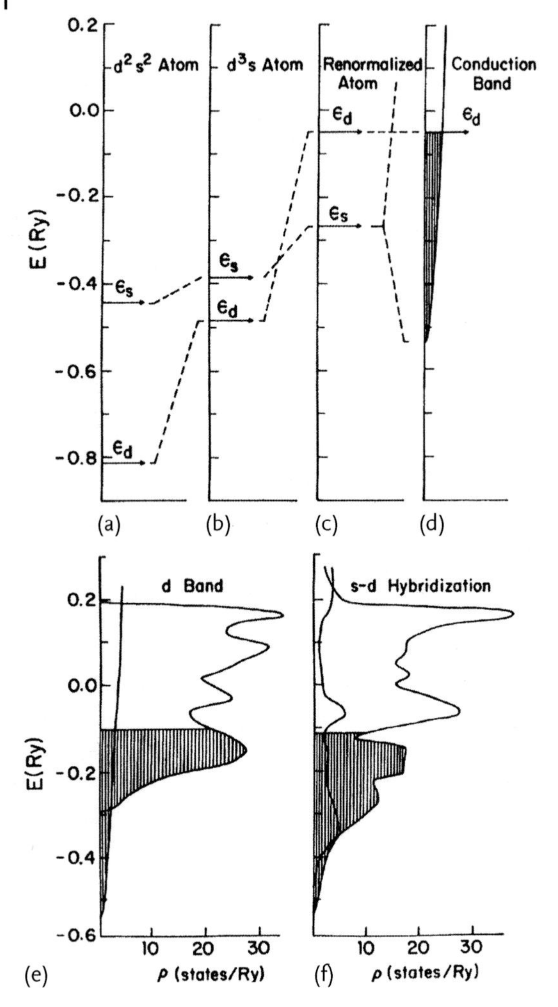


**Figure 13.4** Schematic electronic density of states for titanium at the various stages in the development of the crystalline density of states. $\varepsilon_\mathrm{d}$ and $\varepsilon_\mathrm{s}$ are "centers of gravity" for d band and s band, respectively; (a) is the free-atom state, (b) excited atom, (c) renormalized atom, (d,e) broadening of d level in d band, (f) s–d hybridization (after [52] by permission of American Physical Socity).

s electron, this being the situation close to the nonmagnetic bulk transition metal. This is a positive contribution largest in the middle of the series (or zero for noble metals copper and silver). For example, the average energy of the $3d^9 4s^1$ configuration of nickel relative to the $3d^8 4s^2$ ground state, energy of preparation, is 0.013 Ry = 0.177 eV.

The renormalization energy is the difference between the large repulsive contribution coming from the shift in the "center of gravity" of the d band $E_\mathrm{d}$ (see

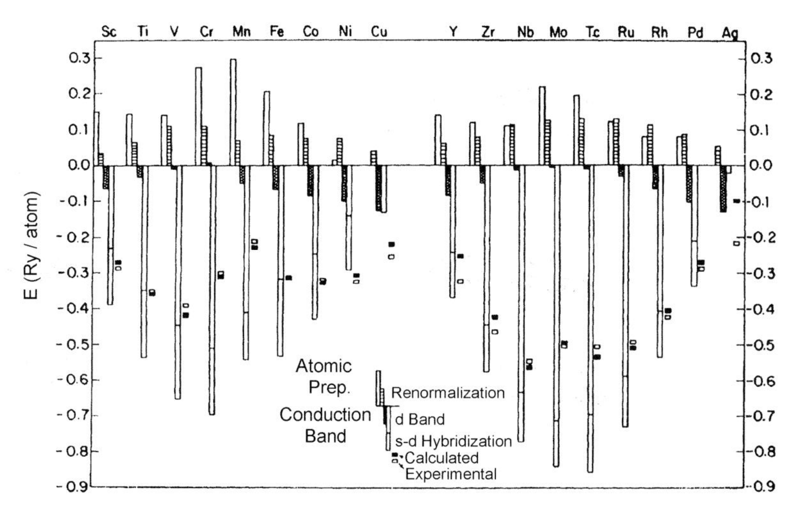


**Figure 13.5** Components of the cohesive energy for the 3d and 4d transition metal rows. For each element the experimental value is denoted by the open box, the calculated value by the filled box. Data of Gellat, Ehrenreich, and Watson ([52] by permission of American Physical Socity.

Figure 13.6) as the atoms bond together and a large double-counting term. This is a small positive contribution of about 1 eV (0.074 Ry) that is shown in Figure 13.5.

The sp band energy is the energy of electrons in the near-free electron conduction band. It is small and negative.

The d bonding energy $U_{\mathrm{dbond}}$ is energy of the d electrons that is measured with respect to the "center of gravity" of the tight binding band. This part of the total energy is given by

$$U_{\mathrm{dbond}} = \int_{E_\mathrm{b}}^{E_\mathrm{F}} (E - E_\mathrm{d}) n_\mathrm{d}(E) dE \tag{13.2}$$

where $n_\mathrm{d}(E)$ is the d band density of states. The energy $U_{\mathrm{dbond}}$ is large and negative. It is of the order of −6.8 eV for manganese in the middle of 3d series and −10.2 eV for molybdenum in the middle of the 4d series. The d bonding energy is analogous to the energy of covalent bonds for the s valent molecules. This contribution to

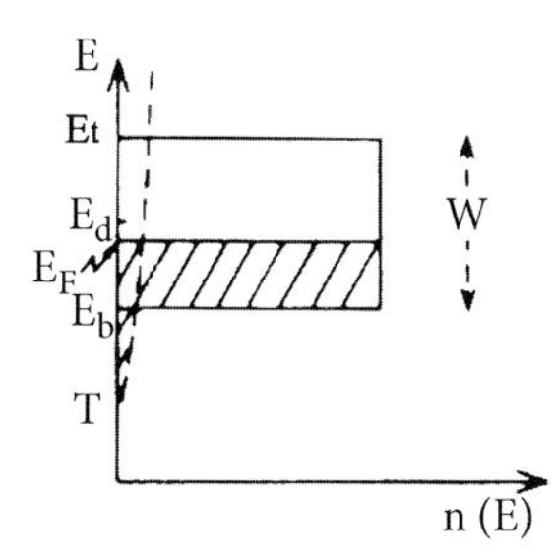


**Figure 13.6** A scheme of a transition metal density of states. The dashed curve is the s,p band, solid curve is the d band. $W$ is the band width. $E_\mathrm{F}$ is the Fermi energy, $E_\mathrm{b}$, $E_\mathrm{d}$, and $E_\mathrm{t}$ mark the bottom, "center of gravity" and top of the d band, respectively. Hybridization of electrons is neglected.

the interatomic bonding vanishes for copper and silver with their nominally-full valence d shells.

The integral (13.2) may be evaluated as [3]

$$U_{\mathrm{dbond}} = -\frac{1}{20} W N_{\mathrm{d}}(10 - N_{\mathrm{d}}) \tag{13.3}$$

where $W$ is the d bandwidth, $N_{\mathrm{d}}$ is number of electrons in the d band, which can hold exactly ten electrons when it is full. This equation displays the parabolic variation of cohesive energy across 3d and 4d series shown in Figure 13.5.

The sp-d hybridization energy is the contribution from mixing between sp and d bands. It is negative taking the values from −1.9 to −3.2 eV.

Figure 13.5 shows an excellent fit between calculated and experimental values of cohesive energy for majority of metals. These values coincide for Ti, Cr, Fe, Co, Mo, Ru. For other metals the data agree within 7%, with the exception of Cu, Y, and Zr. We see from this figure that the atomic preparation energy and the normalization energy are positive. Other contributions to the total energy are negative and ensure the interatomic bonding. The parabolic dependence of cohesive energy on number of d electrons is obvious. The dominating contributions to the cohesive energy result from the formation of d bands and the hybridization of the conduction and d bands. From Figure 13.5 it is clear that these terms are the largest part of the cohesive energy of transition metals. The nearly parabolic shape is the generic feature of the d band contribution.

## 13.2 The Rectangular d Band Model of Cohesion

The simplest model for describing the strength of interatomic bonding in transition metals is based on the evaluation of the bond integrals $\mathrm{dd}\sigma, \mathrm{dd}\pi, \mathrm{dd}\delta$ [3]. Density of states for this model is illustrated in Figure 13.6.

The total energy per atom consists of attraction contribution (13.3) and repulsive contribution,

$$U_{\mathrm{tot}} = U_{\mathrm{dbond}} + U_{\mathrm{rep}} \tag{13.4}$$

where the repulsive contribution is assumed to be pairwise, giving

$$U_{\mathrm{rep}} = \frac{1}{2N} \sum_{i,j} \Phi_{\mathrm{rep}}(R_{i,j}) \tag{13.5}$$

The d orbitals are shown in Figure 3.8. The dominant contribution is $\mathrm{dd}\sigma$. The author of [3] treats dependence of repulsive and attractive forces $U_{\mathrm{dbond}}$ and $U_{\mathrm{rep}}$ on distance in similar way as treatment of molecules. Figure 13.7 illustrates the theoretic predictions and the experimental values across 4d series.

One can see from Figure 13.7 that the rectangular model is able to qualitatively account for the observed trends in cohesive energy, equilibrium nearest-neighbor

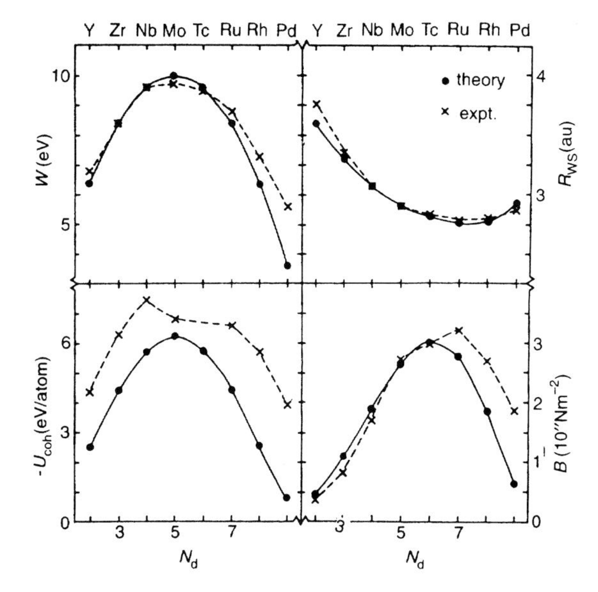


**Figure 13.7** The theoretical (•) and experimental (×) values of the equilibrium bandwidth $W$, Wigner–Seitz radius $R_{WS}$, cohesive energy $U_{coh}$, and bulk modulus $B$ of the 4d transition metals. $N_d$ is the number of electrons in d shell. The "experimental" data for the bandwidth are taken from the local density calculations (after [3] by permission of Oxford University Press).

distance and bulk modulus across the nonmagnetic transition metal series. The parabolic behavior of the cohesive energy reflects the initial filling of the bonding d states, followed by antibonding states. The skewed behavior of the equilibrium nearest-neighbor distance, on the other hand, reflects the competition between the attractive bonding term, which varies parabolically with band filling, and the repulsive overlap term, which at a fixed internuclear separation decreases monotonically across the series as the size of the free atom contracts.

## 13.3 Electronic Structure

The d states of transition metals retain much of their atomic character in the solid.

The promotion energies for transferring a d electron to an s state is presented in Figure 13.8. It can be seen from the solid line in this figure (experiment) that there is a gain in energy $\Delta_{sd}$ for all metals except chromium and copper. The electron shells $3d^{n-2}4s^2$ are more stable than $3d^{n-1}4s^1$ shells.

The results of simulating for transition metals illustrate a large difference between the s and p valence orbitals that are delocalized and the d states that are much more localized. Exchange and correlation of electrons are much more expressed in highly localized orbitals.

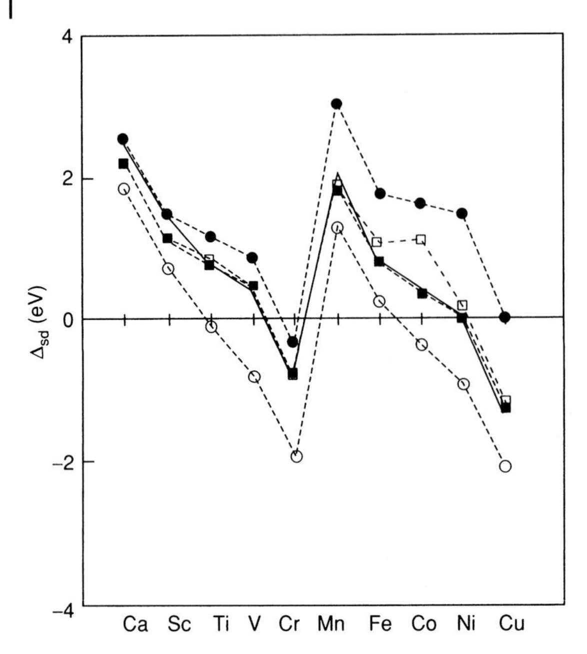


**Figure 13.8** Promotion energies for d → s electrons in the 3d transition metal series. The energies $\Delta_{sd} = E_{total}[3d^{n-1}4s^1] - E_{total}[3d^{n-2}4s^2]$ are shown from experiment (solid line) and from calculated total energy differences with different functionals (after [1]).

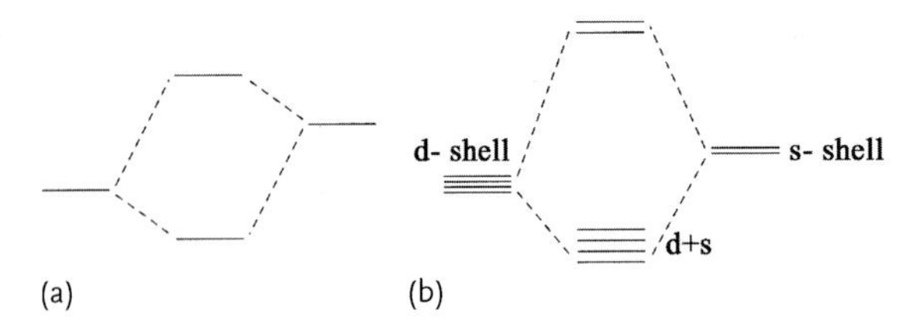


**Figure 13.9** Hybridization of electrons. (a) A pair of electron levels; (b) a twofold and a fourfold nearly degenerate set of levels. In each case "the center of gravity" of the sets of levels is preserved upon mixing. (After [49]).

Although the noble metals such as silver and copper have filled d bands, their bond shares important features in common with the elements to their left in the Mendeleev periodic table.

Hybridization between electron levels is an extended effect that results in a decrease of the total bonding energy. One defines the hybridization as a mixing of electrons that belong to different shells. In Figure 13.9 a two-level situation is plotted. The two levels of energy may be associated with orbitals centered on the adjacent atoms of a diatomic molecule or the levels might reside at the same atomic site in a crystal where they are allowed to mix by symmetry. If the orbitals mix, there will be an orbital with a high energy and the orbital with the lower energy.

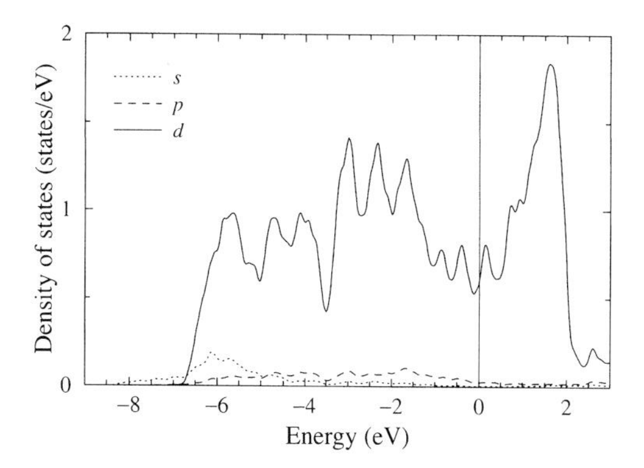


**Figure 13.10** The s, p, and d partial densities of states within the atomic sphere (a muffin-tin approximation) for hcp ruthenium. The outer electronic shell is $^{44}$Ru: [Kr]4d$^7$5s$^1$. Zero at the energy axis corresponds to the Fermi energy (after [49]).

The "center of gravity" of the energies of the two remains fixed and so bonding energy is gained only if the lower bonding level is occupied and the higher-lying antibonding level remains empty. That "the center of gravity" stays fixed is a result. Figure 13.9b presents a slightly more complicated case, that is, a different number of the nearly degenerate states. Again the "center of gravity" holds fixed and the maximum bonding energy is gained if all the lower bonding levels are filled. For many alloys of transition metals the d bands of one constituent lie well below those of the other. An alloy formation will occur at the composition where the low-lying hybridized levels are filled and the higher-lying levels are empty.

The d eigenfunction is of significance in the transition metal formation. Figure 13.10 shows the partial density of states for ruthenium. There are several features of the partial densities of states for this metal, which are typical for the transition metals in general. First, the density of states is dominated by the d component. The onset of the conduction bands has s-like character below the d band, followed by significant d-non-d hybridization at the bottom of the d band.

Self-consistent calculation using the pseudopotential method enables one to determine the electronic structure of a transition metal. The data on the band structure, the density of states, and the charge distribution were obtained for bulk niobium [54]. The results were compared with an experiment that was carried out by the photoemission technique.

Table 13.1 shows the levels of energy for $Nb^{4+}$ ion. One usually assumes that the collective (delocalized) electrons in solids have an energy higher than −6.0 Ry. The electrons of the 4d$^4$ and 5s$^1$ shells in niobium atom are the valence electrons. The data on 5p, 5d, 6s shells in Table 13.1 illustrate the energy of excited states. The agreement between the calculated and measured levels of energy is good.

The histogram density of states for niobium is shown in Figure 13.11. The calculated peak positions in occupied states (−2.5, −1.4, and −0.4 eV) agree with the data of photoemission. Figure 13.12 presents the charge distributions in niobium.

**Table 13.1** Comparison of the calculated energy levels with experimental data for $Nb^{4+}$ ion.

| Level | Calculated energy (Ry) | Experiment (Ry) |
|---|---|---|
| 4d | −3.657 | −3.63 |
| 5s | −2.953 | −2.95 |
| 5p | −2.448 | −2.45 |
| 5d | −1.725 | −1.71 |
| 6s | −1.635 | −1.56 |

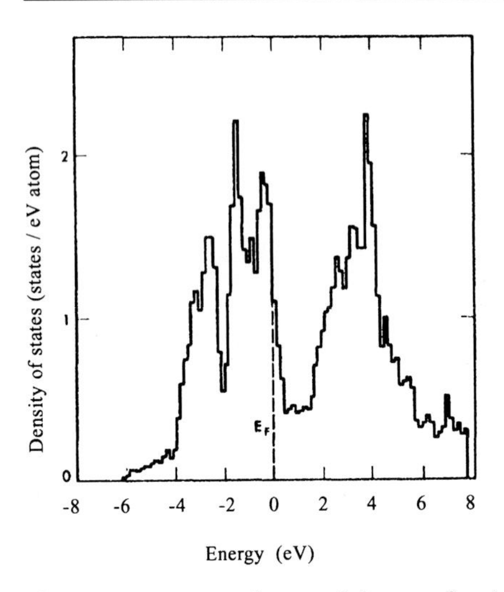


**Figure 13.11** Density of states of electrons for niobium. The zero energy corresponds to the Fermi level (after [54]).

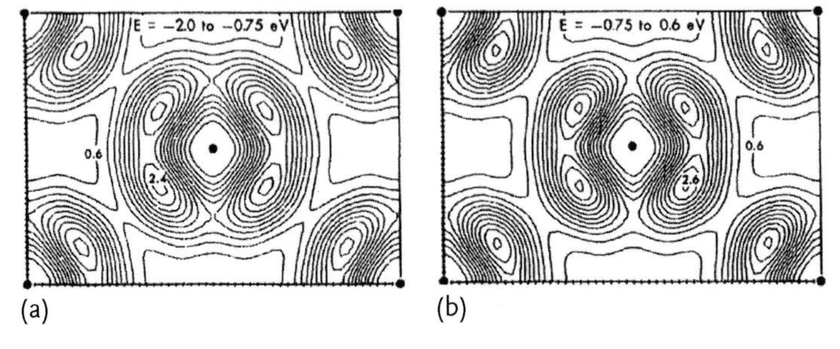


**Figure 13.12** Partial charge density of states for niobium; plane (011). Energy of electrons ranges from −0.20 to −0.75 eV (a) and from −0.75 to 0.60 eV (b) (after [54]).

One can easily see the lobes of the electron clouds that look like bridge-like tension bars along the directions of [111] type.

| Sc | Ti | V | Cr | Mn | Fe | Co | Ni | Cu |
|---|---|---|---|---|---|---|---|---|
| hcp(bcc) | hcp(bcc) | bcc | bcc | $\alpha$ Mn | bcc | hcp | fcc | fcc |
| Y | Zr | Nb | Mo | Tc | Ru | Rh | Pd | Ag |
| hcp(bcc) | hcp(bcc) | bcc | bcc | hcp | hcp | fcc | fcc | fcc |
| La/Lu | Hf | Ta | W | Re | Os | Ir | Pt | Au |
| hcp(bcc) | hcp(bcc) | bcc | bcc | hcp | hcp | fcc | fcc | fcc |

**Figure 13.13** Crystal structures of transition metals.

## 13.4 Crystal Structures

The crystal structures of elemental transition metals form a well-ordered pattern (Figure 13.13).

The top row in Figure 13.13 from Sc to Cu presents the transition metals of 3d series, the middle row presents the metals of 4d series, and the lower row shows the transition metals belonging to 5d series. Every series ends with noble metals copper, silver, and gold, respectively. The d shell of these metals is completely filled.

The reason that the transition elements have hcp or bcc or fcc crystal structures becomes clear when one takes a look at densities of states for the three hypothetic structures of ruthenium (Figure 13.14). It is seen that the occupied one-electron states are pushed down and away from the Fermi level at the expense of the unoccupied states. This stabilizes the hcp structure relative to its fcc and bcc competition. Similarly, there is a bonding–antibonding hollow in the hcp density of states at band fillings appropriate to Tc, Ru, Re, and Os that is more clearly established than that of for the fcc structure, thus stabilizing the hcp structure relative to the fcc for these elements.

The Fermi level falls in the characteristic deep hollow in the bcc density of states for the V and Cr column elements.

As the d shell of transition metals fills by electrons the crystal structure changes as follows (number of electrons in d shell is given in parentheses):

$$\begin{aligned} &\text{hcp}\,(1\text{ or }2) \rightarrow \text{bcc}\,(3, 4\text{ or }5) \rightarrow \text{tcp}(\alpha\text{-Mn})\,(5) \\ &\qquad \rightarrow \text{hcp}\,(6\text{ or }7) \rightarrow \text{fcc}\,(8, 9\text{ or }10)\ . \end{aligned}$$

Manganese, iron, and cobalt have structures different from the other members of their columns, which is generally attributed to magnetism. Nonmagnetic manganese and iron would be expected to be hcp and cobalt would be expected to be fcc crystal lattice.

## 13.5 Binary Intermetallic Phases

The interatomic bond in solid solutions is affected by their composition. Figures 13.15 and 13.16 illustrate an influence of small quantities of Mo or W atoms

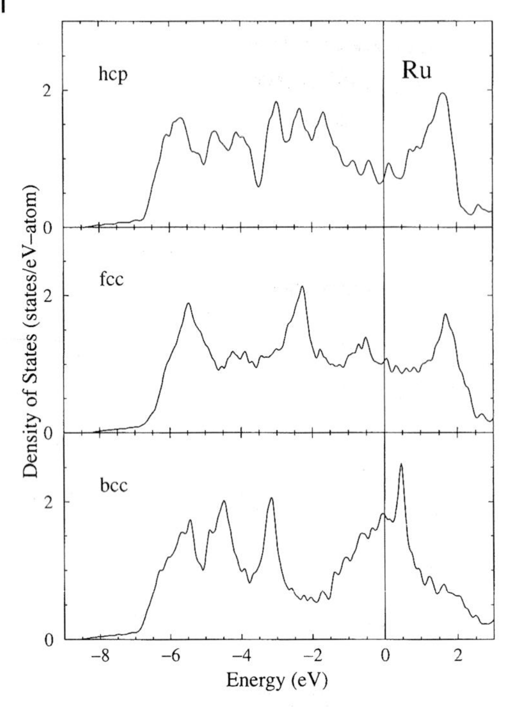


**Figure 13.14** The calculated densities of states for ruthenium in the hcp, fcc, and bcc structures, all at the same volume. Ruthenium has the hcp crystal lattice. The energy zero is the Fermi level (after [49]).

in iron on mean-square amplitude of atomic vibrations. The reduction of the amplitudes in solid solutions is distinct at all temperatures from 673 to 973 K. The influence of tungsten is greater than that of molybdenum. The addition of 7 at.% of tungsten to iron causes the mean-square amplitude to diminish threefold.

The 5d transition elements molybdenum and tungsten supply the crystal lattice of iron with delocalized electrons. This results in strengthening in the interatomic bonding.

Let us consider the formation of intermetallic compounds for zirconium. Table 13.2 and Table 13.3 show the melting temperatures of binary phases of the 4d series elements.

**Table 13.2** The valence electron shells and melting temperatures of some transition metals of 4d series.

| Metal | $^{40}$Zr | $^{44}$Ru | $^{45}$Rh | $^{46}$Pd |
|---|---|---|---|---|
| Valence shell | $4d^2 5s^2$ | $4d^7 5s^1$ | $4d^8 5s^1$ | $4d^{10}$ |
| Melting temperature (°C) | 1830 | 2400 | 1970 | 1553 |

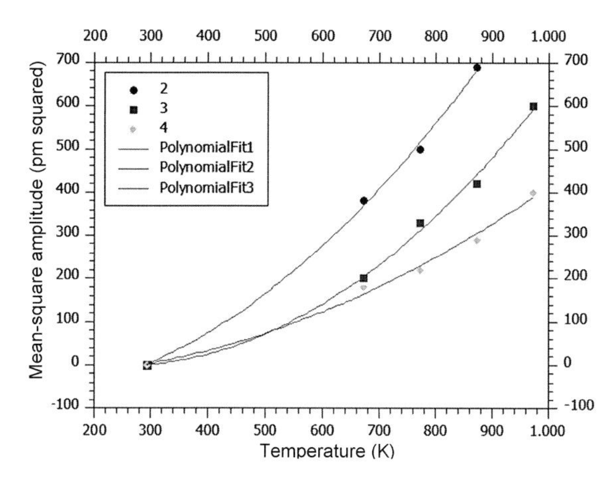


**Figure 13.15** Dependence of mean-square amplitude of the atom vibrations on composition and temperature: 2) $\alpha$-Fe; 3) $\alpha$-Fe+1.2% Mo (wt%); 4) $\alpha$-Fe+3.4% Mo (wt%).

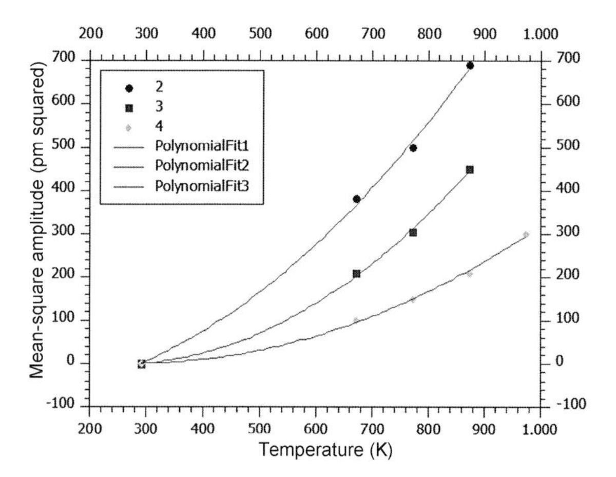


**Figure 13.16** Dependence of mean-square amplitude of the atom vibrations on composition and temperature: 2) $\alpha$-Fe; 3) $\alpha$-Fe+1.7% W; 4) $\alpha$-Fe+2.7% W.

The strength of interatomic bonding is expected to be maximal when the atoms have the maximum number of unlike nearest neighbors. The number of unlike nearest atom pairs A–B goes as B(1 – B), where B is the fraction of the second component, with the maximum at 50 at.% composition. As one can see from Tables 13.2 and 13.3, this is true for RuZr but is not for ZrPd. The $ZrPd_3$ phase have

**Table 13.3** Melting temperatures of intermetallic compounds.

| **Compound** | **ZrRu** | **$ZrRu_2$** | **$ZrRh_3$** | **ZrRh** | **ZrPd** | **$ZrPd_2$** | **$ZrPd_3$** |
|---|---|---|---|---|---|---|---|
| Melting temperature (°C) | 2130 | 2000 | 1900 | 1910 | 1600 | 1600 | 1900 |

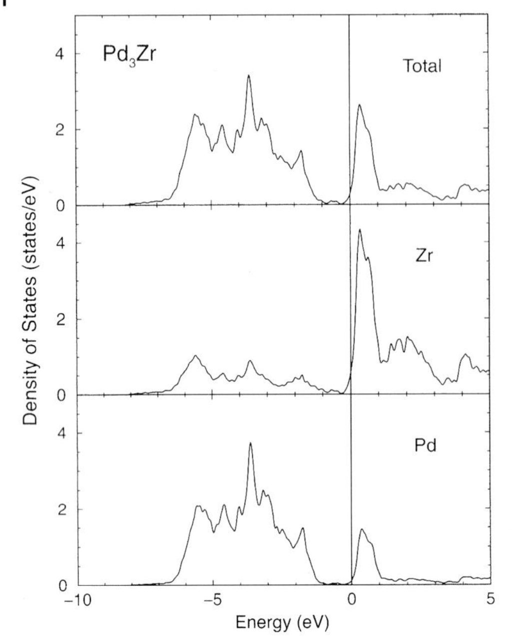


**Figure 13.17** The total (per atom) and local densities of states for $ZrPd_3$ (after [49]).

a greater melting temperature than ZrPd phase and even greater than zirconium and palladium itself.

It is seen in Figure 13.17 that the energy distribution of the d electrons in the compound is palladium in character. The hybridized d electrons of zirconium are accommodated to low energy levels of the $ZrPd_3$. The scope, to which the bonding states are filled and the antibonding are empty, is often critical to the competition among different structures at the same composition or among alloy phases of differing composition.

The Zr-Pd system, whose phase diagram is shown in Figure 13.18, shows that its maximum bonding, as manifested by the melting temperatures, is skewed well away from 50%. If palladium is replaced by ruthenium, each ruthenium atom can accommodate the electrons from one zirconium atom, with the result that the Zr-Ru alloy system has its maximum melting temperature (2130 °C) at a composition of 50 at.% (Figure 13.19).

The heats of formation $\Delta H$ for ordered stoichiometric alloys were determined with self-consistent linear-augmented calculations [53]. For 50 : 50% alloys of titanium, zirconium, and hafnium with the heavier 4d and 5d elements the agreement between the theory and experiment was of the order of the scatter of the experimental data. For instance, heat of formation was found for the chemical compound of RuZr to be equal to −0.75 eV/atom; according to the calorimetric measurements $\Delta H = -0.79$ eV/atom.

Figure 13.19 presents the phase diagram Zr-Ru. The density of states for Zr, Ru, and the ZrRu and $Ru_2Zr$ compounds is shown in Figure 13.20. One can see,

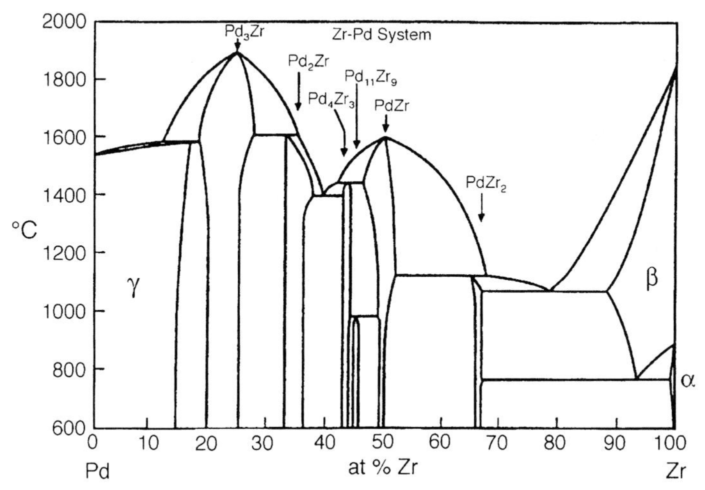


Figure 13.18 The Zr-Pd phase diagram.

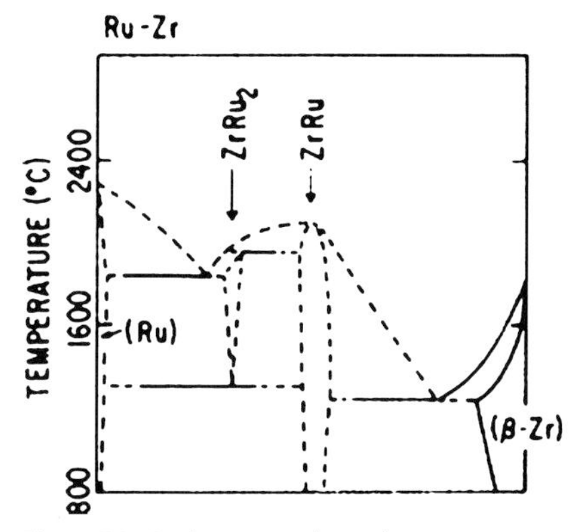


Figure 13.19 The Zr-Ru phase diagram.

comparing two intermetallic phases that in the case of RuZr, a compound with substantial $\Delta H$, the Fermi level falls at the dip having the ruthenium character. In contrast, $Ru_2Zr$ compound has Fermi level falling at the peak of the electron density of states for ruthenium. Accordingly, it has modest values of $\Delta H$ and melting temperature.

There are a number of general features of binary phase diagrams. As a function of composition, a number of ordered compounds exist often in intrinsically different crystal structures. Some structures occur over an extended composition range, while others only over a very narrow range. Some phases exist from low temperatures all the way up to melting temperature, while others exist either only at high or only at low temperatures. Finally, the melting temperature is not uniform across the full composition range, nor is it a linear interpolation between the elemental melting temperatures. For an alloy to be stable, it must be more stable than other phases in a given concentration range.

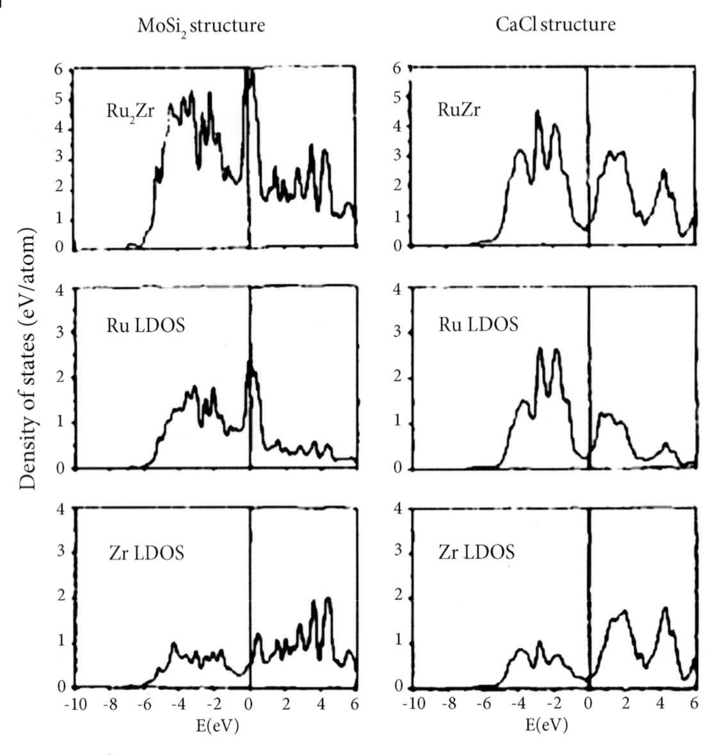


**Figure 13.20** Local density of states (LDOS) for Zr, Ru, RuZr, and $Ru_2Zr$. Note the dip at the curve for RuZr at the Fermi level. Melting temperatures are 2273 and 2403 K for $Ru_2Zr$ and RuZr, respectively.

In Figure 13.21, the heats of formation for a hypothetical A-B alloy are shown schematically. The various ordered phases, $\alpha, \beta, \gamma, \delta$ all have the heats of formation $\Delta H < 0$.[2] This means only that the compound is stable relative to a two-phase mixture of A and B. Each stable phase must lie below the tie line connecting any two other phases. In the figure, the phase stability is dominated by the heat of the $\gamma$ phase. Although the heat for $\delta$ is the second largest, it is still less than a two-phase mixture of $\gamma$ and pure B. On the other hand, $\beta$ is stable relative to a two-phase mixture of $\gamma$ and A, and thus would be stable if only these alloys were considered. However, the presence of the $\alpha$ phase suppresses the $\beta$ phase even though the heat for $\beta$ is more binding than for $\alpha$. Thus, from this simple set of heats, we would expect two stable phases $(\alpha, \gamma)$ and a two-phase region in between. Because of the significantly large binding of $\gamma$, we would expect a peak in the melting temperature, as seen in the Zr-Pd case (Figure 13.18).

2) The heat of formation $\Delta H$ is an enthalpy, not a free energy. Limiting the consideration to $\Delta H$ alone neglects entropic contributions, so this discussion of phase stability is restricted to low temperatures.

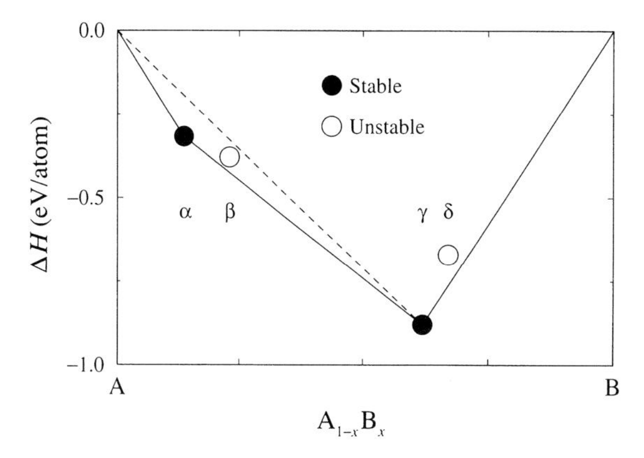


**Figure 13.21** Schematic diagram of the heats of formation for various potential phases $\alpha$, $\beta$, $\gamma$, $\delta$ for a hypothetical $A_{1-x}B_x$ alloy. The tie lines connecting the stable phases are also shown (after [49]).

## 13.6 Vibrational Contribution to Structure

Polymorphism, that is, the ability of atoms of a given element to unite into more than one crystal structure, is very common among the elements [55]. More than one-third of them show polymorphism at ordinary pressure. The free energy curves of two polymorphic structures intersect each other. There must be a difference in heat capacities of these structures. For metals, important contributions to the heat capacity come from the lattice vibration, the conduction electrons and in some cases magnetic excitation or localized electron states.

The free energy due to harmonic lattice vibrations (or equivalently the Debye temperature) is approximately the same for bcc, fcc, and hcp structures but with a significant tendency for the bcc value to be a few percent lower. The more open bcc structure has a transverse phonon mode with a particularly low frequency which causes a more rapid decrease in the free energy with temperature. On cooling, sodium and lithium transform partially from bcc to hcp at very low temperatures ($0.1$–$0.2\,T_{\mathrm{m}}$). Calcium, strontium, beryllium, and thallium transform to a bcc phase at high temperatures ($0.66$–$0.98\,T_{\mathrm{m}}$) when there is a considerable anharmonic contribution to the free energy.

The latent heat at a polymorphic transformation equals to $T\Delta S$ where $\Delta S$ is the entropy difference between two structures. The bcc phase has the higher entropy due to the sharper peak in the $N(E)$ curve (Figure 13.22). Quite generally, if $N(E)$ has a narrow peak in the vicinity of the Fermi level, this peak does not contribute to the heat capacity at high temperatures.

The electrons in simple metals are well described by a free electron gas. Since the atomic volume is changed by only a few percent as a result of polymorphic transformations, the electronic contribution to a free energy does not show any significant structure dependence for these metals. For transition metals, the elec-

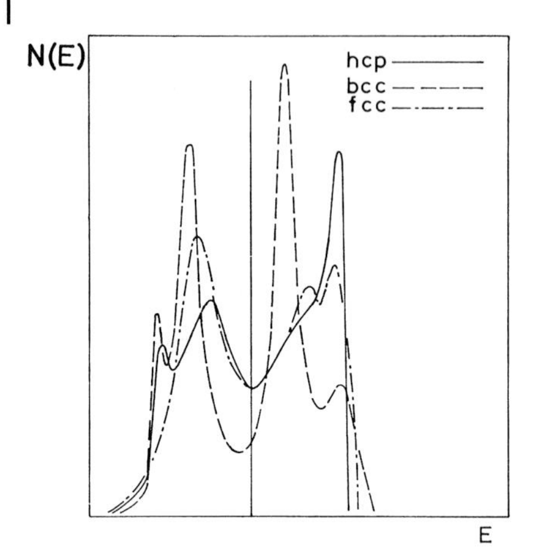


**Figure 13.22** Typical dependence on lattice structure for d band density of states in transition metals. Arbitrary units. The vertical line correspond to the Fermi energy.

tronic contribution to the free energy may be significant. Transition metals, on the other hand, have a d band density of states which may vary considerably with the lattice structure and hence be a reason for polymorphism. For iron, it is obvious that the combined electronic contribution plus magnetic free energy is essential in order to explain its polymorphism. It has been shown [56] that although the electronic contribution has to be taken into account for metals like titanium and zirconium, it is still less important than the vibrational free energy.

The thermodynamical quantity determining the phase stability is the Gibbs free energy $G(p, T)$, that is, the free energy at constant pressure and temperature. At $T = 0$ this reduces to the enthalpy. Thus, $\Delta H$ determines the phase stability at low temperatures, but at higher temperatures entropy effects may play a leading role in determining phase stability [49]. There are a number of contributions to the entropy: electronic, vibrational, and configurational. The vibrational modes of the crystal are an important contribution to the free energy of the system: the magnitude of the vibrational entropy is generally larger than the electronic one, as manifested in the temperature dependence of the specific heat. The vibrational free energy can be written in terms of the phonons as

$$F_{\text{vib}} = k_{\text{B}} T \sum_i \ln\left(2 \sinh \frac{\hbar\omega}{2k_{\text{B}} T}\right) \tag{13.6}$$

where the sum is over all (possibly temperature-dependent) phonon frequencies $\omega_i$.

The vibrational entropy is then given by

$$S_{\text{vib}} = \frac{\partial F_{\text{vib}}}{\partial T} \, . \tag{13.7}$$

The high-temperature stabilization of the bcc lattice is the prototypical example of the importance of vibrational contributions. Friedel proposed that the larger vibrational entropy in the bcc lattice arose from a lower Einstein frequency, as might be expected from the smaller number of nearest neighbors. Basing the estimate of the entropy difference between the structures on the ratio of nearest neighbors (12 in fcc or hcp) and bcc structures gives

$$\Delta S_{\text{bcc-fcc/hcp}} = \frac{3}{2}\ln\frac{3}{2} \approx 0.61 k_B = 5.23 \times 10^{-5}\,\text{eV}\,\text{K}^{-1}\,\text{atom}^{-1}\,. \qquad (13.8)$$

For a temperature of order 1000 K, the contribution $T\,\Delta\,S$ to the free energy is about 0.052 eV/atom$^{-1}$, which is significant on the scale of the energy differences between the bcc and fcc and hcp phases.

# 14
# Semiconductors

The most important aspect of the semiconductor band structure (that is, ranges of levels of the electron energy, Section 6.5) may be summarized as follows: at absolute zero the completely filled band (the valence band) is separated from the empty band (the conduction band) by an energy gap or band gap $E_g$. Therefore the material does not conduct electricity at $T = 0$. At higher temperatures, a variety of processes enable electrons to be excited into the conduction band and empty states to occur in the valence band, thus allowing electrical conduction. These bands are located between the energy of the ground state, in which electrons are tightly bound to the atomic nuclei of the material, and the exited states. The ease with which electrons in the semiconductor can be excited from the valence band to the conduction band depends on the gap between the bands. The size of this energy band gap (roughly 4 eV) serves as a boundary between semiconductors and insulators.

Electrons excited to the conduction band also leave behind electron holes, that is, unoccupied states in the valence band. Both the conduction band electrons and the valence band holes contribute to electrical conductivity. The holes themselves do not actually move, but a neighboring electron can move to fill the hole, leaving a hole at the place it has just come from, and in this way the holes appear to move, and the holes behave as if they were actual positively charged particles.

As the semiconductors proved technologically very important we shall take some time to consider the general properties of their band structure.

Let us consider carbon as an example. It has the valence shell $2s^2 2p^2$ and its atoms form bonding and antibonding states for both s and p orbitals (Figure 14.1); the lowest four states from each atom mix to form $sp^3$ hybrids. As there are two atoms per primitive cell, each with four electrons (Figure 4.4), the lower $sp^3$ states, which overlap to form a band, are completely filled. Thus, diamond is not metallic, as the occupied and unoccupied electron states are separated by a considerable energy gap, $E_g = 5.46$ eV.

Most of the technologically-important semiconductors, such as Ge, Si, GaAs, GaSb, and (Hg,Cd)Te have a face-centered cubic lattice with a two-atom basis. In the case of Si and Ge, the atoms on the A and B sites are identical forming the so-called diamond structure. In the case of binary semiconductors such as GaAs, the A sites are occupied Ga atoms and the B sites As atoms; the crystal is said to have the zinc-blende structure. As stated above, in each case, the underlying lat-

*Interatomic Bonding in Solids: Fundamentals,Simulation,andApplications,* First Edition. Valim Levitin.

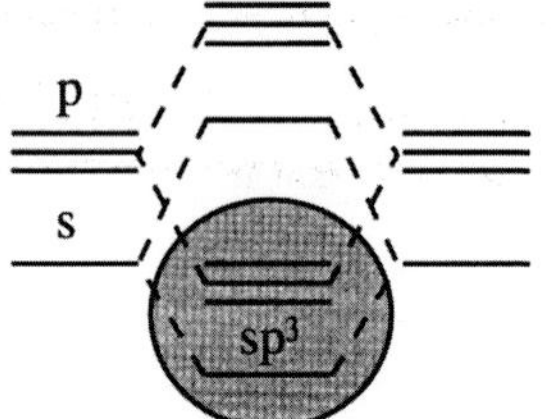


**Figure 14.1** Schematic of the formation of $sp^3$ hybrid bonding states in carbon (diamond, after [4] by permission of Oxford University Press).

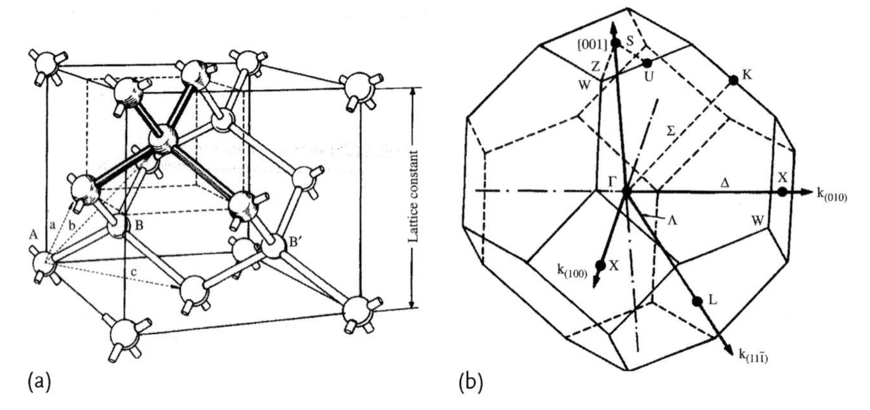


**Figure 14.2** (a) An illustration of the crystal structure of the group IV elements and many of binary semiconductors such as GaAs and CdTe; (b) the first Brillouin zone for the group IV elements and the semiconductors (after [4] by permission of Oxford University Press).

tice is a face-centered cubic and there is a two-atom basis. Thus, both the group IV elements and many of the binary semiconductors have the same shape of first Brillouin zone, which is shown in Figure 14.2.

Figure 14.3 shows development of bands as the atoms are brought together. The larger atomic spacing in tin corresponds to the portion of the diagram where the upper and lower bands overlap, so that tin is a metal; as the atomic spacing decreases, the gap between the band derived from $sp^3$ bonding states and that derived from the antibonding states opens up, resulting in an energy gap between full and empty states.

Figures 14.4 and 14.5 show calculated band structures of silicon and germanium, respectively. At the typical temperatures that will concern us (0–300 K), the only external action (that is, thermal excitation of electrons and holes, optical absorption edges) will occur close to the highest point in the valence band and the lowest points in the conduction band.

Authors of [8] studied electronic and structural properties of silicon and germanium from first-principle models. Results allow one to describe qualitatively the band structure of sp-bonded metals and semiconductors by a few numbers: the values of the spherical atomic-like potentials at a few lowest-reciprocal lattice vectors. By fitting to experimental data, some parameters could be used to describe a

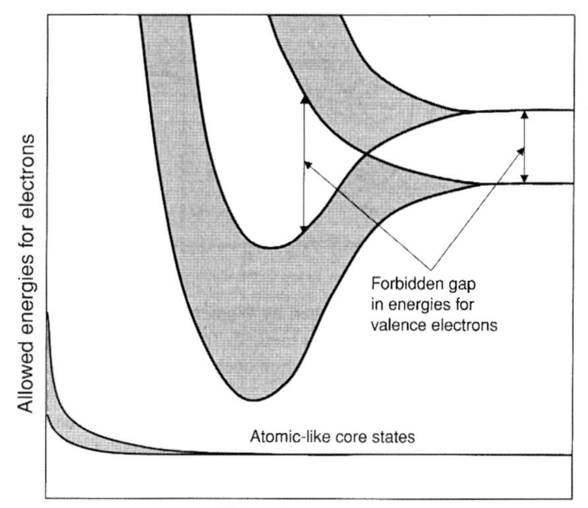


**Figure 14.3** Schematic illustration of energy levels for electrons, showing the evolution from discrete atomic energies to bands of allowed states separated by forbidden gaps, as the atoms are brought together. This leads to the basic division of solids into insulators, where the bands are filled with a gap to the empty states, and metals, where the bands are partially filled with no gap.

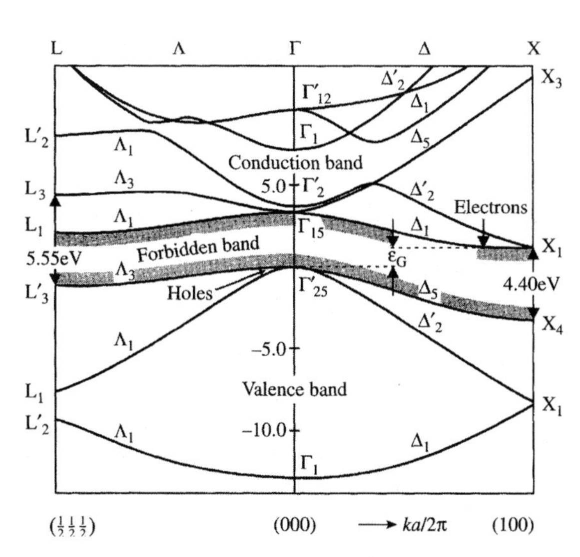


**Figure 14.4** Calculated band structure of Si, projected along [100] and [111] directions. The shading indicates the lowest (unoccupied at $T = 0$) conduction band and the highest (occupied at $T = 0$) valence band, with the forbidden band (data of Cardona and Pollack, after [4] by permission of Oxford University Press).

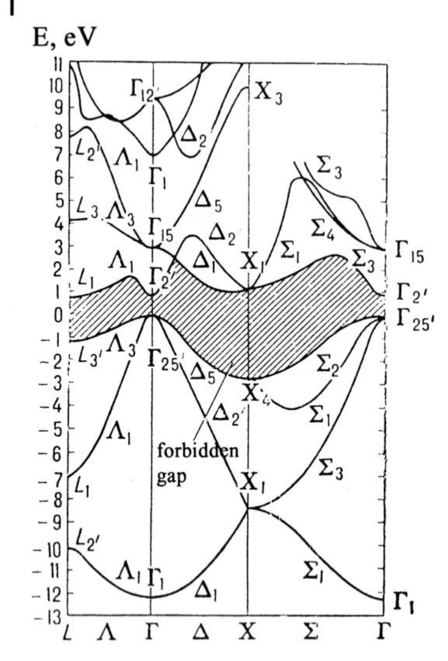


**Figure 14.5** Band structure of Ge (that is, the energy spectrum of electrons and holes) for directions [111] $(\Lambda)$, [100] $(\Delta)$, [110] $(\Sigma)$. The forbidden zone is shaded.

large amount of data related to the band structure, effective mass and band gaps, optical properties, and so on.

Figure 14.6 presents a comparison of experimental and theoretical results for covalent bonds in silicon. It is worth to note that the electron density has been calculated accurately by both used methods. The methods of calculation differ from each other by exchange-correlation functionals (see 8.6–8.8) but produced similar results.

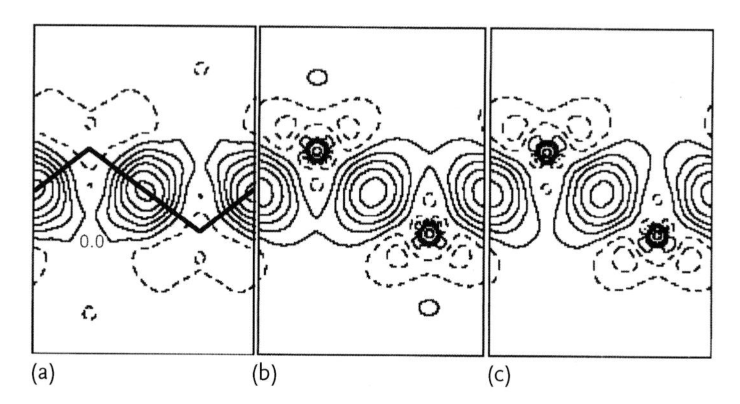


**Figure 14.6** Electron density in silicon. (a) A contour plot of experimental measurements using the electron scattering; (b) theoretical results found using the linear augmented plane wave (LAPW) method and density functionals; (c) calculations using generalized-gradient approximation (GCA). Electron density is presented as the difference of the total density and the sum of spherical atomic densities (after [1] by permission of Cambridge University Press).

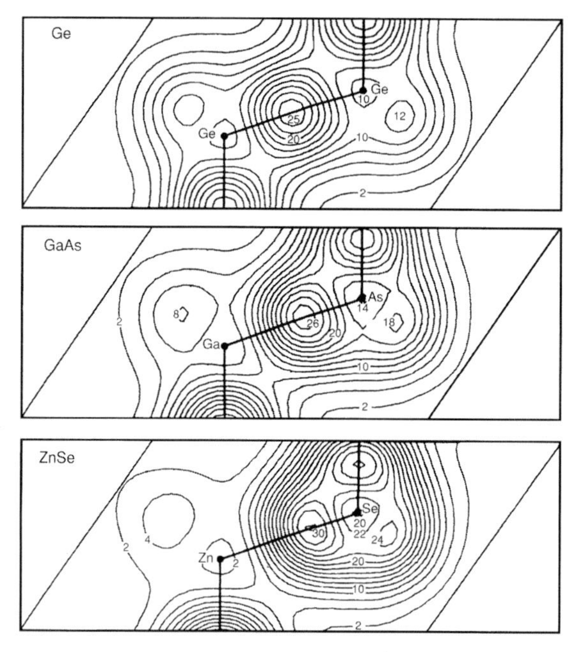


**Figure 14.7** Theoretical calculations of the valence charge density of semiconductors showing the formation of the covalent bond and the progression to more ionic character in the series Ge–GaAs–ZnSe. Data of Chelikovsky and Cohen (after [1]).

A noticeable shift of the negative electric charge from the cation to anion occurs in the semiconductor compounds of elements of groups II and VI. As Figure 14.7 presents, in the case of ZnSe, the electron charge is displaced from Zn atom in the direction of Se atom.

## 14.1 Strength and Fracture

Table 14.1 illustrates properties of semiconductors and the semiconductor compounds of elements of III and V groups. Two values of $\sqrt{u^2}$ for compounds correspond to the mean-square amplitudes of $A^{III}$ and $B^{V}$ elements, respectively.

In the carbon atom, the screening effect of the hybridized sp orbital is very weak. As a result, outer electrons are strongly bounded with the nucleus. Diamond is the insulator. The crystal lattice of silicon and germanium is the same (Figure 14.2). However, with arising of internal shells the screening of external electrons occurs. As a result, a part of valency electrons is delocalized and energy gap decreases up to 1.12 and 0.67 eV for silicon and germanium, respectively. Bulk modulus decreases and mean-square amplitude increases as compared with diamond.

**Table 14.1** Properties of diamond and semiconductors. The values of the lattice parameters $a$, energy gap width $E_g$, and bulk moduli $B$ from *Semiconductors on NSM (National Semiconductor Corporation)*, based on [57]. The values of mean-square amplitudes $\sqrt{u^2}$ are taken from [58–60]. All the data are related to temperature 293 K.

| Material | Formation of electron shells | $a$ (nm) | $E_g$ (eV) | $B$ (GPa) | $\sqrt{u^2}$ (pm) |
|---|---|---|---|---|---|
| C | $2s^1 2p^2$ | 0.357 | 5.46 | 442.0 | 7.5 |
| Si | $3s^1 3p^2$ | 0.543 | 1.12 | 98.0 | 9.7 |
| Ge | $4s^1 4p^2$ | 0.566 | 0.67 | 75.0 | 13.8 |
| GaP | $4p^1 + 3p^3$ | 0.545 | 2.26 | 88.0 | 7.75; 9.54 |
| GaAs | $4p^1 + 4p^3$ | 0.565 | 1.42 | 75.3 | 9.75; 9.22 |
| GaSb | $4p^1 + 5p^3$ | 0.610 | 0.73 | 56.3 | 11.2; 10.4 |
| InAs | $5p^1 + 4p^3$ | 0.606 | 0.35 | 58.0 | 9.24 |
| InSb | $5p^1 + 4p^3$ | 0.648 | 0.17 | 46.6 | 17.4; 12.2 |

When semiconductor compounds are formed, the covalent bond is provided by p electrons. In this case the atom of element $A^{III}$ (Ga, In) supplies one electron and atom of element $B^{V}$ (As, Sb) supplies three electrons.

The less energy gap for the semiconductor compounds the less is bulk modulus and the larger mean-square amplitude.

The bulk modulus decreases and mean-square vibration amplitude increases when going from GaP to InSb. The band gap $E_g$ decreases at the same time. We see that the strength of the interatomic bonding reduces with increasing the sum $n_A + n_B$, where $n_A$ and $n_B$ are the principal quantum numbers of the outer shells for the $A^{III}$ and $B^{V}$ elements, respectively. The sum $n_A + n_B$ equals for C, Si, and Ge to 4, 6, and 8, respectively. It changes from 7 for the semiconductor GaP to 9 for InSb.

Since semiconductors are primarily used in electronic applications, they are not usually thought of as structural materials. Nevertheless, there are important reasons, both technological and scientific, for the study of mechanical properties of semiconductors. The developing field of micro-machines, from microelectromechanical systems (MEMS) to nanotechnology, relies on fabrication techniques developed for electronic devices to make microscopic mechanical system. To a large extent, it is the link between these fabrication techniques, including deposition, masking, and etching, and the materials that has driven the use of semiconductors as structural components. On a more fundamental level, the ability to fabricate extremely pure and nearly defect free samples makes semiconductors excellent model systems for studying the physics of fracture.

Fracture is one possible failure mode of materials under mechanical load. It occurs when a crack grows, eventually entirely through a machine component, causing it to fail. In brittle fracture the crack is sharp, and its geometry causes a concentration of stress at the tip. In a continuum description of the solid and in the limit of an infinitively sharp crack the stress concentration becomes singular, and the stress field diverges at the crack tip. This stress concentration makes the material ahead of an existing crack most susceptible to failure, and causes the behavior of the material to be dominated by preexisting cracks. Since the amount of stress concentration, that is the coefficient of the singular term, is correlated with the length of the crack, brittle materials tend to break catastrophically.

Griffith set up an energy balance equation, comparing the amount of elastic energy released by crack extension with the amount of surface energy needed to generate the newly exposed crack surface. The Griffith criterion for brittle fracture is expressed as

$$G \geq 2\gamma_{\mathrm{S}} \tag{14.1}$$

where $G$ is rate of the elastic energy release and $\gamma_{\mathrm{S}}$ is the surface energy. The elastic energy release rate is generally directly proportional to stress squared. The Griffith approach ignores any atomistic details of process of the bonding-breaking.

Because the Griffith criterion is based on a conservation energy argument it probably gives, in most cases, a lower level for the critical load. It is difficult to validate the Griffith criterion because of the lack of accurate, independent ways of experimentally measuring the surface energy. Since semiconductors seem to be ideal brittle materials, a simulation of fracture can be used to test the Griffith criterion.

Another failure mode for a material under mechanical load is the plastic strain. The stress concentration at the tip of a crack increases the probability that dislocations will nucleate and move at the head of the crack tip. However, unlike in brittle fracture, plasticity dissipates a lot of energy, thus reducing the stress concentration by blunting the crack. This type of ductile behavior, typical in metals, leads to robust structural materials: the initiation of failure does not necessarily extend catastrophically through the entire specimen, and a lot of energy is dissipated during the process of the material strain.

Because both brittle and ductile failure of materials are controlled by processes on atomic scale such as bonding rupture and dislocation nucleation, atomistic simulation is almost the only tool that can provide us with an "atomic resolution view" of what is happening at the crack tip during fracture. One question that we can address is whether the Griffith criterion for brittle fracture is valid given the discrete, atomistic nature of matter. Another question is the microscopic mechanism that occurs at the crack up as the material fractures or deforms plastically. We can also point out the technological importance, such as the development of new, stronger materials, or the prevention of the failure of existing materials.

We consider the fatigue damage in detail in Chapter 17.

All semiconductors, elemental or compound, essentially consist of a network of atoms joined by covalent or mixed covalent–ionic bond. These covalent bonds typically involve $sp^3$ hybrid orbitals that favor tetrahedral coordination, leading to open lattices such as diamond structure (for elemental semiconductors) or its two-component analog, the zinc-blende structure (Figure 14.2a). The covalent bonding is stiff with respect to deformation of the angles between the bonding direction,

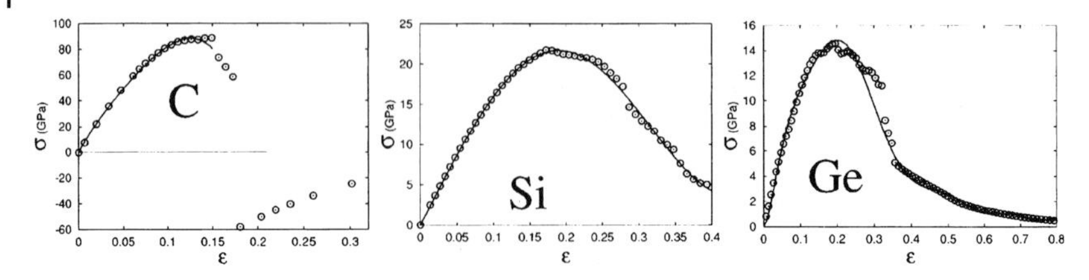


**Figure 14.8** Plot of the calculated stress as a function of applied tensile strain for diamond and two semiconductors (after [61]).

leading to a strong resistance to shear stress. The directionality of the bondings leads to a large energy needed to form dislocations that allow a plasticity. This suppression of the dislocation generation makes most semiconductors brittle, at least at low temperatures.

In a real material, the fracture process is a complex one and it may span a wide range of length scales. An elastic field extends over an entire macroscopic sample, is focused, through singularity, at the crack tip. The propagation of the crack is affected by many factors, including the crystal lattice, stress field, and defects already present.

Thus, a number of approaches are used to simplify the problem. These can be roughly classified as idealized models and direct simulations, both either quasi-static or dynamic.

A simple approach for the calculation of fracture properties neglects the complexities of fracture mechanics and calculates instead an "ideal strength." In a simplified picture, this is the peak stress that a uniform system sustains as a result of applied strain, typically uniaxial tension or simple shear. The ideal strength is relatively easy to compute, even using accurate first-principles approach. It requires that the energy and stress of a bulk system (that is a small unit cell with periodic boundary conditions) be computed as a function of strain for a range of applied strains. Positions of the different atoms in a complex lattice have to be relaxed at each applied strain. Figure 14.8 shows an example of this technique applied to three elemental semiconductors. The stress in the simulated system at a range of applied stress shows a linear rise in the elastic regime, followed by inelastic behavior and finally a maximum of the stress $\sigma_{\max}$ the material can support. The $\sigma_{\max}$ values equal to 90, 22, and 15 MPa for diamond, silicon, and germanium, respectively. These values correlate with bulk moduli (Table 14.1).

It is possible to use energy and force calculations to follow the trajectory of a system of atoms. One specifies the position and velocities of all atoms at zero time. The Newton equations of motion are integrated using a numerical algorithm. This type of simulation can give us the most direct view of the fracture process. The greatest difficulty in carrying out direct numerical simulations is in developing a method for computing the energy of the system. The method must be accurate enough to take into account the important characteristic physical features while remaining fast enough to be practical.

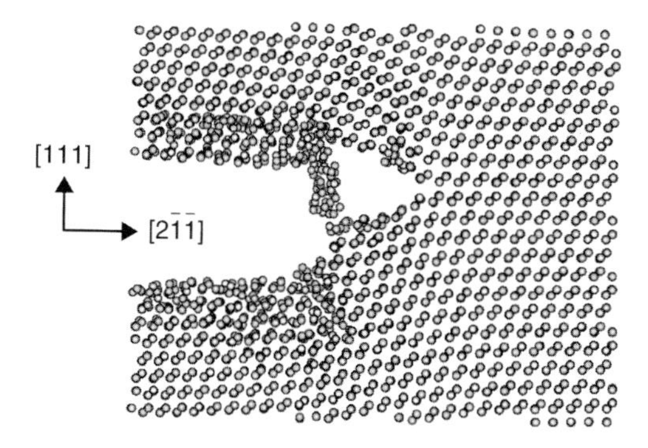


**Figure 14.9** Image of crack propagation through silicon at high loading, showing surface steps and dislocations (after [61]).

Three different approaches have been tried. One approach includes very accurate first-principles methods which are however applicable to small systems containing few tens of atoms. A second approach comprises methods of simulations based on interatomic potentials, which can be easily applied to systems containing $10^4$ atoms or more, but has serious problems with accuracy that can led to qualitatively wrong results. The third approach combines the previous two using an accurate method near the tip of the crack, and an interatomic potential far from the crack.

## 14.2 Fracture Processes in Silicon

Silicon is a suitable material for study of the perfectly brittle failure.

Two important conclusions may be drawn from the Griffith criterion (14.1)

- crystal lattice planes with low surface energies are energetically favored as cleavage planes;
- a given cleavage plane has a single unique value of $G$.

A perfectly brittle crack in a crystal is therefore expected to choose a cleavage plane with low surface energy and propagate on this plane with equal ease in all directions.

The *ab initio* methods make it possible for one to analyze in detail the bonding-breaking process at crack tip in silicon. These methods can accurately describe the nonlinear forces acting on atoms at the crack tip.

The starting configuration of atoms for a two-dimensional model is shown in Figure 14.10. The stability of the crack propagation in various directions on the (111) and (110) cleavage planes is different [62]. The total energy of the system and interatomic forces were determined within the local density approximation to density functional theory. The calculations show that the bonding-breaking process at a crack tip can have very different character.

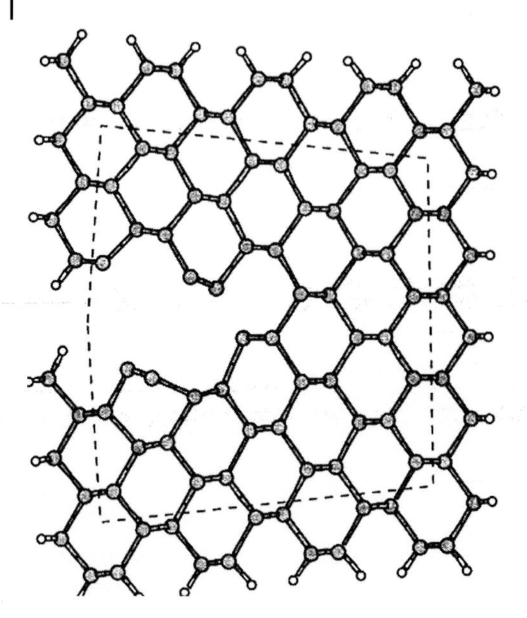

**Figure 14.10** Illustration of the starting atom configuration. The atoms outside the dashed line are fixed during the relaxation process, while the inner atoms are included in the atomic relaxation (after [62]).

The rupture of a bonding can be continuous or discontinuous. The discontinuous process is clearly connected to atomistic relaxations and rearrangements of only a few atoms around the crack tip. This can be partly understood as a result of some load sharing between the crack tip bonding and the neighboring bonding. The discontinuous process results in a rather large lattice trapping as compared to the continuous bonding-breaking event.

This difference in the bonding-breaking behavior introduces an anisotropy to the propagation direction of a crack on the same cleavage plane. If the crack is driven on a $\{110\}$ plane in a $\langle 001 \rangle$ direction the bonding-breaking process is discontinuous and is associated with pronounced relaxations of surrounding atoms. The bondings break continuously and cracks propagate easily on $\{111\}$ and $\{110\}$ planes provided crack propagation proceeds in the $\langle \bar{1}10 \rangle$ direction.

## 14.3 Graphene

Material that should not exist

*Andre Geim and Konstantin Novoselov*

Many scientists believed that two-dimensional crystals, which are one atom thick, could not exist. The theory pointed out that a divergent contribution of thermal vibrations in a low-dimensional crystal lattice should lead to such displacements of atoms that they become comparable to interatomic distances at any finite temperature. The films should become unstable and segregate into islands or decompose.

"Graphene is a rapidly rising star of the horizon of materials science and condensed-matter physics. This strictly two-dimensional material exhibits high crystal and electronic quality, and, despite its short history, has revealed a cornucopia of new physics and potential applications ... Graphene is the name given to a flat monolayer of carbon atoms tightly packed into a two-dimensional (2D) honeycomb lattice, and is a basic building block for graphic materials of all other dimensionalities" [64].

The problem to produce the graphene was solved by A. Geim and K. Novoselov. By gently rubbing or pressing a freshly cleaved crystal on an oxidized wafer graphene flakes with the correct thickness of oxide, single atomic layers are visible under an optical microscopy due to thin film interference effects.

As two-dimensional crystals are unusual and quite a new type of a material we spare a few lines to a method of their making. The authors of [65] used a simple but effective procedure. A fresh surface of a layered crystal was rubbed against another surface (virtually any solid surface is suitable), which left a variety of flakes attached to it. Unexpectedly, among the resulting flakes the authors always found single layers. Two-dimensional crystallites become visible on top of an oxidized silicon wafer, because even a monolayer adds up sufficiently to the optical path of reflected light so that the interference color changes with respect to the one of an empty substrate (phase contrast), Figure 14.11. The whole procedure takes literally half an hour to implement and identify probable two-dimensional crystallites. Their further analysis was done by an atomic force microscopy (AFM), for which single-layer crystals were selected as those exhibiting an apparent thickness of approximately the interlayer distance in the corresponding three-dimensional crystals. The critical step that allowed authors to find two-dimensional crystallites is the discovered possibility of their tentative identification in an optical microscope when they are placed on top of an oxidized silicon wafer.

It turned out that the graphene or its bilayer have simple electronic spectra: they are both zero-gap semiconductors. They have one type of electron and one type of hole as carriers. The electronic structure rapidly evolves with the number of layers, approaching the three-dimensional limit of graphite at 10 layers. For three or more layers, the spectra become increasingly complicated: several charge carrier appear, and the conduction and valence bands start notably overlapping. This allows single-, double-, and few- (3 to $< 10$) layer graphene to be distinguished as three different types of graphenes (two-dimensional crystals). Thicker structures should be considered as thin films of graphite.

Figure 14.12 illustrates the electric conductivity of the selected two-dimensional materials. The data were obtained using a field-effect-transistor-like microscopic device. The authors have measured the carrier mobilities and their density.

Mechanical properties of graphene have been measured by methods of static deflection using an atomic force microscope [66]. Spring constants ranging from 1

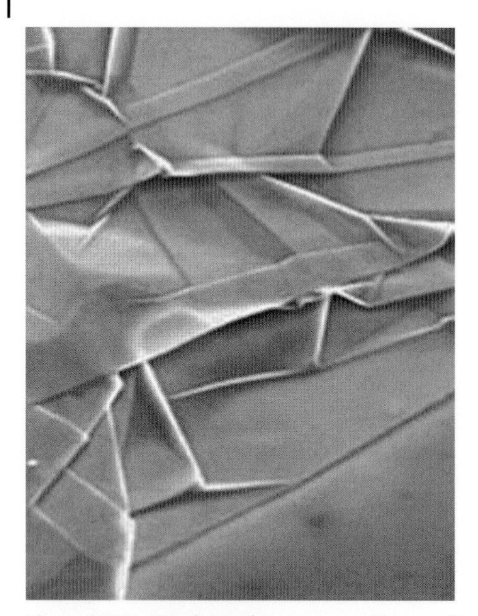

**Figure 14.11** Graphene thrown on a silicon wafer (after [63]).

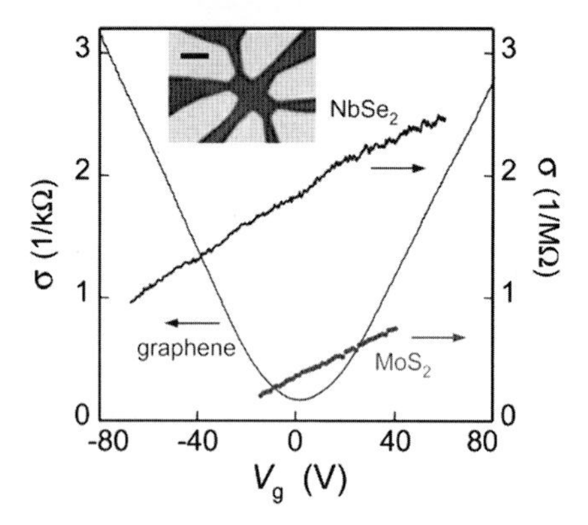


**Figure 14.12** Changes in electrical conductivity $\sigma$ of two-dimensional crystals $NbSe_2$, $MoS_2$ and graphene. Authors instrument (inset) is an optical image of two-dimensional $NbSe_2$ on top of an oxidized silicon wafer (used as a gate electrode) with a set of gold contacts (after [65]).

to 5 N/m were observed for suspended graphene sheets less than 10 nm thick. The Young modulus was found to be 500 GPa.

The authors of [67] measured the elastic properties and intrinsic breaking strength of free-standing monolayer graphene membranes by nanoindentation in an atomic force microscope. The found quantities correspond to a Young modulus of 1000 GPa. The authors believe that "these experiments establish graphene as the strongest material ever measured, and show that atomically perfect nanoscale materials can be mechanically tested to deformations well beyond the linear regime."

The mechanical behavior of graphene under both tension and compression has also been investigated [68]. In tension, the material seems to sustain strain up to

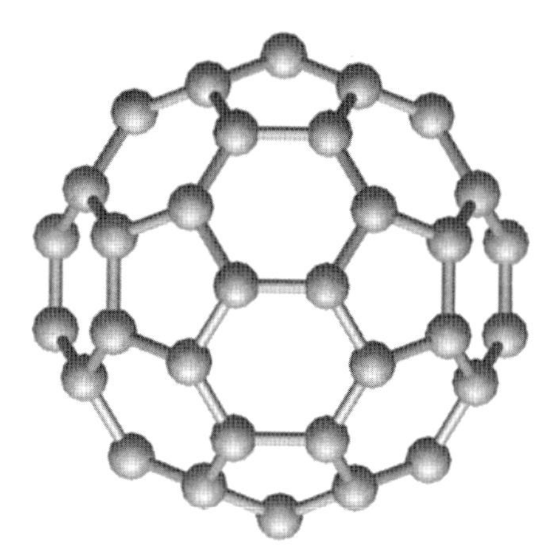

**Figure 14.13** The structure of $C_{60}$. The symmetric molecule has the shape of a football.

1.3%, whereas in compression, there is an indication of flake buckling at about 0.7% strain.

Figure 14.13 presents an example of a structure of the $C_{60}$ molecule that belongs to a so-called fullerene family. The structure is composed of stacked graphene sheets. Each carbon atom is covalently bonded with three others.

**Applications of graphene.** In 2005, the Nobel prize winners A. Geim and K. Novoselov suggested the main field of application of this specific material.

- The most immediate is the application of graphene in composite materials. Conductive plastics at less than one percent of graphene filling have revealed their availability.
- An enticing possibility is use of graphene in electric batteries. A large surface-to-volume ratio and high electric conductivity provided by graphene powder can lead to improvements in efficiency of batteries.
- It is likely that graphene power can offer distinct emission properties.
- Graphene can become an excellent material for solid-state gas sensors.
- Graphene is a potential material for superconductive field-effect transistors.
- Microprocessors are unlikely to appear for the next 20 years.

## 14.4 Nanomaterials

The theoretical and experimental research of nanomaterials and nanostructures are the most dynamic new areas of material science.

A nanomaterial is defined as a natural, incidental, or manufactured material containing particles, in an unbound state or as an aggregate or as an agglomerate and where, for 50% or more of the particles in the number size distribution, one or

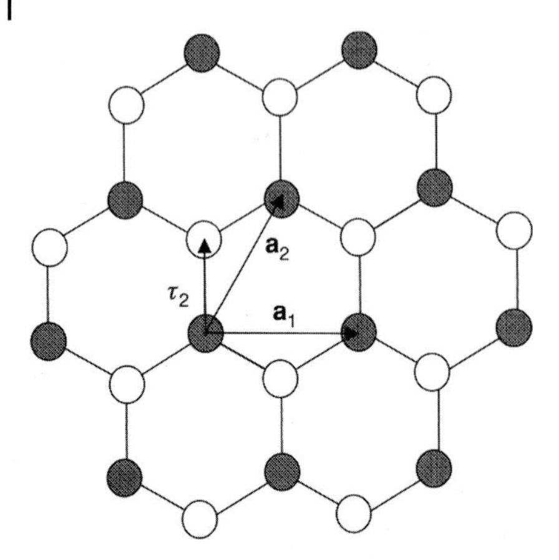


**Figure 14.14** The structure of graphite in a single (001) plane. The primitive cell contains $6/3 = 2$ atoms. The basis vectors $\boldsymbol{a}_1$ and $\boldsymbol{a}_2$ are shown.

more external dimensions is in the size range 1–100 nm.[1] The nanomaterials are small enough that the properties can be tuned by varying the size and the shape of the particles. Physical properties alter with the transition from macroscopic systems to ones of a size measurable in nanometers.

Nanotubes of carbon are made from graphene-like sheets rolled into a tube. Carbon nanotubes are considered to be the strongest and stiffest materials yet discovered in terms of tensile strength and elastic modulus, respectively. This strength results from the covalent $sp^2$ bonds formed between the individual carbon atoms.

In a perfect single-wall tube, all carbon atoms are equivalent and each atom is at a junction of three hexagons. There are various ways a sheet of graphene can be rolled into a variety of materials that are semiconductors or metals. A tube is defined by rolling the plane of graphene to bring the equivalent points together. The tube axis is perpendicular to vector $\boldsymbol{c} = m\boldsymbol{a}_1 + n\boldsymbol{a}_2$. The basis vectors $\boldsymbol{a}_1$ and $\boldsymbol{a}_2$ are shown in Figure 14.14.

In 2000, a multiwalled carbon nanotube was tested to have a tensile strength of 63 GPa.

Further studies, conducted in 2008, revealed that individual carbon tube shells have strengths of up to 100 GPa, which is in good agreement with quantum atomistic models. Since carbon nanotubes have a low density for a solid of 1.3–1.4 g cm$^{-3}$, its specific strength of up to 48 000 kN m kg$^{-1}$ is the best of known materials, compared to 154 kN m kg$^{-1}$ for a high-carbon steel.

Figure 14.15 presents a rolling of a graphene sheet in order to form a nanotube. A magnitude of the $\boldsymbol{a}_1$ vector determines the diameter of the nanotube, whereas a magnitude of the $\boldsymbol{a}_2$ vector determines the lead of helix.

When many carbon atoms are brought into a crystal, each bonding or antibonding orbital acquires a dispersion to become an energy band. The bands which corre-

1) The definition of the European Commission of 18 October 2011.

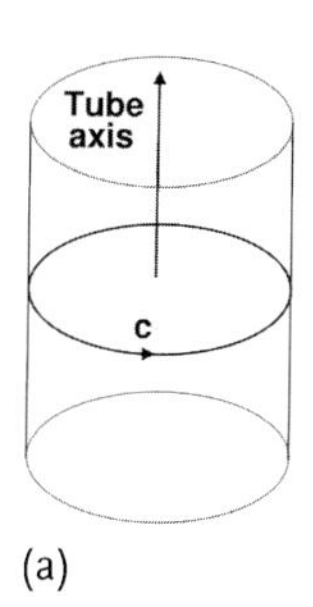


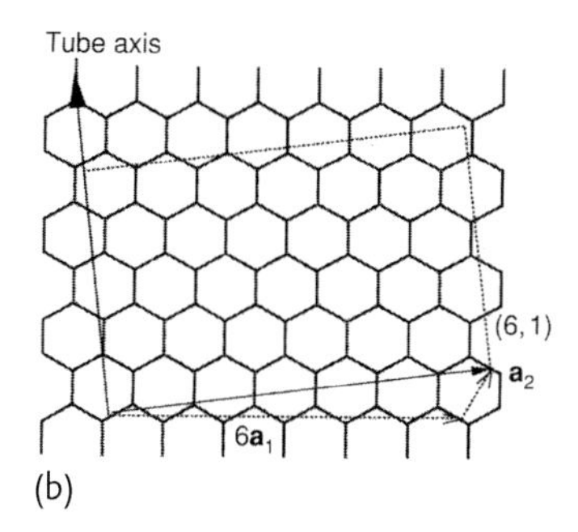


(a) (b)

**Figure 14.15** Rolling of a graphene to form a nanotube (a); (b) shows a sheet for rolling with the 6 $\boldsymbol{a}_1$ and 1 $\boldsymbol{a}_2$ vectors.

spond to the dispersion of bonding and antibonding molecular orbitals constructed from $p_z$ orbitals on two carbon atoms are called $\pi$ and $\pi^*$ bands.

Because the lattice is two-dimensional, all translations commute with reflection in the plane of the lattice, so any electron (or vibrational) eigenstate can be chosen to be either even or odd under this reflection. For this reason, the single-particle electron states are rigorously separated into two classes, called $\sigma$ and $\pi$. The even $\sigma$ states are derived from carbon s and $p_x$, $p_y$ orbitals (that is, their hybridized $sp^2$ orbitals), while the odd $\pi$ states are derived from carbon $p_z$ orbitals. These latter are cylindrically symmetric in the $x-y$ plane, lie near the Fermi level (half-filled) and are the electrically active states of interest in low energy experimental probing of graphene.

Figure 14.16 presents different types of the nanotubes.

The average value of the Young module for multiwalled nanotubes of diameter from 26 to 76 nm was found to be $1280 \pm 590$ GPa [69]. The authors believe that "the carbon nanotubes have great potential for applications requiring high-modulus, high-strength materials. However, detail characterization of the strength of carbon nanotubes in tension is still lacking … Current progress in nanotechnology is very rapid and no doubt it will be possible in the future."

The perspectives of applications of the nanomaterials are immense. Plastic articles produced using the compounded base material exhibit higher thermal stability, high thermal conductivity, high dimensional stability (low coefficient of thermal expansion), high fracture toughness, and high wear resistance. Due to the extremely small particle size and uniformity, they do not adversely affect mold flow ability or influence the set point. The ultra-high-purity materials have been obtained: the nanotechnology is currently capable of delivering purification factors in excess of 10 million, and ultimately will deliver chemicals having impurity levels in the parts per quadrillion (ppq) range. Industry is currently developing a rapid consolidation process which will allow for the production of fully dense, very large structures ($> 1\,m^2$) having uniform nanostructure.

Single-walled carbon nanotubes (SWNTs) obtained in the Rice University, Houston, were investigated in high resolution mode (0.335 nm) using a Philips 420 transmission electron microscope. Digital images were captured at the atomic lev-

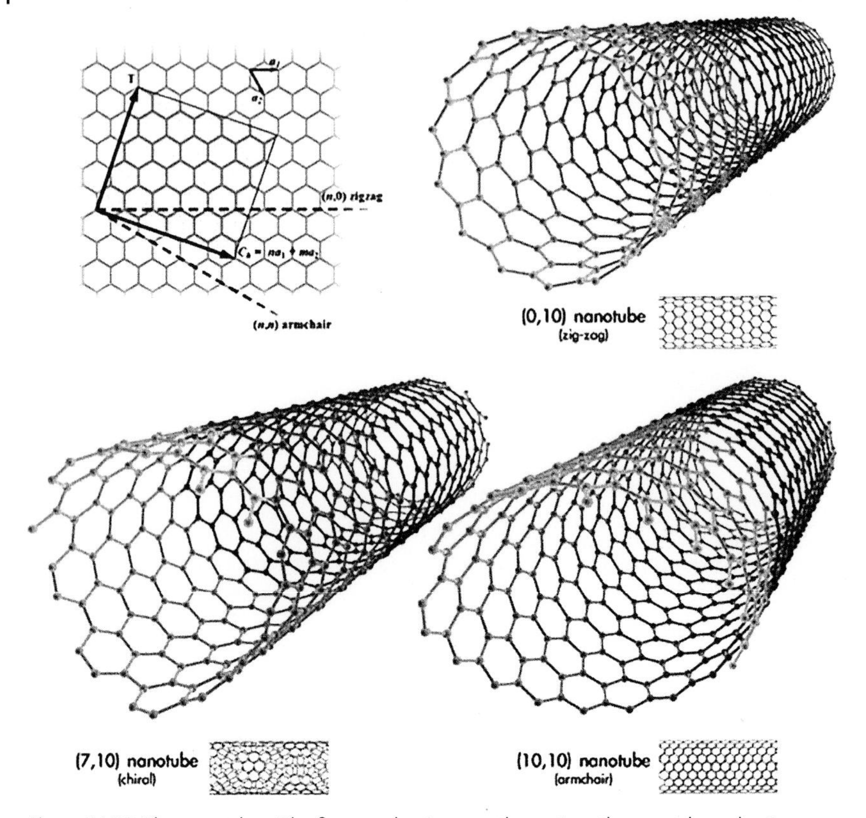


Figure 14.16 The nanotubes. The first number in parentheses is $n$, the second number is $m$.

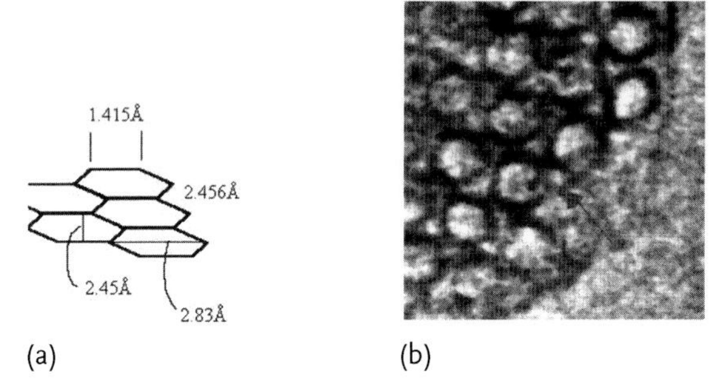


**Figure 14.17** (a) Spacing within a crystal lattice of graphite (1 Å = 0.1 nm). (b) The image is assumed to be a view of a growing bundle single-walled carbon nanotube. The pattern is obtained in a high resolution mode transmission electron microscope. Final magnification is 32 700 000. Note the symmetry of the dark outline of the tube to the left of the arrow and the tube to its right. These may be linear chains of carbon adding to the bundle. Data of Spires and Brown, Jr.

el, demonstrating structural characteristics of the nanotubes. The used technique allowed one to easily measure on the high resolution monitor and record the atomic distances less than 0.2 nm. In Figure 14.17 the picture of a nanotube is shown. A dark chain of carbon, seen more clearly in Figure 14.17, is attached to the tip of one of the tubes in the bundle.

# 15 Molecular and Ionic Crystals

A molecular solid consists of individually distinguishable molecules. Atoms inside a molecule are glued by the strong and short-range bondings. These bondings are realized by interatomic forces that we have considered in preceding chapters. Contrary to atoms, the ensembles of molecules are attached together by the weak and long-range intermolecular forces. Generally, the bonding between molecules provides the hydrogen bond and the van der Waals forces.

The relatively weak intermolecular forces can be easily disrupted by thermal vibrations. The molecular crystals differ from ionic and covalent crystals in lower melting points, lower densities, and relatively lower strength.

According to empirical evidence, the Lennard-Jones potential (11.26) describes the interaction between molecules well. The Lennard-Jones potential does not depend on a bonding direction, therefore the molecular solids tend to have the largest possible number of neighbors. Solid molecular crystals often have the face-centered cubic lattice or close hexagonal crystal lattice.

Examples of molecular crystals are ice, solid noble gases, oxygen, nitrogen, halogens ($F_2$, $Cl_2$, $Br_2$, $I_2$), sulfur as an $S_8$ ring, phosphorus as a $P_4$ tetrahedron, and the hydrocarbons and organic compounds.

The molecular crystals are important for science and widely used in industry.

## 15.1 Interaction of Dipoles: The van der Waals Bond

The van der Waals forces act between dipoles of molecules. These forces are referred to as dipole–dipole, dipole–induced and dispersion forces.

In a dipole molecule a center of the positive charge $+q$ is separated from a center of a negative charge $-q$. The dipole moment $p$ of the molecule is given by

$$p = q \cdot d \,, \tag{15.1}$$

where $d$ is the distance between the charge centers. Only asymmetrical molecules have constant dipole moments. Dipole moment is often expressed in CGS system in an unit named after Debye. One Debye equals to $3.336 \times 10^{-30}$ C m in SI system of units.

*Interatomic Bonding in Solids: Fundamentals,Simulation,andApplications*, First Edition. Valim Levitin.

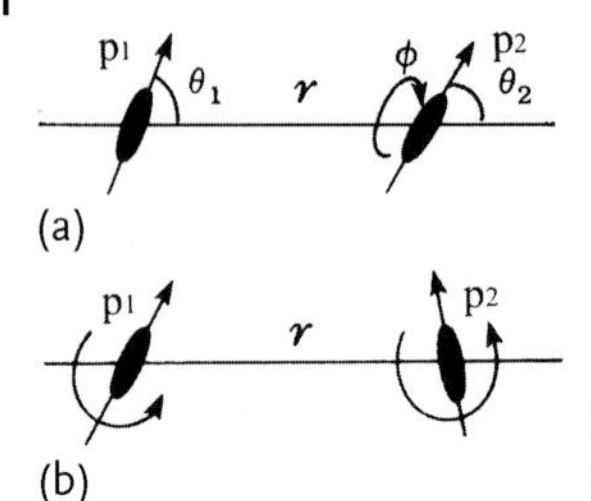


**Figure 15.1** Interaction of dipoles. (a) The dipoles are fixed; (b) the dipoles are free to rotate.

In the solid, successive molecules have electric dipole moments which alternate in orientation so as to produce successive attractions.

A net interaction between two dipole molecules can be expressed in polar coordinates $\theta_1$, $\theta_2$, and $\phi$ as

$$E_{\mathrm{di}}(r, \theta_1, \theta_2, \phi) = -\frac{1}{4\pi\varepsilon_0}\frac{p_1 p_2}{r^3}(2\cos\theta_1\cos\theta_2 - \sin\theta_1\sin\theta_2\cos\phi)\ , \quad (15.2)$$

where $p_1$ and $p_2$ are the dipole moments of molecules. The interaction energy between molecules decreases with distance as $r^{-3}$.

If two dipole molecules align parallel to each other head to tail at the distance $r$ apart the energy of their interaction is given by

$$E_{\mathrm{di}}\left(r, \frac{\pi}{2}, \frac{\pi}{2}, \pi\right) = -\frac{1}{4\pi\varepsilon_0}\frac{p_1 p_2}{r^3}\ . \quad (15.3)$$

When two dipoles are lying in line the energy $E_{\mathrm{di}}(r, 0, 0, \phi)$ halves. Table 15.1 presents a quantitative estimation of the equilibrium interdipole separation at room temperature calculated according to formula (15.3). The energy of heat vibrations equals to $k_B T$, that is $4.046 \times 10^{-19}\,\mathrm{J} = 0.0252\,\mathrm{eV}$. The distances in Table 15.1 are of the order of molecular separation in solids. This means that the dipole–dipole interaction is strong enough to bind molecules in a solid.

Figure 15.1a shows the interaction of fixed molecules. The dependence of energy on distance is changed if two molecules can rotate freely, Figure 15.1b. The angle-average interaction for two free dipole molecules is usually referred to as Kessom

**Table 15.1** Equilibrium distance $r$ between dipole molecules at $T = 293$ K in vacuum. Arrows show the relative position of molecules.

| Molecule | Dipole momentum (*D*) | $r_{\uparrow\downarrow}$ (nm) | $r \rightarrow\rightarrow$ (nm) |
|---|---|---|---|
| HCl | 1.08 | 0.31 | 0.24 |
| $NH_3$ | 1.47 | 0.38 | 0.30 |
| $C_2H_5OH$ | 1.70 | 0.42 | 0.33 |
| $H_2O$ | 1.85 | 0.44 | 0.35 |
| NaCl | 8.5 | 1.21 | 0.96 |
| CsCl | 10.4 | 1.38 | 1.10 |

interaction. The averaging of the energy over all orientations leads to

$$E_{\mathrm{di}} = \frac{1}{(4\pi\varepsilon_0)^2}\frac{p_1 p_2}{3k_B T r^6} . \qquad (15.4)$$

The electric field of intensity $E$ induces in a nonpolar molecule a dipole moment,

$$p_{\mathrm{ind}} = \alpha E , \qquad (15.5)$$

where $p_{\mathrm{ind}}$ is the induced dipole moment, $\alpha$ is the polarizability. The polarizability of the nonpolar molecules is caused by a displacement of the electric charge. The electric field of a polar molecule is given by

$$E = \frac{1}{4\pi\varepsilon_0}\frac{2p_1}{r^3} . \qquad (15.6)$$

The interaction energy is therefore

$$E_{\mathrm{bond}} = -\left(\frac{1}{4\pi\varepsilon_0}\right)^2 \frac{\alpha p^2}{r^6} . \qquad (15.7)$$

In a general case, the bonding energy for two different molecules can be expressed as

$$E_{\mathrm{bond}} = -\left(\frac{1}{4\pi\varepsilon_0}\right)^2 \frac{p_1^2\alpha_2 + p_2^2\alpha_1}{r^6} , \qquad (15.8)$$

where $\alpha_1, \alpha_2$ and $p_1, p_2$ are polarizabilities and dipole moments of molecules, respectively.

Equation (15.8) that claims inverse six power dependence energy on separation is referred to as the Debye interaction between molecules.

Dispersion forces also known as the London forces act between molecules and atoms that are even neutral. The dispersion forces are long-range forces. Their range of action varies from 0.2 to 10 nm.

Thus, the physical mechanism of intermolecular bonds that involves the van der Waals attraction is an interaction between electric dipoles of the molecules. Because of the fluctuating quantum mechanical behavior of the electrons in a molecule, all molecules have a fluctuating dipole moment, even though for many of them symmetry consideration requires that it fluctuates about an average value of zero. At a time when a molecule has a certain instantaneous dipole moment, its electric field will induce the dipole moment as a result of the charge redistribution in a nearby molecule.

The interaction energy is proportional to the mean-square of the inducing electric dipole moment. The resulting attraction is weak, the binding energies being of the order of $10^{-2}$ eV and the force varying as the inverse sixth power of the intermolecular separation. Because the bonding is weak, solidification takes place only at very low temperatures where the disruptive effect of thermal action is small. The melting point of solid hydrogen is 14 K, for example. The relatively weak intermolecular bonding makes molecular solids easy to deform and compress, and the absence of free electrons makes them very poor conductors of electricity or heat.

## 15.2
## The Hydrogen Bond

The hydrogen bond is formed by a hydrogen atom $^1$H: $1s^1$, which is already bonded to an atom A with high electronegativity.[1] The same hydrogen atom attracts as acceptor another atom B, likewise with high electronegativity. Thus, the hydrogen bond is a donor–acceptor interaction involving a hydrogen atom. This attraction is realized between proton and a partly negatively charged atom. The scheme of the hydrogen bond can be represented as

$$A\text{–}H \cdots B\,.$$

Here A is an electronegative atom,[2] more often nitrogen N (external shell is $2p^3$), oxygen O ($2p^4$), fluorine F ($2p^5$), sulfur S ($3p^4$) or phosphor P ($3p^5$). The hydrogen bond is shown by the dotted line. To interact with the donor A–H bonding, the acceptor B must have lone-pair electrons or polarizable $\pi$ electrons.

The hydrogen atom, which has formed the covalent bond $A\text{–}H^+$ with nitrogen or oxygen or fluorine (it shares its electron with A atom), gains a partial positive charge. This partial positive charge, indeed, the partial charge of proton, attracts an electronegative atom B that belongs to another molecule. This molecule can be the same as is schematically shown as A–H···A. An example of this is the hydrogen bond between molecules of water O–H···O.

This type of bonding is known to be of importance in chemistry and biology. As early as in 1960 Linus Pauling wrote: "Although the hydrogen bond is not a strong bond ..., it has great significance in determining the properties of substances ... I believe that as the methods of structural chemistry are further applied to physiological problems it will be found that the significance of the hydrogen bond for physiology is greater than that of any other single structural feature [70]."

Water, ammonia, and hydrogen fluoride are examples of materials that have hydrogen bonds. The hydrogen bond is responsible for the binding of the molecules in water and for the tetrahedral bonding of molecules in ice. Each $H_2O$ molecule acts as a donor for two hydrogen bonds and as an acceptor for two others, with the acceptor sites located in directions tetrahedrally opposite the covalent O–H bonds. The hydrogen bond of water has a strong directional property, so that a linear bond O–H···O with an angle of 180° is the strongest. Figure 15.2 illustrates the location of atoms and molecules in a tetrahedral structure.

The strength of hydrogen bond varies from 10 to 40 kJ/mol (0.10–0.41 eV/atom).[3] This is much less than, for instance, ionic bond of NaCl at 764 kJ/mol (7.95 eV/atom).

A representative value of the O–H···O size that has been calculated from the *ab initio* method is of the order of 0.19 nm. The experimental value of the covalent

1) L. Pauling [70] defines the electronegativity as "the power of an atom in a molecule to attract electrons to itself."
2) We remember that the filled stable p shell contains 6 electrons.
3) 1 eV/atom = 96.4853 kJ/mol; 1 kJ/mol = 0.010 36 eV/atom.

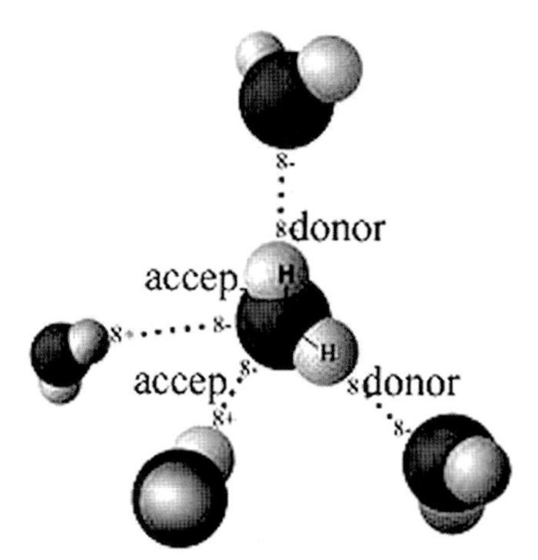


**Figure 15.2** Space bonds between molecules of water. The dotted lines show hydrogen bonds. Each molecule of water acts as a donor for two hydrogen bonds and an acceptor for two others.

bond H—O distance in $H_2O$ molecule, which has been found from the rotation–vibration spectra, equals to 0.0957 nm. The angle H–O–H is 104.52°.

Figure 15.3 illustrates the electron density in an isolated water molecule. The electron density near the oxygen atom equals to 0.84 electrons per $(a.u.)^3$ as compared with 0.36 electrons per $(a.u.)^3$ near the hydrogen atoms. The H—O bond is the polar covalent bond, and the oxygen atom is more electronegative (3.45) than the hydrogen atom (2.05 according to the Pauling scale [70]).

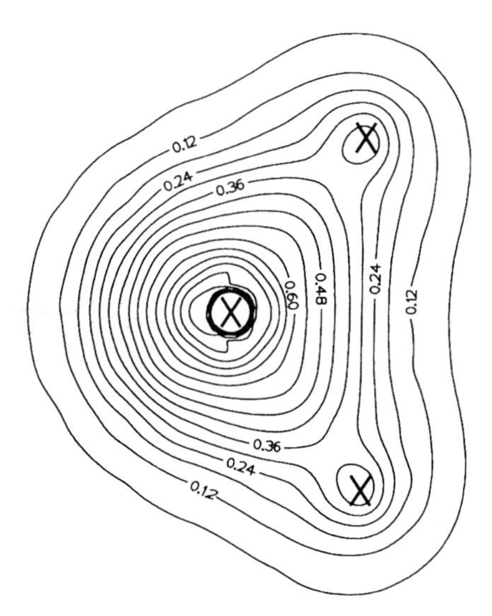


**Figure 15.3** Electron density in an isolated water molecule. The units on the contour lines are electrons per $(a.u.)^3$. The calculated *ab initio* data of M.I.J. Probert. The negative charge is shifted to the oxygen atom (after [72]).

Many condensed liquid and solid systems exist due to hydrogen bonds. The molecules $NH_3$, $OH_2$, FH, $PH_3$, $SH_2$, ClH can serve as the proton acceptors. For visualization purposes the proton acceptor is here indicated as the first.

It is apparent that the strength of the hydrogen bond depends upon the electronegativity of an A atom, dipole moment of a molecule and the ionization potential of the proton acceptor.

The author of [73] deduced a physical model of the hydrogen bond from *ab initio* molecular orbital wave functions. The characteristic of the model are as follows: the dipole moment of the A–H bond $\mu_{\text{A–H}}$; the difference between the first ionization potential of the electron donor and the noble gas in its row of the Mendeleev Table $\Delta I$; the length $R$ of the hydrogen bonding lone pair. A number of characteristic of the intermolecular strength can be described in terms of these quantities.

The energy of the dimer formation is one of the criteria of the strength of the hydrogen bond between two molecules. The dimerization is defined as the chemical compound of two identical molecules. The energy of dimerization may be expressed as

$$E_{\text{D}} = -K\frac{\mu_{\text{A–H}} \cdot \Delta I}{R} , \tag{15.9}$$

where $R$ is the heavy-atom internuclear separation.

We may estimate the strength of the hydrogen bond by analyzing values of energy of dimerization (Table 15.2). We see from this table that even for the hydrofluoric acid the bond energy equals to 0.3 eV/atom. For water we have almost the same value. Fluorine and oxygen are by far the most electronegative atoms having 4.0 and 3.5 relative units, respectively.

The purely electrostatic interpretation of the hydrogen bond is not sufficient [71]. The other component of the hydrogen bond energy is identified as the quantum charge-transfer contribution, delocalization, and dispersion. Some authors decompose the total bonding energy into electrostatic, polarization, exchange repulsion, and coupling.

**Table 15.2** Energy of dimerization. The data after [73]; error of experimental results is from 8 to 28%.

| Dimer | Theory | | Experiment | |
|---|---|---|---|---|
| | (kJ/mol) | (eV/molecule) | (kJ/mol) | (eV/molecule) |
| $(FH)_2$ | – | – | −29.1 | −0.30 |
| $(OH_2)_2$ | −21.7 | −0.23 | −28.0 | −0.29 |
| $(NH_3)_2$ | −16.7 | −0.17 | −18.2 | −0.19 |
| $(ClH)_2$ | −6.3 | −0.07 | −8.9 | −0.09 |
| $(SH_2)_2$ | −5.0 | −0.05 | −7.1 | −0.07 |

## 15.3 Structure and Strength of Ice

Ice has a special place among the solids that surround a human. The oceans cover 70% of the earth surface, and the sea ice covers about 10% of the ocean surface either seasonally or eternally. 10% of the land mass is covered with ice to depth of up to several kilometers. Ice is a factor of the global climate. It plays the leading role in polar marine transport activities. Icing occurs at airplanes and electrical transmission lines. Scientists of different specialties investigate the nature, distribution, and properties of ice.

Ice can crystallize forming 12 different structures. At atmospheric pressure the freezing water transforms into ice termed Ih that has a hexagonal crystal lattice. The Ic ice with the cubic crystal lattice is formed at low temperatures, that is at 143 K (–130 °C). Phases of ice other than the Ih phases are produced by the application of high pressures.

The structure of usual ice Ih is shown in Figure 15.4a. The X-ray investigation has revealed that the oxygen atoms form hexagonal crystal lattice similar to that of wurtzite ZnS.

Each $H_2O$ molecule has four nearest neighbors arranged near the vertices of a regular tetrahedron centered about the molecule of interest. The oxygen atom of each molecule is strongly covalently bonded to two hydrogen atoms, while the

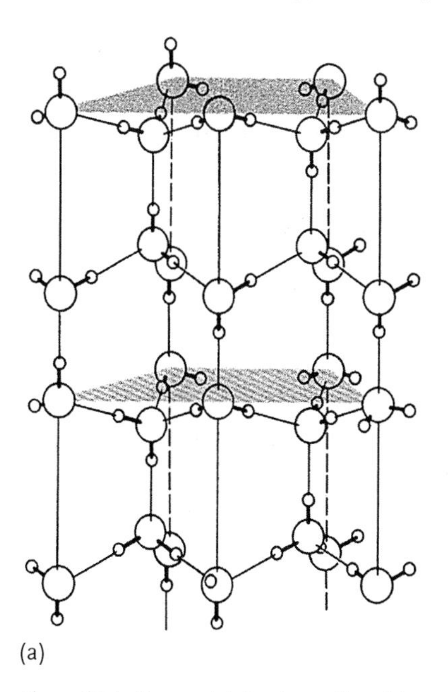

(a)

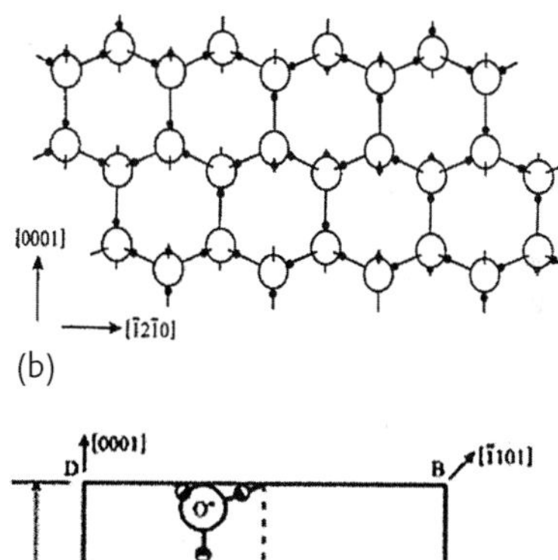


(b)

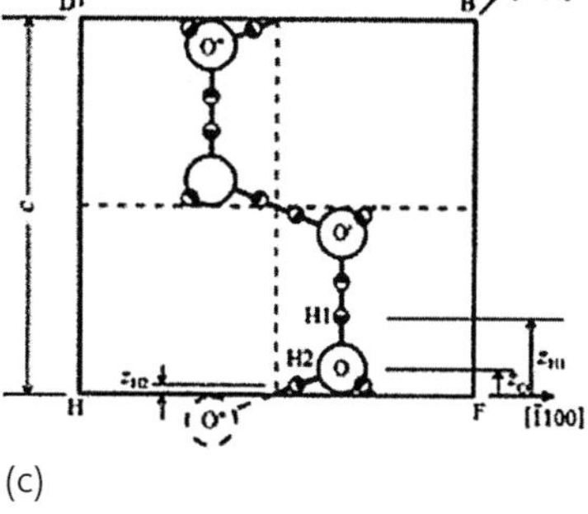


(c)

**Figure 15.4** The crystal structure of ice. (a) Hexagonal crystal lattice of Ih ice; the large circles are the oxygen atoms, the small circles are the hydrogen atoms. (b) A layer of the ice structure projected on the (10$\bar{1}$0) plane. The hydrogen atoms are shown by small black circles. (c) Unit cell of ice projected on (11$\bar{2}$0) plane (after [70, 72]).

molecules are weakly hydrogen bonded to each other. One proton realizes this bond between two adjacent oxygen atoms. Each oxygen atom is surrounded tetrahedrally by four other oxygen atoms at a distance of 0.276 nm.

When projected onto the plane perpendicular to the *c*-axis, the molecular stacking sequence is ... ABBAABBA...

In the structure presented in Figure 15.4a the oxygen atoms lie in layers perpendicular to the c-axis. These layers are built up of puckered hexagonal rings that are stacked in the sequence ... ABABAB... Two layers A are shown as shaded in the figure. The puckering of the rings forming these layers consists of displacements of the O atoms out of plane following the sequence ... up-down-up-down-up... The structure contains also hexagonal rings in planes parallel to c-axes, as it is seen in Figure 15.4b.

The crystal belongs to $P6_3/mmc$ space group of symmetry.[4)]

The crystal lattice parameters of ice depend on temperature. At 253 K (–20 °C) parameter $a = 0.4519$ nm and $b = 0.7357$ nm. The ratio $c/a = 1.628$ is sufficiently close to the characteristic quantity of the hexagonal unit cell. For ice it is almost independent of temperature.

Each $H_2O$ molecule is linked by hydrogen bonds to four others at the corners of a regular tetrahedron, offering protons to two neighbors and accepting hydrogen bonds from two others. A proton lies about 98.5 pm from one of oxygen atoms, and O–O distance equals to 250 pm. There is no long-range order in the orientations of the molecules. Many properties of ice are related to this disorder. The ice Ih unit cell is relatively open with the packing factor below 0.34, and this accounts for ordinary ice being less density than water.

"Each water molecule is so oriented that its two hydrogen atoms are directed toward two of the four surrounding oxygen atoms, that only one hydrogen atom lies along oxygen–oxygen line, and that under ordinary conditions the interaction of nonadjacent molecules is such as not stabilize appreciable any one of many configurations satisfying these conditions with reference to the others." "Thus we assume that an ice crystal can exist in any one of a large number of configurations, each corresponding to certain orientations of the water molecules." "It can change from one configuration to another by rotation of some of molecules or by motion of some of the hydrogen nuclei, each moving 0.76 Å from a position 1.00 Å from one oxygen atom to the similar position near the other bonded atom ... The proton will tend to jump in this way in groups ... [70]" Thus, the author supposes that at temperatures higher than 200 K (–73 °C) the molecules of ice have a certain degree of freedom.

We can calculate a number of molecules in the crystal unit cell from experimental values of density and lattice parameters. The density of ice can be expressed as

$$\rho_{\text{ice}} = \frac{N_{\text{molecules/cell}} \cdot \mu_{\text{water}}}{V_{\text{u.c.}} \cdot N_{\text{A}}}, \tag{15.10}$$

4) No 194 according to International Tables for Crystallography.

**Table 15.3** Interatomic distances and mean-square amplitudes of atomic vibrations in the Ih ice. Experimental data of X-ray investigation of Kuhs and Lehman (after [72]).

| Parameter | 15 K | 123 K | 223–233 K |
|---|---|---|---|
| O–O (pm) | 275.0 | – | 276.0 |
| O–H (pm) | 100.7 | – | 100.2 |
| ∠ O–O–O (deg) | 109.47 | – | 109.47 |
| $\sqrt{u^2{}_O}$ (pm) | 9.5 | 15.8 | 21.2 |
| $\sqrt{u^2{}_{H\parallel}}$ (pm) | 11.4 | 16.7 | 21.8 |
| $\sqrt{u^2{}_{H\perp}}$ (pm) | 16.7 | 20.6 | 24.3 |

where $V_{u.c.}$ is the volume of unit cell, $\mu_{water} = 18 \times 10^{-3}\,kg\,m^{-3}$ is molar mass of water, $N_A = 6.022 \times 10^{23}$ molecules mol$^{-1}$ is the Avogadro number. At temperature 273 K (0 °C) $\rho_{ice} = 916.68\,kg\,m^{-3}$. The unit cell parameters measured at 265 K (–8 °C) were found to be $a = 0.45214$ nm and $c = 0.73616$ nm. Substituting these values in (15.10) we obtain the number of the water molecules in unit cell equal to 11.9910. Thus, the hexagonal unit cell of ice Ih contains 12 molecules of water.

The interatomic distances for ice are presented in Table 15.3. The mean-square amplitudes of atomic vibrations are also shown. At temperatures 223–233 K (−50 to −40°C) amplitude of proton vibration reaches up to 22–24% of interatomic O–H distance. The amplitudes of transverse vibrations is greater than those of longitudinal ones. Large values of amplitudes like these are not characteristic features for metallic and ionic crystals. This fact confirms a partially disordered structure of ice.

The average cohesive energy in crystal lattice of ice was found to be equal to −0.306 eV per bond. This energy consists of the hydrogen bond, the van der Waals interaction (that is attraction), and the repulsion of the atomic shells. The intermolecular bond in ice is stronger than the single hydrogen bond in the water dimer (−0.24 eV). An O–O distance of 0.2750 nm in ice is less than distance 0.2976 in dimer $(H_2O)_2$. These facts indicate polarization of the ice molecules in the electric field of their neighbors.

The elastic properties of ice are characterized by moderate anisotropy. At temperatures near the melting point, the Young modulus of single crystals varies by less than 30%, from 12 GPa along the direction parallel to the $c$-axis to 8.6 GPa along direction inclined to both the $c$- and $a$-axes. Along directions within the basal plane the Young modulus is 10 GPa. For randomly oriented polycrystals at 268 K (–5 °C) typical values of the Young modulus and the Poisson ratio are 9.0 MPa and 0.33, respectively [74].

## 15.4 Solid Noble Gases

The atoms of noble gases contain the stable outer electron shells, neon has the ($2p^6$) shell, argon ($3p^6$), krypton ($4p^6$), and xenon the ($5p^6$) shell. These gases are known to be inert because they usually do not enter into chemical reactions.

At low temperatures they solidify. The solid noble gases have the cubic face-centered crystal lattice. The molecules of solid noble gases are glued by very weak fluctuating dipole forces. The atoms are distorted from the stable configuration and this creates a potential that holds atoms of solid together. This attractive potential depends on the interatomic distance $r$ as the inverse sixth power. The repulsive force depends inversely on the distance to the twelfth power. An empirical formula for potential is known as Lennard-Jones potential (11.26). It is appropriate to use this equation in the dimensionally-handy form

$$U_{\mathrm{LJ}}(r) = 4\epsilon \left[ \left(\frac{\sigma}{r}\right)^{12} - \left(\frac{\sigma}{r}\right)^{6} \right] , \tag{15.11}$$

where $\epsilon = A^2/4B$, $\sigma = (B/A)^{1/6}$. Constant $\epsilon$ is measured in eV and $\sigma$ in pm. For solid gases, this equation has rather an illustrative meaning than (can be used as) a calculation tool. The interaction potential is attractive at a large separation and varies as $1/r^6$. The potential reveals a strong repulsion of neighboring atoms at a small separation.

The constants $\epsilon$ and $\sigma$ have been measured for the low-density noble gases. It is possible to estimate some properties of solid gases using only these values and the potential (15.11) [7].

The atoms interact with each other pairwise. Thus, the energy of interaction of an $i$ atom in origin with all the others is given by

$$E(r) = \sum_j \phi(r_{ij}) . \tag{15.12}$$

Total energy per atom is expressed as

$$E_{\mathrm{at}}(r) = \frac{1}{2} \sum_{i,j;i \neq j} \phi(r_{ij}) . \tag{15.13}$$

The coefficient 1/2 is needed because the energy of interaction between atoms $i$ and $j$ was counted twice. We may use the Lennard-Jones potential as the function $\phi(r_{ij})$.

It is convenient to denote the interatomic spacings as

$$r_{ij} = r \cdot q_{ij} , \tag{15.14}$$

where $r$ is the nearest-neighbor separation, $q_{ij}$ is a dimensionless value. In other words, we measure interatomic distances $r_{ij}$ in units of the nearest-neighbor spacing. Inverse quantities are expressed as

$$\frac{1}{r_{ij}} = \frac{1}{r} \cdot \frac{1}{q_{ij}} . \tag{15.15}$$

Combining (15.15), (15.13) and (15.11) we find the energy of crystal lattice per atom,

$$E_{\text{at}} = 2\epsilon \left[ A_{12} \left( \frac{\sigma}{r} \right)^{12} - A_6 \left( \frac{\sigma}{r} \right)^6 \right] , \tag{15.16}$$

where

$$A_{12} = \sum_{i,j;i \neq j} \frac{1}{q_{ij}^{12}} ; A_6 = \sum_{i,j;i \neq j} \frac{1}{q_{ij}^6} . \tag{15.17}$$

$A_n$ ($n = 12$ or 6) is the sum of the inverse $n$th powers of the distances from an origin in crystal lattice to all others, where the unit of distance is taken to be the distance between nearest neighbors. By definition, $q_{ij} = 1$, when a vector joining nearest neighbors. For face-centered cubic lattice the sum is

$$\sum_{ij;i \neq j} \frac{1}{q_{ij}} = \frac{12}{1^n} + \frac{6}{\sqrt{2}^n} + \frac{24}{\sqrt{3}^n} + \frac{12}{2^n} + \frac{12}{\sqrt{5}^n} + \ldots \tag{15.18}$$

The exponent $n$ is an important quantity in the empirical description of the interatomic interactions. This quantity uniquely determines the dependence of energy on distance, $E(r) \sim r^{-n}$. Series of the type of (15.18) have been calculated and tabulated (see [7]). For $n \leq 3$ these series diverge. As $n \to \infty A_n$ approaches the number of nearest neighbors, which is 12 for the face-centered crystal lattice.

The lattice sums $A_n$ equal to 12.13 and 14.45 for $n = 12$ and 6, respectively.

In order to find nearest distance between atoms one calculates derivative of (15.11) with respect to $r$ finding $\partial U_{\text{LJ}}(r)/\partial r = 0$. This gives the equilibrium value of interatomic distance $r_0$ in the solid noble gases,

$$r_0 = \left( \frac{2A_{12}}{A_6} \right)^{\frac{1}{6}} = 1.09\sigma . \tag{15.19}$$

Substituting the equilibrium interatomic distance (15.19) into equation for energy per atom (15.16) we find the cohesive energy,

$$E_{\text{coh}} = -\frac{\epsilon A_6^2}{2A_{12}} = -8.6\epsilon . \tag{15.20}$$

The calculated bulk modulus is given by [7]

$$B = \frac{75\epsilon}{\sigma^3} . \tag{15.21}$$

**Table 15.4** Theoretic and experimental parameters of the solid noble gases. Data of [7].

| Gas | $r_0$ (nm) | | $E_{coh}$ (meV/atom) | | $B$ (MPa) | |
|---|---|---|---|---|---|---|
| | Theory | Experiment | Theory | Experiment | Theory | Experiment |
| Ne | 0.299 | 0.313 | −27 | −20 | 1.81 | 1.10 |
| Ar | 0.371 | 0.375 | −89 | −80 | 3.18 | 2.70 |
| Kr | 0.398 | 0.399 | −120 | −110 | 3.46 | 3.50 |
| Xe | 0.434 | 0.433 | −172 | −170 | 3.81 | 3.60 |

Table 15.4 presents the comparison of theoretic and experimental values of nearest-neighbor distance $r_0$, cohesive energy $E_{coh}$ and bulk modulus of the solid noble gases.

We can see a very good agreement between theoretic and experimental data on the nearest interatomic spacing. The more atomic mass, the bigger the $r_0$. If one compares theoretic and experimental values of cohesive energies the agreement again is found to be good. $E_{coh}$ ranges from –0.02 to –0.17 eV/atom (compare with −0.829 to −1.592 eV/atom for simple metals). Agreement for bulk modulus is satisfactory for krypton and xenon, but the theoretical value is about 17% larger for argon and 64% larger for neon than experimental ones.

## 15.5 Cohesive Energy Calculation for Noble Gas Solids

The density functional theory and the Kohn–Sham approach provide an efficient method for calculation properties for a wide range of materials. These methods are considered in Chapters 8–10.

The basis of the Kohn–Sham functional is an approximation to exchange-correlation energy of electrons. The approximations should include all many-body contributions to energy that is beyond the Hartree theory. The most common choices for the exchange-correlation energy functionals are local density approximation and generalized density approximation (Section 8.6). However, application of these functionals reveals a poor fitness for treatment systems that are bonded by the van der Waals forces.

The face-centered cubic crystals neon, argon, and xenon are the most prominent examples of the van der Waals bonded systems.

The first-principle method is being developed for systems with long-range dispersion forces. There are two ways to include dispersion forces in first-principle calculations. A semiempirical van der Waals interaction can be taken into account in *ab initio* calculations. It is realized by using the Lenard-Jones potential of the form (11.26). The second approach is based on the adiabatic connection fluctuation-dissipation theorem. This theory includes seamless long-range dispersion forces

**Table 15.5** Comparison of the first-principle results and experimental data of lattice constants and cohesive energies for solid noble gases. Data of [75].

| Solid | Lattice constant (nm) | | Cohesive energy (meV/atom) | |
|---|---|---|---|---|
| | Calculated | Experiment | Calculated | Experiment |
| Ne | 0.450 | 0.435 | −17.0 | −27.3 |
| Ar | 0.530 | 0.523 | −83.0 | −88.9 |
| Kr | 0.570 | 0.561 | −112.0 | −122.5 |

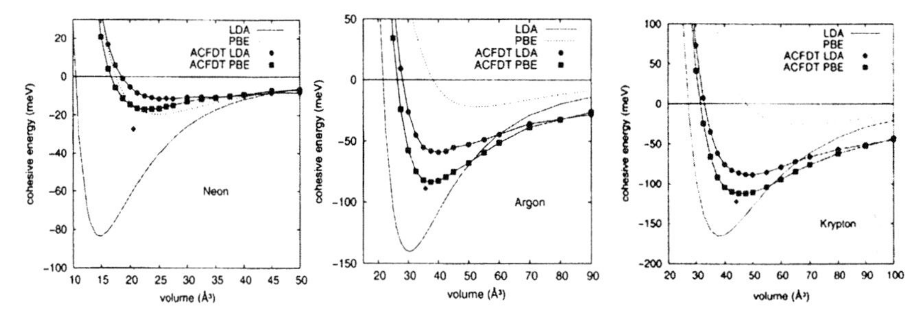


**Figure 15.5** Cohesive energies of solid noble gases as a function of the unit cell volume. LDA is the local density approximation; PBE is approximation by Perdew, Burke, and Ernzerhof [16]; ACDFT is adiabatic connection fluctuation-dissipation theorem. The experimental values are shown by black diamonds (after [75]).

and at the same time remains reasonably accurate for overlapping electron densities.

The advanced methods allow one to calculate equilibrium lattice constants and cohesive energies of the noble gas crystals.

In Table 15.5 the predicted equilibrium lattice constants and cohesive energies of the solid neon, argon, and krypton are compared with experiments. The results of the first-principle calculations are satisfactory, except for a value of cohesive energy for neon. As one can see, the theory overestimates the equilibrium lattice constant. Nevertheless, the updated theory [75] provides the adequate description of van der Waals bonded systems.

Figure 15.5 shows the dependence of calculated cohesive energy on volume of the unit cell. Local density approximation of the exchange-correlation energy of electrons fails for these molecular systems. Minima at curves based on improved theory (Figure 15.5, small squares) correspond well to experimental quantities (diamonds) for argon and krypton (error of the order of 9%). For neon the error of the cohesive energy calculation equals to 39%.

These examples illustrate the fact that the density functional theory and the first-principle technique are in a continuous progressive development and can be applied for study of molecular crystals.

We note that the experimental data for $r_0$ and $E_{\text{coh}}$, which are cited by authors [7] and [75] (Tables 15.4 and 15.5), differ appreciably. A certain scatter of experimental data that have been obtained by different research is a common and expected occurrence. We can state, however, that both theories under consideration give reasonable results of properties of solid noble gases.

## 15.6
## Organic Molecular Crystals

The first group of molecular crystals are the nonpolar A–H···B molecules. They are uncharged, and only relatively weak nondirectional dispersion forces operate. The molecules tend to attain an effective packing arrangement to minimize the repulsion forces between them.

The organic crystals constructed from nonpolar molecules often form structures packed face-to-face in stacks. Other neutral organic molecules compose the face-centered cubic crystal lattice. Examples of these structures are illustrated in Figures 15.6 and 15.7.

The second group can be presented as $A–H^{+}···B^{-}$. In this case the dipole–dipole and dipole–induced forces operate. The charges change the electronic environment around the bond (polarization), so as to increase attraction or repulsion.

The energy of interaction of two dipole molecules varies with distance as $r^{-3}$ (15.3). An example of a dipole interaction is shown in Figure 15.8. This interaction leads to the formation of chains of molecules with strong attraction between the cyanogen end of one molecule and the iodine end of the next one. The chains conjugate into layers. The molecular dipoles all have the same direction in a given layer and in opposite directions in successive layers.

A partial transition of electron charge from a donor molecule to an acceptor occurs in the third group of structures. This transition can affect intramolecular and intermolecular distances as well as the orientation of the molecules in the crystal. A relative orientation of neighbor molecules is such that maximizes the overlap of donor and acceptor orbitals.

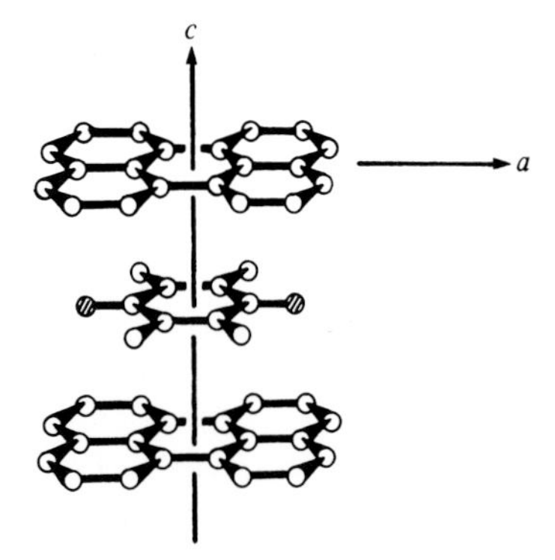


**Figure 15.6** The layer structure of perylene, which is a polycyclic hydrocarbon. Its chemical formula is $C_{20}H_{12}$. Melting temperature of the compound equals to 550.5 K (277.5 °C) (after [76] by permission of Cambridge University Press).

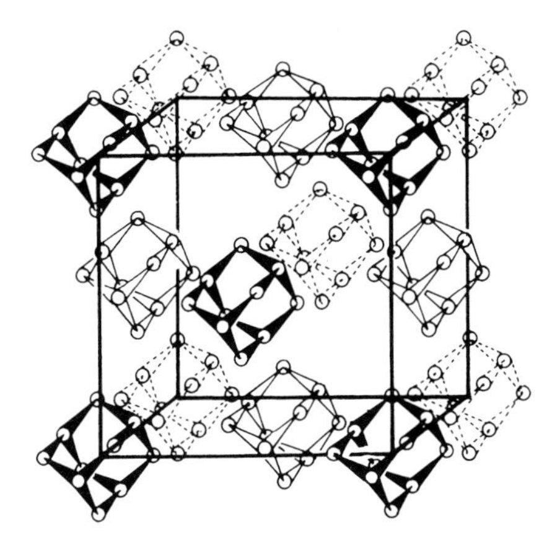

**Figure 15.7** The face-centered cubic structure of adamantane that is a chemical compound with a formula $C_{10}H_{16}$. The carbon atoms are located in the adamantane crystal like in diamond. Melting temperature of the compound equals to 543 K (270 °C) (after [76] by permission of Cambridge University Press).

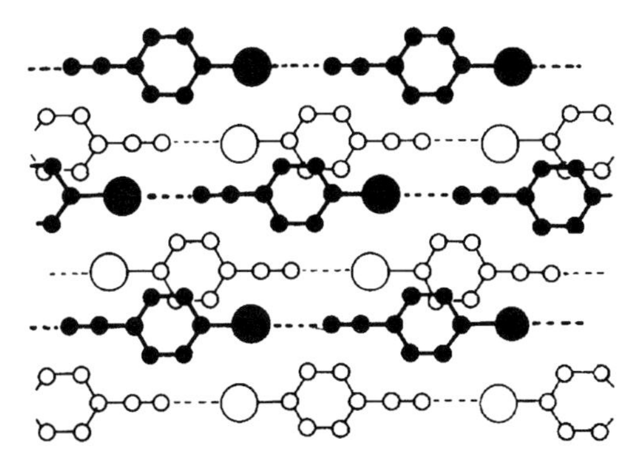

**Figure 15.8** The structure of *p*-Iodobenzonitrile. The molecular formula is $C_7H_4IN$. The melting temperature equals to 314.5 K (41.5 °C). Two small circles at the benzene rings denote the CN ends of the molecules, big circles denote the I atoms, attraction between them is shown by dotted lines (after [76] by permission of Cambridge University Press).

Figure 15.9a presents the edge-to-face $C^{-}{-}H^{+}\cdots\pi$ interaction of benzene rings. Rotation of the molecule about the single bonding occurs. The bonding arises due to the dispersion interaction and the Coulomb polarization. The face-to-face $\pi\cdots\pi$ bonding is shown in Figure 15.9b. This bonding is accepted to have several high-lying occupied donor orbitals or low-lying vacant acceptor orbitals of different symmetry, but is nevertheless capable of contributing significantly to the total charge-transfer interaction. The $\pi\cdots\pi$ interaction often leads to an offset of the benzene rings. The offset allows a hydrogen atom in C–H group to be located close to the center of the $\pi$ bonding, thereby increasing the electrostatic attraction.

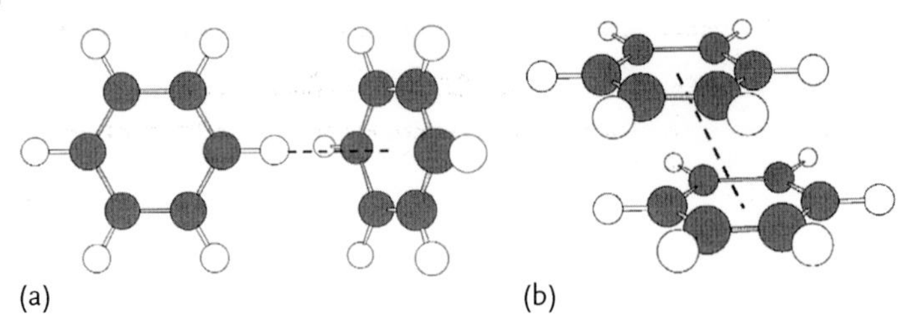


**Figure 15.9** Two types of the $\pi$ interaction between molecules: (a) the $C^{-}-H^{+}\cdots\pi$ bond; (b) the $\pi\cdots\pi$ bond (after [77]).

## 15.7 Molecule-Based Networks

It is appropriate to consider a type of molecules, which are bonded by intermolecular forces in branching networks, as a specific class of molecular crystals. Being a result of synthetic processes, these molecules represent a bulk material with desired properties as catalytic potential, porosity, magnetism, color. These solids are composed of specific spatial blocks. The building blocks are able to connect to each other by directional intermolecular bondings. The authors of [77] designate these structures as three-dimensional or 3D net. The molecules in such crystal may form "infinite" patterns (Figure 15.10).

It is of interest that different types of bonding occurs in a net within intermolecular interactions. Figure 15.11 illustrates the possible bondings along a path of a net. A scheme that represents the structure built from molecular blocks allows one to understand the construction of a solid compound with predefined properties.

The structures of the three-dimensional crystals are represented by means of schemes built from molecular blocks. It is reasonable to draw a bonding retracing curve as shown in Figure 15.11: path of covalent bonds, path of coordinative

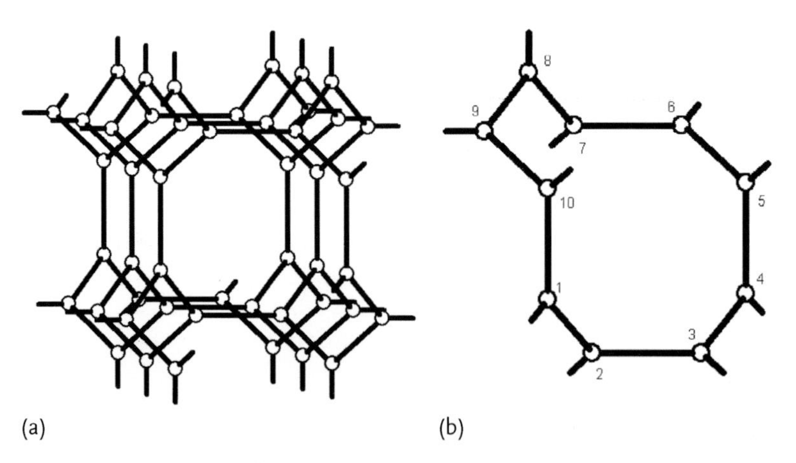


**Figure 15.10** An example of a net known as "Laves net" or "$SrSi_2$" net; (a) the net has nodes of more than one type. (b) A 10-ring contained in this net (after [77]).

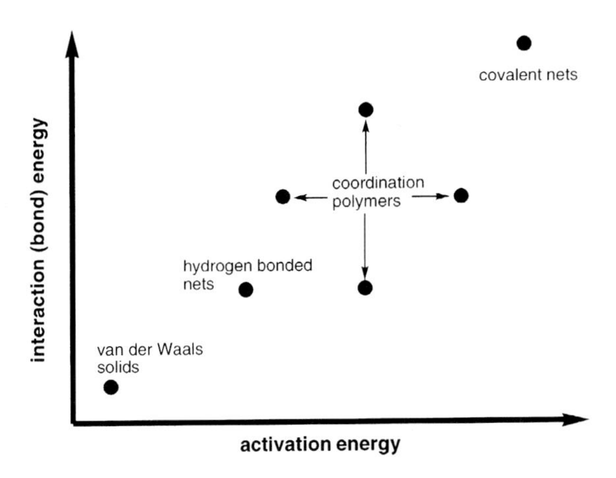


**Figure 15.11** A scheme of the bonding energy for nets and the activation energy for network link formation (after [77]).

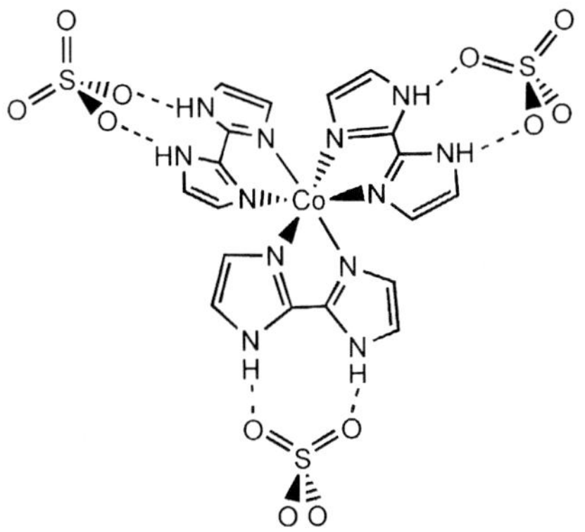

**Figure 15.12** Structure of an organic three-dimensional crystal. Seen are the covalent bonds S=O, hydrogen bonds (dotted lines) and charged bonds (triangles) (after [77]).

bonds, hydrogen, and van der Waals bonds. "A path along the net should follow the strongest directional intermolecular interactions (or bonds) and be significantly stronger than all other directional short-range intermolecular interactions [77]."

A scheme of the three-dimensional network with bondings of a different nature is shown in Figure 15.12.

Table 15.6 illustrates the strength of feasible bonds of a hydrogen atom with other atoms in organic crystals. This classification is proposed by the author of [71]. As expected, the covalent bond reveals the largest value of the bonding strength and the least value of the bonding length. The essence of cohesion in this case is quite similar to bonding in molecules (3.8). The overlap of the $^{1}H: 1s^{1}$ eigenfunction with an eigenfunction of another atom results in decrease of the total electron energy. An equilibrium value of interatomic distance of the order of 0.15 nm and

**Table 15.6** A classification of bonds of a hydrogen atom. $E$ is the energy of the covalent A–H bond or the hydrogen H···B bond, $l$ and $\theta$ are the length and the angle of the bond, respectively.

| Bonding type | Feature | $E$ (kJ/mol) | $E$ (eV/atom) | $l$ (nm) | $\theta$ (deg) |
|---|---|---|---|---|---|
| Covalent | Strong | −59 to −167 | −0.61 to −1.74 | 0.12–0.15 | 175–180 |
| Hydrogen (electrostatic) | Moderate | −16.7 to −62.7 | −0.17 to −0.65 | 0.15–0.22 | 130–180 |
| Hydrogen (dispersion) | Weak | > −16.7 | > −0.17 | 0.22–0.32 | 90–150 |

the flat angle are because of attraction and the core-core repulsion under the Pauli principle.

An induced dipole charge creates a relatively weak bonding of the H···B type.

## 15.8 Ionic Compounds

In ionic solids the particles are charged ions that are bonded together by the Coulomb forces.

For a positive ion $i$ and a negative ion $j$ separated by a distance $r_{ij}$ the potential of interaction is given by

$$E_{ij}(r) = -\frac{1}{4\pi\varepsilon_0}\frac{e^2}{r_{ij}} + \frac{C}{r_{ij}^n} . \tag{15.22}$$

The first term on the right side of (15.22) characterizes the Coulomb attraction of the ion pair. The Coulomb interaction is a long-range one since it changes as $r_{ij}^{-1}$. The second term with $n > 1$ characterizes the repulsion of electron shells at closest approach according to the Pauli principle. In classical approximation, the constants $C$ and $n$ are considered as unknown quantities to be determined from the lattice and compressibility data.[5]

In order to calculate the energy of the NaCl-type lattice we summarize interactions over all $ij$ pairs of ions. To take into account the double counting of interatomic interactions we introduce coefficient 1/2. Energy per ion is given by

$$E_{\text{ion}}(r) = \frac{1}{2}\sum_{i,j;i\neq j} E_{ij}(r) . \tag{15.23}$$

It is appropriate to introduce dimensionless values $p_{ij}$,

$$r_{ij} = p_{ij} r , \tag{15.24}$$

5) The values in (15.22) are in the SI units. It is seen that the unit of the $C$ constant is J m$^n$.

where $r$ is the distance between the neighboring atoms in the crystal. Then

$$\frac{1}{r_{ij}} = \frac{1}{p_{ij}}\frac{1}{r}\,; \quad \frac{1}{r_{ij}^n} = \frac{1}{p_{ij}^n}\frac{1}{r^n}\,. \tag{15.25}$$

Substituting (15.22) and (15.25) to (15.23) we arrive at

$$E_{\text{ion}}(r) = \frac{1}{2}\left[-\frac{1}{4\pi\varepsilon_0}\frac{1}{\sum_{i,j;i\neq j} p_{ij}}\frac{e^2}{r} + \frac{C}{\sum_{i,j;i\neq j} p_{ij}^n}\frac{1}{r^n}\right]. \tag{15.26}$$

Denote

$$\alpha = \frac{1}{\sum_{i,j;i\neq j} p_{ij}}\,; \quad A_n = \frac{1}{\sum_{i,j;i\neq j} p_{ij}^n}\,, \tag{15.27}$$

where $\alpha$ is the Madelung constant. The energy of an ionic crystal per ion can be expressed finally as

$$E_{\text{ion}}(r) = \frac{1}{2}\left[-\frac{1}{4\pi\varepsilon_0}\frac{\alpha e^2}{r} + \frac{A_n C}{r^n}\right]. \tag{15.28}$$

The Madelung constant plays the leading part in theory of ionic crystals.

In the crystal lattice of NaCl the nearest neighbors of the $Cl^-$ ion are 6 $Na^+$ ions at a distance of $r$ ($p = 1$). Further 12 $Cl^-$ ions are situated at a distance $r\sqrt{2}$ ($p = \sqrt{2}$), further 8 $Na^+$ ions are at a distance $r\sqrt{3}$ ($p = \sqrt{3}$), further again 6 $Cl^-$ ions are at a distance $2r$ ($p = 2$)... The summation gives[6)]

$$\alpha = \frac{6}{1} - \frac{12}{\sqrt{2}} + \frac{8}{\sqrt{3}} - \frac{6}{2} + \ldots$$

An exact calculation of the Madelung constants gives $\alpha = 1.7476, 1.7627, 1.6381$ for NaCl, CsCl, ZnS, respectively.

Taking derivative from (15.28) in order to find the equilibrium interatomic distance $r_0$ we arrive at

$$\frac{\partial E_{\text{ion}}}{\partial r} = \frac{1}{4\pi\varepsilon_0}\frac{\alpha e^2}{r_0^2} - \frac{nCA_n}{r_0^{n+1}} = 0\,. \tag{15.29}$$

Substituting $A_n$ from (15.29) to (15.26) we obtain

$$E_{\text{ion}} = E_{\text{coh}} = -\frac{1}{4\pi\varepsilon_0}\frac{\alpha e^2}{r_0}\left(1 - \frac{1}{n}\right). \tag{15.30}$$

An estimation of the $n$ value from the compressibility data gives $n = 9.4$ [78], $\alpha = 1.747\,558$. Further, we have for NaCl from X-ray data the lattice spacing $r_0 = 2.815 \times 10^{-10}$ m. Electron charge $e = 1.60 \times 10^{-19}$ C, $\varepsilon_0 = 8.85 \times$

6) With such calculation we may omit the coefficient 1/2 in (15.26).

**Table 15.7** The properties of some alkali halides. $r_0$ is the distance between neighbor atoms; $B$ is bulk modulus; $E_{coh}$ is the cohesive energy.

| Compound | Experiment | | | Theory | |
|---|---|---|---|---|---|
| | $r_0$ (nm) | $B$ (GPa) | $E_{coh}$ (kcal/mol) | $E_{coh}$ (kcal/mol) | $E_{coh}$ (eV/ion) |
| LiF | 0.201 | 6.71 | −241.9 | −240.1 | −10.42 |
| LiCl | 0.256 | 2.98 | −198.1 | −199.2 | −8.65 |
| LiBr | 0.275 | 2.38 | −188.3 | −189.3 | −8.22 |
| LiI | 0.300 | 1.72 | −177.1 | −181.1 | −7.86 |
| NaF | 0.231 | 4.65 | −214.6 | −213.4 | −9.26 |
| NaCl | 0.282 | 2.40 | −182.8 | −183.1 | −7.95 |
| NaBr | 0.299 | 1.99 | −173.3 | −174.6 | −7.58 |
| NaI | 0.324 | 1.51 | −166.4 | −163.9 | −7.11 |
| KF | 0.267 | 3.05 | −190.1 | −189.7 | −8.25 |
| KCl | 0.315 | 1.75 | −164.4 | −165.4 | −7.18 |
| KBr | 0.330 | 1.48 | −156.2 | −159.3 | −6.91 |
| KI | 0.353 | 1.17 | −151.5 | −150.8 | −6.54 |

$10^{-12}\,\mathrm{C\,V^{-1}\,m^{-1}}$. Substituting these quantities to (15.30) we obtain cohesive energy $E_{coh\,NaCl} = -768.755\,\mathrm{kJ/mol} = -183.732\,\mathrm{kcal/mol} = -7.95\,\mathrm{eV/ion}$. Experimental value equals to $-182.8\,\mathrm{kcal/mol}$. The fit is good.

Our calculation is based on a semiempirical classical theory because we use (15.22) and experimental values for $r_0$ and $n$. A more rigorous theory, however, would not improve the result much. That is because the major part of cohesive energy of ionic compounds is due to the Coulomb interaction of the ions which can be considered as electric charges at lattice sites.

Table 15.7 presents theoretic and measured properties of alkali halides. We can see that the ionic bond is a relatively strong one. The cohesive energy ranged from $-6.5$ to $-10.5\,\mathrm{eV/ion}$. In a series, for instance, of NaF–NaCl–NaBr–NaI (valence shells of halogen atoms are $2p^5$, $3p^5$, $4p^5$, $5p^5$, respectively) the interatomic distance increases, whereas bulk modulus and cohesive energy decreases. This regularity is in accordance with the Pauling electronegativity scale. L. Pauling characterizes the qualitative property of an atom to attract electrons to itself by numbers of 4.0, 3.0, 2.8, and 2.5 for atoms F, Cl, Br, and I, respectively.

# 16
# High-Temperature Creep

Problem of strength of metallic materials at high temperatures is of great scientific and practical importance. The durability of gas-turbine engines, steam pipelines, reactors, aeroplanes, and aerospace vehicles depends directly on an ability of their parts and units to withstand changes in shape. On the other hand, a significant mobility of the crystal lattice defects and of atoms plays an important role in the behavior of materials under applied stresses at high temperatures and is also of great interest for materials science as well as for practical applications.

In Figure 16.1 one can see a dependence of strain upon time $\varepsilon(t)$ when applied stress $\sigma$ remains constant. In general, the creep curve contains four stages: an incubation, primary, steady-state, and tertiary stages. The steady-state stage is the most important characteristic for metals, because it takes up the greater part of durability of the specimen. Correspondingly, the minimum strain rate during steady-state stage $\dot{\varepsilon}$ is an important value because it determines the lifetime of the specimen. The tertiary stage is associated with a proportionality of the creep strain rate and the accumulated strain. It is observed to a certain extent in creep resistant materials. A tertiary stage is followed by a rupture. Figure 16.1 illustrate the macroscopic level of strength of metals at high temperatures.

Energies of the vacancy generation $E_v$ and vacancy diffusion $U_v$ in crystal lattice turn out to be a controlling factor of a material resistance to strain at high temperatures. These factors are affected by strength of interatomic bonding.

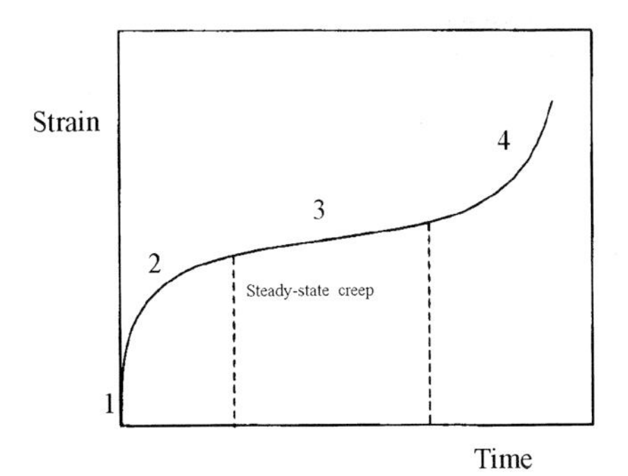


**Figure 16.1** The typical curve of creep.

*Interatomic Bonding in Solids: Fundamentals,Simulation,andApplications*, First Edition. Valim Levitin.

I would like to show compactly in this chapter stages of a physical research:

- experimental investigation;
- determination of parameters that have a significant influence on the process;
- construction of a physical model;
- derivation of equations corresponding to the model and their solution;
- comparison of calculated results with the experimental data.

## 16.1
## Experimental Data: Evolution of Structural Parameters

A theory of strength of metals at high temperatures has to be based on experimental data. An experimental technique was developed for *in situ* X-ray investigations of metals and alloys directly during creep process [79]. The structural peculiarities of the high-temperature strained metals were also studied by transmission electron microscopy (TEM). Our goal was to obtain data that enable us to observe the events on atomic and microscopic levels.

*In situ* studies allow conclusions to be drawn concerning the effect of stresses at high temperatures on the evolution of structure in metals. The X-ray reflections change in shape after the stress is applied. As a rule, the integral intensity of reflections grows and the angle width of the intensity curve-base increases appreciably. The increase in the diffracted irradiation energy indicates that the reflecting structural elements become smaller.

In Figure 16.2 the lower curve $\varepsilon(t)$ is typical for the creep. The primary and the steady-state stages of deformation are seen. Variations in subgrain sizes $D$ (upper curves) and in their misorientations $\eta$ are presented in the same graph. Here each type of symbols corresponds to one of crystallites of the same specimen.

The initial mean size of subgrains $D$ is equal to 3 μm, in the primary stage of deformation it decreases to 0.8 μm and then is almost unchanged during the steady-state stage. The misorientation angle $\eta$ increases from 2 to 5–7 mrad. The change in $\eta$ is observed at the primary stage.

An electron micrograph of a subgrain and sub-boundaries in crept nickel is shown in Figure 16.3. One can see a clean area in the center of the picture, that is, subgrain or cell surrounded by dislocation aggregations. The cell walls separate relatively dislocation-free regions from each other. Subgrains are also seen at the borders of the picture. The substructure is formed inside of crystallites during the first, transitive stage of creep. The origin of the steady-state deformation stage coincides with the end of the substructure formation.

Some regularities are revealed as the result of systematic examinations of the bright- and dark-field image pictures and diffraction patterns of a large number of specimens.

Most of the dislocations in specimen after high-temperature tests are associated in sub-boundaries. The parallel sub-boundary dislocations are situated at equal distance from each other. It follows from results of the Burgers vector determinations

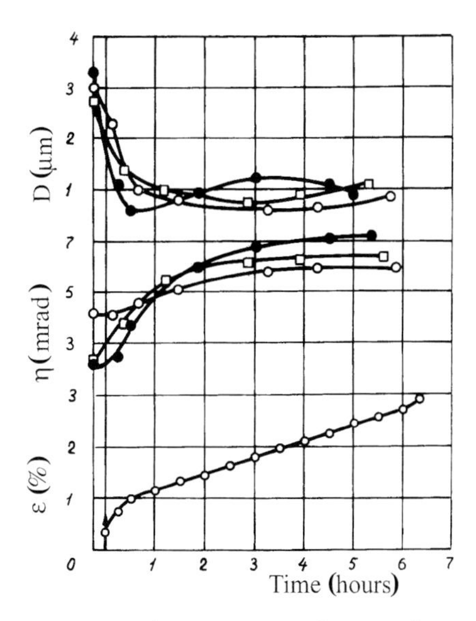


Figure 16.2 The creep curve and structural parameters as a function of time for nickel. Tests at temperature $T = 673$ K (0.39 $T_m$); stress $\sigma = 130$ MPa ($1.7 \times 10^{-3}\mu$).

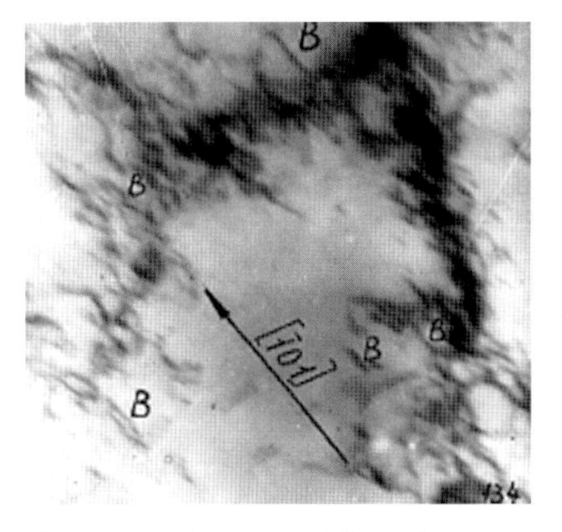

Figure 16.3 Subgrain in nickel tested at temperature 1073 K, stress 20 MPa. Screw dislocations along the [$\bar{1}$01] direction are denoted as *B*. 33 000 ×.

and of the repeating structural configuration that the parallel sub-boundary dislocations have the same sign. Two intersected dislocation systems are often observed inside sub-boundaries .These systems form small-angle boundary.

Figure 16.4 illustrates the dislocation sub-boundary in alpha-iron. Two systems of dislocations, which intersect each other at the right angle, are observed. Dislocation lines are parallel to face diagonals, that is, along crystalline directions [110] and [1$\bar{1}$0]. One of systems is inclined noticeably to the foil plane. This is the cause of an oscillating contrast in the dislocation images. Atomic displacements are directed along the body diagonals of the elementary body-centered cubic cell.

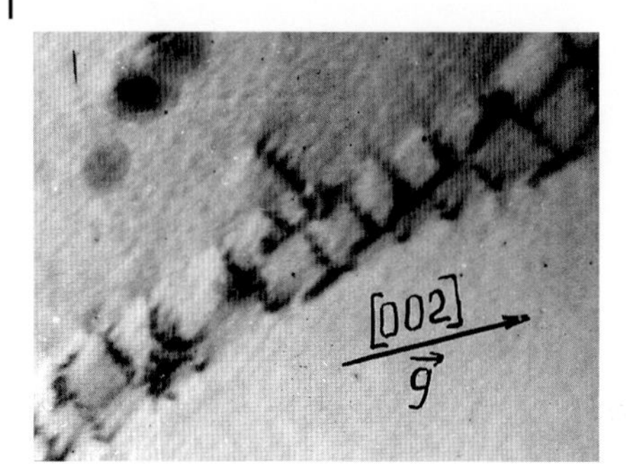


**Figure 16.4** Transmission electron micrograph showing the dislocation sub-boundary in alpha-iron, which is formed in the steady-state creep. $T = 813$ K and $\sigma = 49.0$ MPa. The first dislocation system is directed along [110] with the Burgers vector $\boldsymbol{b}_1 = \boldsymbol{a}/2[111]$, the second along with $\boldsymbol{b}_2 = \boldsymbol{a}/2[\bar{1}\bar{1}1]$. The plane of the foil is $(2\bar{1}0)$. 84 000 ×.

**Table 16.1** Comparison of distances between dislocations in sub-boundaries $\lambda$ and average distances $z_0$ between jogs in mobile dislocations.

| Metal | $T$ (K) | $\sigma$ (MPa) | $\lambda$ (nm) X-ray data | $\lambda$ (nm) TEM data | $z_0$ (nm) TEM data |
|---|---|---|---|---|---|
| Ni | 1073 | 14.0 | $45^{\pm 3}$ | $42^{\pm 5}$ | $48^{\pm 4}$ |
| | 1073 | 20.0 | $38^{\pm 3}$ | $34^{\pm 4}$ | $45^{\pm 7}$ |
| Nb | 1233 | 39.2 | $74^{\pm 6}$ | $67^{\pm 13}$ | – |
| | 1370 | 29.4 | $107^{\pm 10}$ | $109^{\pm 14}$ | $120^{\pm 10}$ |
| V | 1096 | 24.5 | $80^{\pm 8}$ | – | $90^{\pm 8}$ |
| | 1096 | 34.3 | $67^{\pm 6}$ | – | $60^{\pm 7}$ |
| Cu | 746 | 7.8 | $87^{\pm 6}$ | $83^{\pm 6}$ | $83^{\pm 7}$ |

The regular dislocation networks as low-angle sub-boundaries are found to be typical for the high-temperature tested metals.

The distance $\lambda$ between parallel dislocations of the same sign in a small-angle boundary can be represented (when $\eta \ll 1, \tan\eta \simeq \eta$) by an expression of the form

$$\lambda = \frac{b}{\eta}, \tag{16.1}$$

where $b$ is the magnitude of the Burgers vector. Measuring results of $\lambda$ values by means of two methods are shown in Table 16.1.

The sub-boundaries that have been formed seem to be sources of slipping dislocations. The process of generation of mobile dislocations by sub-boundaries is readily affected by the applied stress. The TEM technique allows one to observe the beginning of a dislocation emission. The creation of dislocations occurs as if the sub-boundary blows the dislocations loops. These loops broaden gradually and

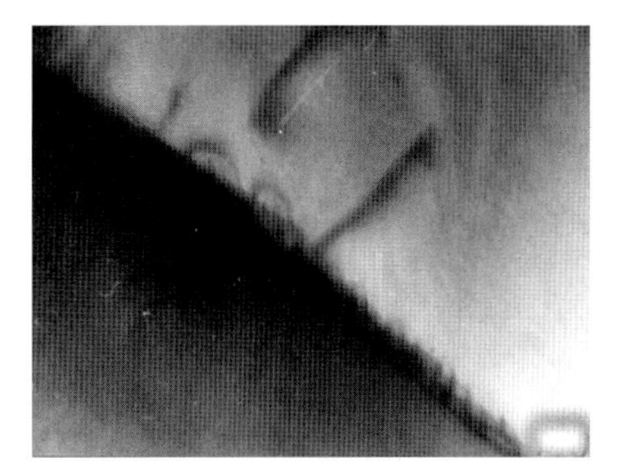

**Figure 16.5** Emission of dislocation loops from sub-boundary in alpha-iron. Tests at 813 K and $\sigma = 49$ MPa; 60 000 ×.

move farther inside subgrains. The sub-boundary in alpha-iron that generates dislocations is shown in Figure 16.5. The subsequent dislocation semiloops are blown by the ordered boundary.

At the same time, the sub-boundaries act as obstacles for moving dislocations. One can often observe a sequence of dislocation lines which are pressed to the sub-boundary and these can enter the sub-boundary.

Some dislocations, which are observed in specimen after high-temperature deformation, are not associated in sub-boundaries. They are located inside subgrains. Bends and kinks at these dislocations attract one's attention. They give an impression that certain points of mobile dislocations are pinned up.

There are good reasons to assume that kinks and bends that have been described by us are jogs. A jog is known to be a segment of a screw dislocation, which does not lie in its plane of slipping. In fact, the jog is a segment of the edge extra-plane and therefore it can move with the gliding screw dislocation only by the emission or absorption of point defects (vacancies or interstitial atoms). During movement the jog brakes the dislocation and lags behind.

We compared the average distances $\bar{\lambda}$ between sub-boundary dislocations, determined with the X-ray method, and the spacings $\bar{z}_0$ between jogs in mobile dislocations measured with the aid of electron microscopy. We revealed that the two values are close to each other. The data of comparison are shown in Table 16.1.

We obtained the same data for Ni, Fe, Cu, V, Nb with the measurement error of the order of 10%. In my opinion, the experimental result that was obtained

$$\bar{z}_0 \approx \bar{\lambda} \tag{16.2}$$

is of great importance in our understanding of the physical mechanism of high-temperature deformation.

Closed dislocation loops as well as helicoids are observed very often in structures of high-temperature tested metals. Dark-field analysis makes it possible for one to determine the sign and the type of loops. Undoubtedly, the loops have been found to be of a vacancy type. Helicoids are known to be formed with screw dislocations usually, under conditions of supersaturation with point crystalline defects.

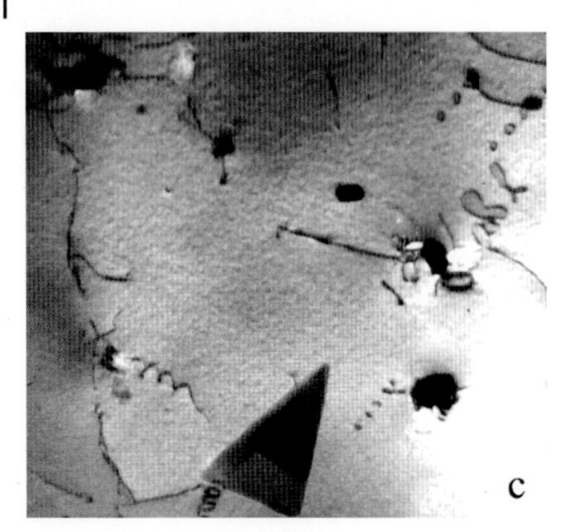


**Figure 16.6** Transmission electron micrograph showing vacancy loops and helicoids in niobium tested at $T = 1508$ K, $\sigma = 17.3$ MPa; 39 000 ×.

In Figure 16.6 a very interesting effect may be observed. Three chains of loops have been left behind two segments of screw dislocations. These moved in the slip plane (110). Dislocations have the Burgers vector $\boldsymbol{b} = \boldsymbol{a}\sqrt{2}/2[\bar{1}10]$; the loops are of the vacancy type. One can also see helicoids.

## 16.2 Physical Model

The dislocation density increases at the beginning of the plastic strain. In the primary stage of deformation some of the generated dislocations form discrete arrangements. They enter into low-angle sub-boundaries. The interaction of dislocations having the same sign is facilitated by the high-temperature conditions and applied stress. These conditions make it easy for dislocations to move and to the edge components of dislocations to climb. The immediate cause of formation of the dislocation walls is the interaction between dislocations of the same sign that results in a decrease in the internal energy of the system.

The well-formed subgrains are sources, emitters of mobile dislocations, which contribute to strain. Emission of mobile dislocations from sub-boundaries leads to formation of jogs in them. It is the dislocation sub-boundary that generates jogs in mobile dislocations. The screw components of emitted dislocations "keep their origin in their memory." They contain the equidistant one-signed jogs.

A jog is a section of dislocation, which does not lie in the slip plane. The jog cannot move without generation of point defects, that is, vacancies. The jogged dislocation can slip if a steady diffusion of vacancies occurs from it. The nonconservative slipping of jogged dislocations is dependent on the material redistribution. The shorter the distance between jogs, the lower the dislocation velocity. Hence, it is the diffusion process that controls the velocity of the deforming dislocations. It is the generation and diffusion of the lattice vacancies that determine the strain rate.

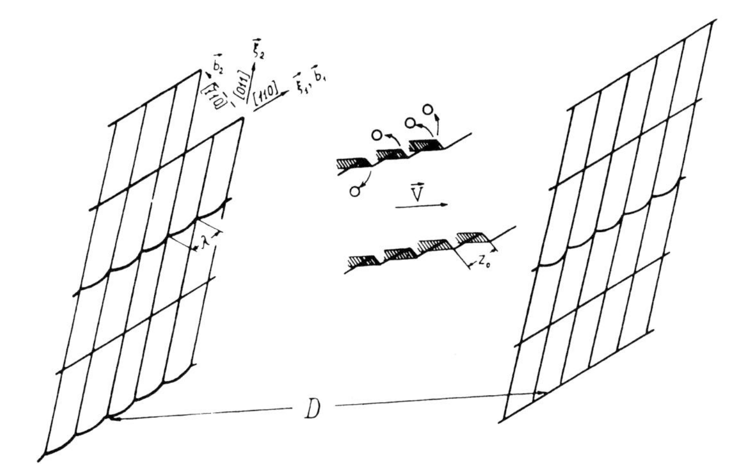

**Figure 16.7** The physical model of the steady-state strain at high temperature. The sub-boundaries are built of two systems: the pure screw one and the 60°-dislocation system. $\lambda$ is the spacing between dislocations in sub-boundaries; $D$ is the distance between sub-boundaries, that is, the size of the sub-grain. Emission of mobile dislocations from sub-boundaries is shown. $z_0$ is the spacing between jogs at mobile dislocations; the edge extra-planes are shaded. The jogs at the mobile screw components have the same sign and generate vacancies.

The interstitial-producing jogs are practically immobile because the energy of formation of the interstitial atoms is several orders greater than that of the vacancies.

The distance between dislocations in sub-boundaries decreases during the primary stage of deformation and the role of sub-boundaries as obstacles for slipping dislocations increases.

At the end of the substructure formation the dislocation arrangement is ordered and a steady-state stage begins. During this stage a dislocation emission from sub-boundaries takes place. The emitted dislocations are replaced in sub-boundaries with new dislocations, which move under the effect of applied stress. Having entered a sub-boundary a new dislocation is absorbed by it. Jogs of the same sign appear along the screw components of the emitted dislocations. The distance between jogs at mobile dislocations is equal to the distance between the immobile sub-boundary dislocations, see (16.2).

Figure 16.7 illustrates the described model of the steady-state creep. It can be seen that atoms are required in order to complete the extra-planes during the motion of the jogs. Vacancies are generated in the vicinity of jogs since atoms are consumed in completing the extra-planes. Consequently, the relay-like motion of the vacancy-emitted jogged dislocations from one sub-boundary to another is the distinguishing feature of the high-temperature strain.

## 16.3 Equations to the Model

The rate of the dislocation slip is given by

$$\dot{\gamma} = bVN\,, \tag{16.3}$$

where $b$ is the Burgers vector, $V$ is average velocity of dislocations, $N$ is the density of deforming dislocations.

The velocity of screw components with vacancy-producing jogs is given by the following expression [80]

$$V_{\mathrm{p}} = \frac{\pi \nu r_0 z_0}{bF(\alpha)} \exp\left(-\frac{E_{\mathrm{v}} + U_{\mathrm{v}} + \varepsilon_0}{k_{\mathrm{B}}T}\right)\left[\exp\left(\frac{\sigma_{yz} b^2 z_0 + \varepsilon_0}{k_{\mathrm{B}}T}\right) - 1\right], \tag{16.4}$$

where $\nu$ is the Debye frequency, $r_0$ is the radius of the dislocation "tube," $E_{\mathrm{v}}$ is energy of the vacancy generation, $U_{\mathrm{v}}$ is energy of the vacancy diffusion, $\varepsilon_0$ is the energy of bonding a dislocation and a vacancy, $\sigma_{yz}$ is the component of the stress tensor, $\alpha = V_{\mathrm{p}} r_0/2D_{\mathrm{v}}$, $D_{\mathrm{v}} = E_{\mathrm{v}} + U_{\mathrm{v}}$, $F(\alpha)$ is a weak function, $k_{\mathrm{B}}$ is the Boltzmann constant, and $T$ is temperature.

We conclude that the velocity of deforming dislocations depends on interatomic bonding through the energies of generation and diffusion of vacancies in a crystal lattice, $E_{\mathrm{v}}$ and $U_{\mathrm{v}}$, respectively. This implies that the strength of interatomic bonding determines the creep resistance of a solid.

The velocity of dislocations depends exponentially on the stress. One can see that the exponent in (16.4) contains the sum $E_{\mathrm{v}} + U_{\mathrm{v}}$. This implies that the effective energy of the dislocation motion is close to the energy activation of diffusion. The activation volume in (16.4) equals to $b^2 z_0$.

Figure 16.8 shows velocity of jogged dislocations producing vacancies versus temperature, stress, and the distance between jogs. The data have been calculated according to (16.4).

We have measured the density of dislocations $N$, which are not associated in sub-boundaries, and the results are presented in Table 16.2.

Now we can calculate the rate of the steady-state high-temperature deformation $\dot{\varepsilon}$. Three groups of physical parameters are needed:

- External parameters: temperature $T$ and stress $\sigma$.

**Table 16.2** The density of dislocations inside the subgrains.

| Metal | $T$ (K) | $\sigma$ (MPa) | $N$ ($10^{11}$ m$^{-2}$) | Metal | $T$ (K) | $\sigma$ (MPa) | $N$ ($10^{11}$ m$^{-2}$) |
|---|---|---|---|---|---|---|---|
| Ni | 1023 | 9.8 | 1.3 | Nb | 1508 | 17.2 | 2.9 |
| | | 40.0 | 6.3 | | | 28.4 | 5.3 |
| Cu | 678 | 8.8 | 2.2 | V | 1096 | 29.4 | 1.6 |
| | | 20.6 | 9.6 | | | 34.3 | 1.4 |

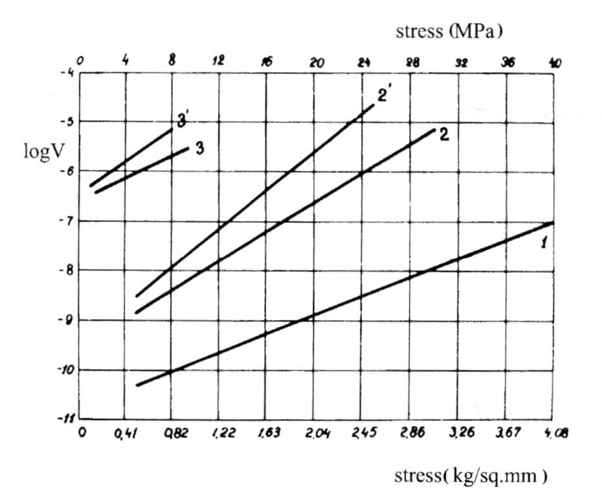


**Figure 16.8** The effect of temperature, stress, and distance between jogs on the velocity of screw dislocations in $\alpha$-iron: 1 – 773 K, $z_0$ = 35 nm; 2 – 813 K, $z_0$ = 57 nm; 2′ – 813 K, $z_0$ = 75 nm; 3 – 973 K, $z_0$ = 52 nm; 3′ – 973 K, $z_0$ = 75 nm.

- Diffusion parameters: the energy of the vacancy generation $E_v$ and the energy of the vacancy diffusion $U_v$.
- Structural parameters: the subgrain size $D$ and the distance between dislocations in sub-boundaries $\lambda = b/\eta$.

These parameters and constants are the input data for calculations. For $\alpha$-iron $E_v = 2.56 \times 10^{-19}\,\mathrm{J\,atom^{-1}}$ $(1.60\,\mathrm{eV\,atom^{-1}})$. $U_v = 1.92 \times 10^{-19}\,\mathrm{J\,atom^{-1}}$ $(1.20\,\mathrm{eV\,atom^{-1}})$ [81, 82].

## 16.4 Comparison with the Experimental Data

The ability of the physical model to correctly and quantitatively represent the macroscopic creep behavior is shown in Figures 16.9 and 16.10.

The theoretical curves are in close agreement with the experimentally observed values of the steady-state creep rate in $\alpha$-iron at 813, 873, 923 and 973 K.

It can be seen that the theory does not represent the experimental results at 773 K.

We obtained analogical results for other metals and for solid solutions. For example, the measured rate of the steady-state creep $\dot{\varepsilon}$ equals to $13.9 \times 10^{-7}\,\mathrm{s^{-1}}$ for alloy Ni + 9.9 at.% Al at 873 K and $\sigma = 136$ MPa. In the same conditions, the calculated $\dot{\varepsilon}$ value equals to $13.0 \times 10^{-7}\,\mathrm{s^{-1}}$.

The present theory is understood to be valid within certain limitations. It has a scope of application. When temperature is relatively low, the dislocation climb is depressed and hence the regular sub-boundaries cannot be formed. The lower limit of

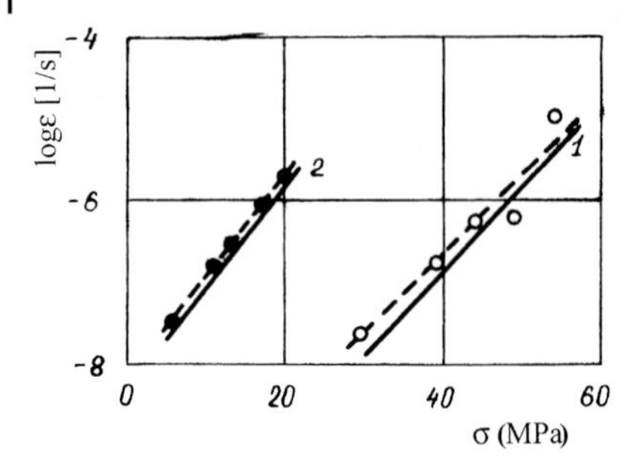


**Figure 16.9** The steady-state creep rate of $\alpha$-iron as a function of stress. Comparison of the experimental data (symbols and dotted lines) and computed results (solid curves); 1 — 813 K, 2 — 923 K.

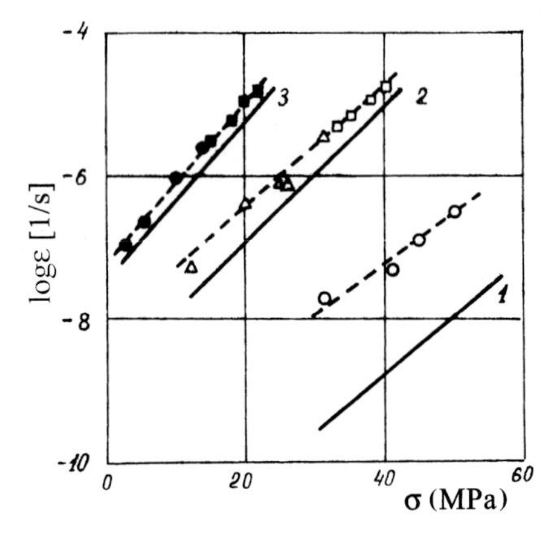


**Figure 16.10** The steady-state creep rate of $\alpha$-iron as a function of stress. Comparison of the experimental data (symbols and dotted lines) and computed results (solid curves); 1 — 773 K; 2 — 873 K; 3 — 973 K.

the sufficient climb rate is about 0.40 or 0.45 $T_m$. It is necessary to emphasize that an adequate understanding of creep processes in these ranges of temperatures and stresses is of great practical importance. Most of heat-resistant metals, steels and alloys operate at temperatures between 0.40 and 0.75 $T_m$. The low-temperature deformation is controlled by another processes, for example, by the overcoming of the Peierls stress in a crystal lattice. The upper limit of temperature is (0.70–0.75) $T_m$. Diffusion creep takes place (the mechanism of Herring–Nabarro) at higher temperatures and at relatively lower stresses.

The stable sub-boundaries are of major significance in the process of high-temperature strain for pure metals and solid solutions. The upper stress limit of the sub-boundary stability depends upon metal properties and temperature. The lower the shear modulus $\mu$ and the higher temperature, the lower the limit. An inactivated emission of dislocations from sub-boundaries occurs when the applied external stress is higher than about $2 \times 10^{-3}\mu$. Then the sub-boundaries break up.

# 17
# Fatigue of Metals

A metal which is subjected to alternating stresses fails at a much lower stress than that required to cause a fracture on the single application of a load. Failures occurring under conditions of oscillating loading are called fatigue failures. The basic mechanism of fatigue fracture is the origination on the surface of a crack which slowly spreads. Fatigue fracture is generally observed after a considerable period of time.

Fatigue failure is known to be particularly insidious because it occurs without any obvious warning. Today it is often stated that fatigue accounts for at least 90% of all service failures which are due to mechanical causes.

Fatigue becomes progressively more prevalent as technology develops greater amount of equipment, such as turbines, aircrafts, automobiles, compressors, pumps, bridges, which are subjected to repeated loading and vibration.

The fatigue damage propagates at several levels: atomic, microscopic, and macroscopic (see Table 1.1). Development of the fatigue failure is determined by processes on atomic scale. We shall see further that it is connected with parameters of interatomic bonding, such as energy of the vacancy formation $E_v$ and the surface energy $\gamma$. It appears that strength of interatomic bonding in structural materials determines our standard of living and our safety.

Fatigue results in a brittle-appearing fracture, with no gross deformation at the fracture. On a macroscopic scale the fracture surface is usually normal to the direction of the principal tensile stress. The crack extension depends upon material, amplitudes of alternating stress, frequency of stress, temperature of operation. The process of fatigue failure under a load of a constant amplitude typically includes several stages:

- cyclic plastic deformation, with formation of stable cyclic dislocation substructures;
- formation of a crack embryo on the surface in a region of the localized shear stress;
- sharpening of the crack front and onset of propagation (Figure 17.1);
- propagation of the crack until fatigue fracture of the specimen or the component.

*Interatomic Bonding in Solids: Fundamentals,Simulation,andApplications*, First Edition. Valim Levitin.

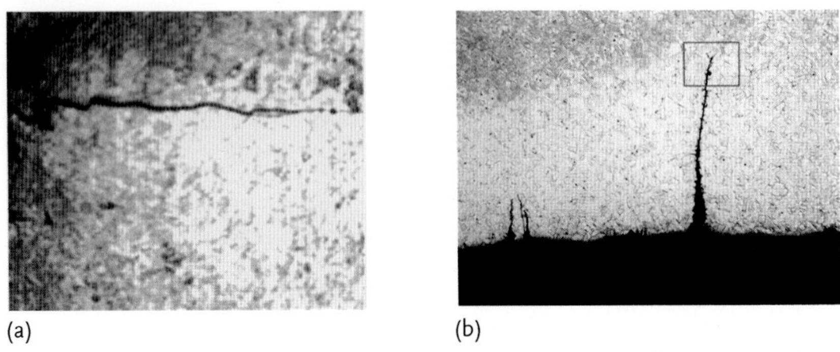


**Figure 17.1** Fatigue cracks in an aluminum alloy (a) and in stainless steel (b).

## 17.1 Crack Initiation

The movement of mobile dislocations plays a significant role in the fatigue failure preparation. If a specimen is subjected to alternating stresses the generation of dislocations is known to occur. There is a threshold stress $\tau_s$ at which this process begins [81]

$$\tau_s = \frac{\mu_s b(n_{pu}\rho)^{\frac{1}{2}}}{2\pi}, \tag{17.1}$$

where $\mu_s$ is the shear modulus; $b$ is the Burgers vector; $n_{pu}$ is the number of dislocations in the pile-up; $\rho$ is the density of dislocations. In metals and alloys the initial value of $\rho$ is of the order of $10^{10}$–$10^{12}$ m$^{-2}$. The dislocation density increases during the deformation. Rough estimate gives a ratio of $\tau_s/\mu_s = 10^{-4}$–$10^{-3}$.

A force is known to attract dislocations to the surface of a specimen. Energy required to produce a certain amount of slip inside a solid is about twice that required to produce the same amount of slip at the surface. Motion of the dislocations caused by cycling loading occurs long before the yield strength is reached.

The structure of a fatigued nickel-based superalloy is presented in Figure 17.2. The deforming dislocations already begin to move after six cycles of the alternating load.

The slip of the majority of dislocations is concentrated in strained metals in slip bands. Figure 17.3 illustrates the clearly-marked formation of surface steps as a result of a material strain. The hollow initiation on the surface is also seen. Surface microscopic cracks start easily in slip bands.

However, the process of the crack nucleus extension is in fact reversible up to some point in time.[1)]

1) Preparatory processes of fatigue and their reversibility can be discovered by the work function technique because of formation of surface steps. The electronic work function can be defined as a smallest amount of energy required to remove an electron from the surface to a point outside the metal with zero kinetic energy. The generation of surface steps early in the fatigue process results in a distinct decrease in work function [83].

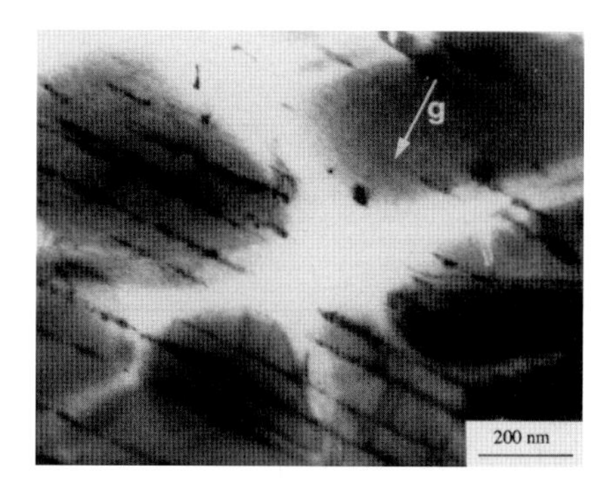


**Figure 17.2** Transmission electron micrograph showing the dislocation structure of a superalloy after six cycles with a shear strain of $1.1 \times 10^{-3}$. Vector $g \parallel \langle 111 \rangle$.

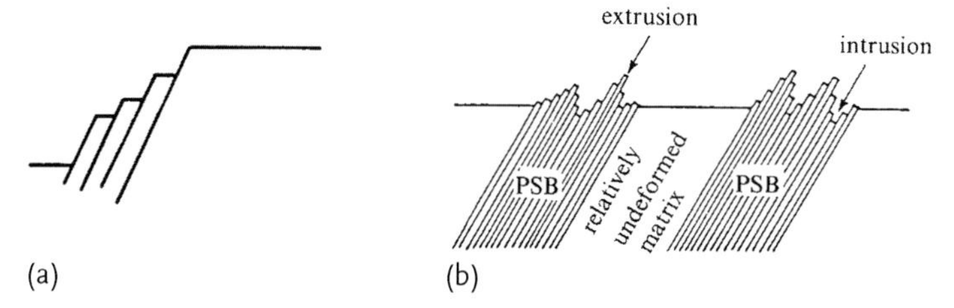


**Figure 17.3** Slip bands at the metal surface: (a) a series of steps resembling a staircase pattern produced by monotonic plastic strain; (b) the rough surface consisting of hills and valleys produced by cyclic plastic strain. PSB are the persistent slip bands at the cycling surface of a specimen.

The author of [84] worked out a mathematical model simulating the development of discrete structures in fatigued metals. The fatigue crack initiation is interpreted in the model as a state of dislocation structures, where the Gibbs free energy is the sum of the elastic strain energy, the potential energy of the applied load and the surface energy of the crack. The essence of the model is the conversion of dislocation pile-ups in parallel planes into the crack.

The total strain energy of dislocation pile-ups $F_1$ accumulates during alternating loading. The energy $F_2$ is released as a result of the relaxation and opening of a crack. The free energy changes in going from the state of dislocation dipole accumulation to that of a crack of size $a$. This change is given by

$$\Delta F = F_1 + F_2 - 2\gamma S \,, \tag{17.2}$$

where $2S$ is the crack embryo area and $\gamma$ is the surface energy. The plus sign in $F_1 + F_2$ means that the system gains the elastic energy as a consequence of transformation. The minus sign in $-2\gamma S$ means that the system expends the energy to create two new faces of the crack. As a result, the increment of the free energy $\Delta F$ first increases with the number of loading cycles, then it achieves a maximum and begins to decrease.

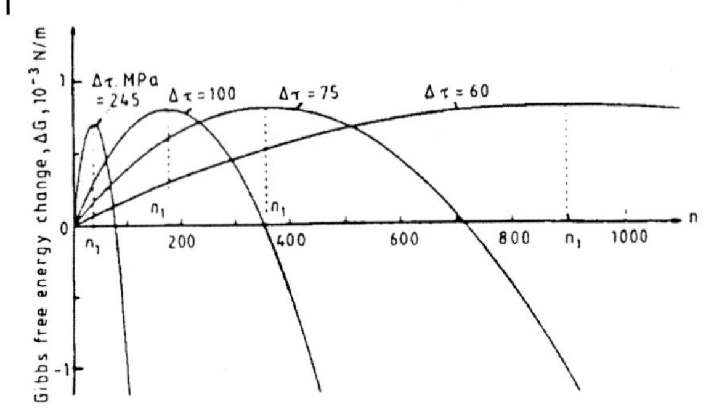


**Figure 17.4** Variation in the Gibbs free energy with the number of loading cycles $n$ and stress amplitude $\Delta\tau$ (after [84]).

The surface area of the critical crack embryo is determined from the equality:

$$2\gamma S = F_1 + F_2 . \tag{17.3}$$

A crack embryo grows if

$$2\gamma S < F_1 + F_2 . \tag{17.4}$$

The closure of the crack embryo occurs if

$$2\gamma S \geq F_1 + F_2 . \tag{17.5}$$

Inequalities (17.4) and (17.5) define the transition from the crack initiation period to the crack growth period.

The variation in the free energy with the number of loading cycles is presented in Figure 17.4. The graph shows that the more the stress amplitude the smaller the size of the crack embryo. The critical crack length is achieved in 170 cycles under an amplitude of 100 MPa, but under a stress amplitude of 75 MPa, 350 cycles are necessary. There is a critical number of cycles at which the free energy has a maximum and where the system becomes unstable, that is, when

$$\frac{\partial(\Delta F)}{\partial n} = 0 . \tag{17.6}$$

Solving (17.6) the author arrives at

$$(\sigma_a \sin 2\alpha - 4\sigma_{th}) N_c \approx \frac{4.6\pi\gamma}{l\zeta_1}\left(\frac{2-f}{f}\right) , \tag{17.7}$$

where $\sigma_a$ is the stress amplitude; $\alpha$ is the angle of inclination of the slip plane to the surface; $\sigma_{th}$ is the threshold stress; $N_c$ is the critical number of cycles at which the crack embryo begin to grow; $\gamma$ is the surface energy; $l$ is the length of the dislocation pile-up; $\zeta_1 = (h/l)^{4/3}$; $h$ is the distance between neighbouring slip planes, and $f$ is a factor of irreversibility of slip.

## 17.2 Periods of Fatigue-Crack Propagation

It is important to stress that there is an initial time of fatigue when only a surface crack is initiated but the crack embryo cannot yet grow. It is preferable to term this time as an incubation period of fatigue.

We introduce the concept of the critical crack length, $a_c \equiv a_{incub}$ that corresponds to the critical value of cycles, $N_c \equiv N_{incub}$. If the number of cycles $N < N_{incub}$ the crack does not grow. If the number of the loading cycles $N \geq N_{incub}$ the crack length begins to increase.

It is appropriate to divide the total number of cycles to the fatigue fracture $N_f$ into three terms:

$$N_f = N_{incub} + N_{gr} + N_{inst} \,, \tag{17.8}$$

where $N_{incub}$ is the number of cycles when the crack length is less than the critical size, the crack does not grow, and the process is still reversible (an incubation period); $N_{gr}$ is the number of cycles when the fatigue crack grows in length (a period of growth); $N_{inst}$ corresponds to the instant fracture, when the cross-section of a specimen or a component has already decreased due to the crack (a period of fracture). Generally, $N_{inst} \ll (N_{incub} + N_{gr})$.

The higher the amplitude of alternating stress $\sigma_a$ the smaller the critical crack size. At large $\sigma_a$ $N_{incub} \ll N_{gr}$. With this condition the initiation of a crack is a very fast process, and the crack begins to grow immediately. On the contrary, under a relatively small applied stress amplitude $\sigma_a$ the incubation period takes a considerable part of the specimen life. The number of cycles until specimen fracture becomes very large and sometimes does not occur.

In Figure 17.5 points C present the experimental dependence of $N_f$ on $\sigma_a$ and number of loading cycles. Points B show data of $N_{incub}$ calculated by us according to (17.7). Reasonable values for the constants were assumed as follows: $\mu = 3.58 \times 10^4$ MPa for a titanium-based alloy, $\sigma_{th} = 3 \times 10^{-3}\mu$, $\alpha = \pi/4$, $\gamma = 1.885\,\mathrm{J\,m^{-2}}$, $f = 0.5$, $l = 100\,\mu\mathrm{m}$, $h = 1.6\,\mathrm{nm}$ [84].

It follows from Figure 17.5 that the incubation period lasts for $10^5$ cycles under a relatively low stress amplitude of 450 MPa, whereas the period of the crack growth continues up to $5 \times 10^7$ cycles. Consequently, under this condition $N_{incub}$ is two orders greater than $N_{gr}$. Under a stress amplitude of 900 MPa the short incubation period $N_{incub} = 4300$ cycles gives place at once to the crack growth period and the fracture of the specimen occurs after a further 5000 cycles.

Figure 17.6 illustrates the relation between periods of fatigue-crack propagation for a titanium-based alloy and for stainless steel. One can see again that the smaller the stress amplitude, the longer the incubation period of fatigue. The crack embryo can exist as long as $2 \times 10^5$ cycles, curve B Figure 17.6.

We have derived [83] the equation for the length of the fatigue-crack embryo,

$$a_{incub} = \frac{0.508\gamma E}{(1-\nu^2)(\sigma_a)^2} \,, \tag{17.9}$$

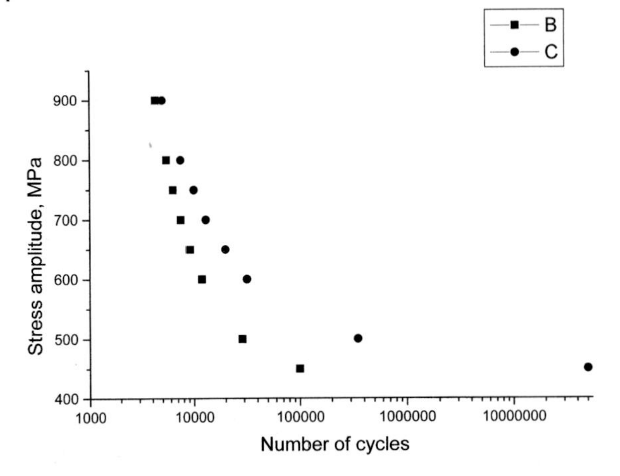


**Figure 17.5** Effect of the stress amplitude on the number of cycles to the beginning of crack growth and to the fracture of the specimen. C – number of cycles to fracture $N_f$ for the titanium-based alloy, test data of [85]; B – the calculated number of cycles during the "incubation" period $N_{incub}$ for the same alloy.

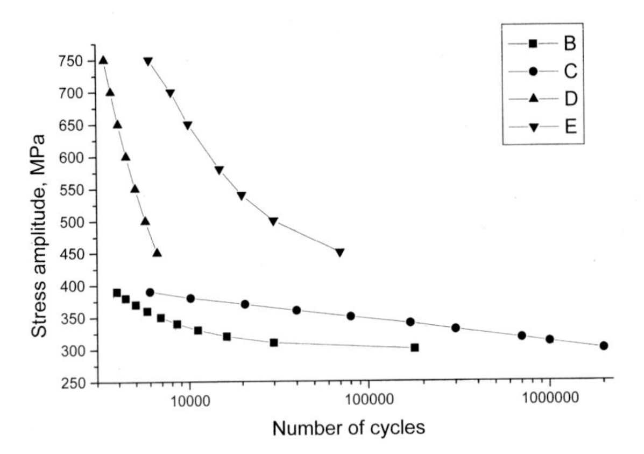


**Figure 17.6** Dependence of the incubation period, and of the crack-growth period on the stress amplitude for two materials: C – the curve of the fatigue fracture (*S–N* curve) for 316L stainless steel, test data [86]; B – calculated values of $N_{incub}$ for the same steel; E – the curve of fracture for the Ti + 6Al + 4V alloy, test data [87]; D – calculated values of $N_{incub}$ for the same alloy.

where $\gamma$ is the surface energy of a material, $E$ is the Young modulus, $\nu$ is the Poisson coefficient, and $\sigma_a$ is the stress amplitude. Equation (17.9) states that the length of the crack embryo is affected by strength of interatomic bonding through the surface energy and the Young modulus. The greater the crack embryo $a_{incub}$ the later fatigue crack starts its dangerous propagation from the surface inside material. The length of the crack embryo that can grow is inversely proportional to the stress amplitude squared.

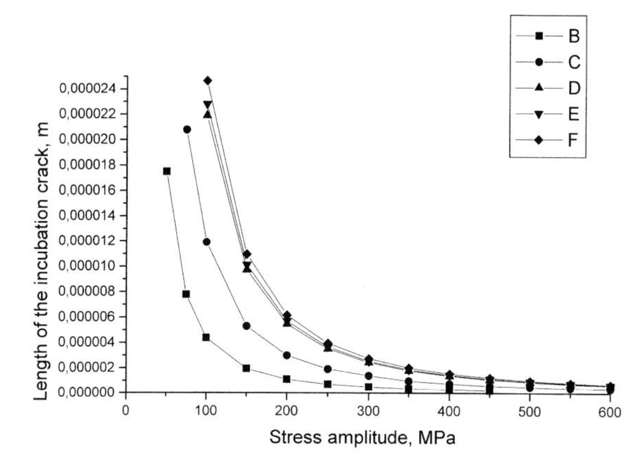


**Figure 17.7** The length of the incubation crack as a function of the stress amplitude: B – an Al-based alloy; C – a Ti-based alloy; D – a low-alloy steel; E – a superalloy; F – a stainless steel. See Table 17.1 for details.

Figure 17.7 illustrates the effect of the applied stress amplitude on the length of the crack embryo for five materials. The $a_{\text{incub}}$ values were calculated according to (17.9). The criteria of interatomic bonding, values of the Young moduli [88] and the surface energies [89], are presented in Table 17.1. The composition of alloys is shown in the same table.

It is obvious from curves in Figure 17.7 that, under stress amplitudes of more than 300 MPa, the length $a_{\text{incub}}$ is of the order of 1–2 μm for Al- and Ti-based alloys. The crack begins to grow almost at once. Other alloys give the $a_{\text{incub}}$ value of 4–5 μm and the crack growth occurs later.

For steels and superalloys under a stress amplitude of from 150 to 300 MPa, the embryo length is found to be 4–10 μm. The values of $N_{\text{incub}}$ and $N_{\text{f}}$ increase correspondingly in comparison with Al- and Ti-based alloys. The stress amplitude from 100 to 150 MPa for these alloys does not lead to the crack-embryo propagation. As

**Table 17.1** The Young modulus $E$ and the surface energy $\gamma$ for some alloys.

| Alloy | Composition (wt%) | E (GPa) | $\gamma$ (J m$^{-2}$) |
|---|---|---|---|
| Al-based | Al + 4.4Cu + 1.5Mg + 0.6Mn +0.5Si + 0.5Fe | 73 | 1.075 |
| Ti-based; Ti metal 811 | Ti + 9Al + 1Mo + 1V | 124 | 1.690 |
| Fe-based; low-alloy steel | Fe + 0.16C + 0.9Mn + 0.55Ni +0.50Cr + 0.20Mo | 208 | 1.900 |
| Superalloy Nimonic 263 | Ni + 20Cr + 20Co + 6Mo +2Ti + 0.7Fe + 0.5Al | 224 | 1.810 |
| Stainless steel AISI 17-7 | Fe + 0.09C + 17Cr + 7.1Ni +1Mn + 1Al + 1Si | 204 | 2.150 |

we should expect, Al-based alloys have a lower fatigue strength than steel, because under the same stress amplitude their $a_{incub}$ is several times less than that of steel.

The nanostructuring of surface of the materials can retard the generation and propagation of fatigue cracks considerably.

## 17.3 Fatigue Failure at Atomic Level

We have seen in Section 17.1 that the basis of the crack origination is the deformation process. The foundation of the fatigue crack propagation is the rapture of interatomic bonding.

In Figure 17.8 the dependence of the crack length on the number of cycles for low-carbon steel is presented. One can see that the rate of the crack growth increases in the course of time. Figure 17.8 also shows that crack initiation and propagation in cold-worked specimens are significantly retarded and the crack growth rates are smaller compared to those in non-cold-worked specimens.

The fatigue crack originates at the surface, then the crack embryo begins to grow. In my opinion, a movement of vacancies to the tip of the crack plays a dominant role in crack growth.

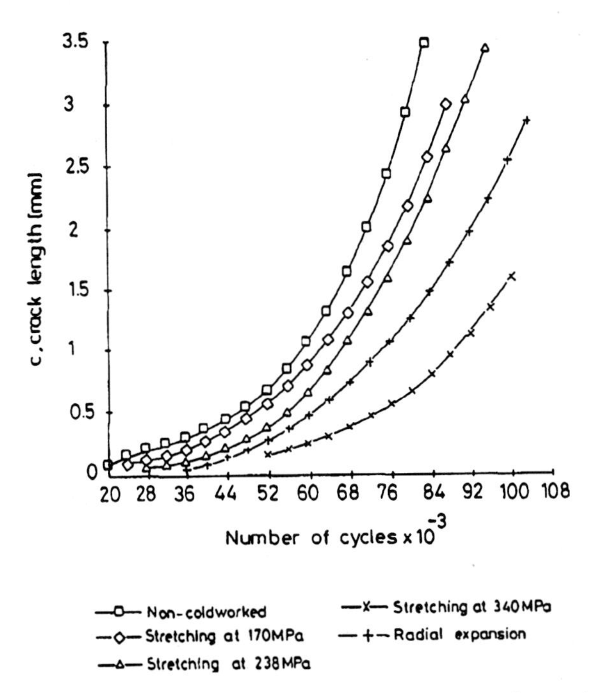


**Figure 17.8** Experimental crack length versus number of test cycles. Amplitude of loading equals to 170 MPa. Frequency of loading is equal to 23 Hz. Some specimens were subjected first to various cold-working processes (data of [90]).

Dislocations move under the influence of loading in intersecting slip planes. Jogs are generated at the intersection of screw components of mobile dislocations. A jog is a segment of dislocation, which does not lie in the slip plane. The jog cannot move without the generation of vacancies. The interstitial-producing jogs are practically immobile because the energy of formation of the interstitial atoms is several orders greater than that of vacancies. Thus, the slip of the jogged dislocation components plays an important role in the increase of vacancy concentration.

There are other sources of vacancies. They are also generated because of the annihilation of edge dislocation segments of opposite signs. The concentration of vacancies produced within slip bands is larger than that in the vicinity of the free surface. The gradient of the vacancy concentration creates a driving force for the diffusion of vacancies within the slip bands to the free surface. The activation energy of diffusion along dislocation pipes is far less than that in the bulk.

One would expect a sufficient time to be available for vacancy diffusion even at room temperature since fatigue involves a sufficiently long period of time for crack initiation and propagation.

An equilibrium concentration of vacancies in the crystal lattice is given by

$$c_0 = \exp\left(-\frac{E_\text{v}}{k_\text{B} T}\right) , \tag{17.10}$$

where $c_0$ is the fraction of lattice sites that are occupied by vacancies, $k_\text{B}$ is the Boltzmann constant.

The relative number of vacancies which are formed at the intersection of dislocations $c_\text{in}$ for face-centered crystals is given by [81]

$$c_\text{in} \approx 1 \times 10^{-4} \varepsilon^2 , \tag{17.11}$$

where $\varepsilon$ is the strain.

Figure 17.9 illustrates fluxes of the vacancies in a specimen subjected to the alternating tensile – compressive stresses. Vacancies near the crack are attracted by the crack tip during the tensile phase of cycling.

The stress distribution near a crack tip is presented schematically in Figure 17.10. The nearer the crack tip the larger a value of stress. The value of the stress gradient at the tip,

$$\text{grad}\sigma = \left(\frac{d\sigma}{dx}\right)_{x=0} \tag{17.12}$$

is of great importance.

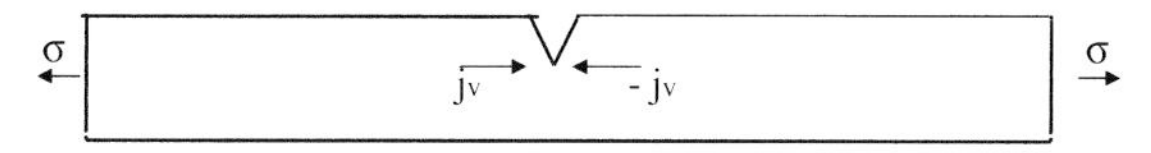


**Figure 17.9** A specimen in the field of the cycling stresses $\sigma$ during the fatigue test. The vectors $j_\text{v}$ and $-j_\text{v}$ show fluxes of vacancies in the direction of the crack tip.

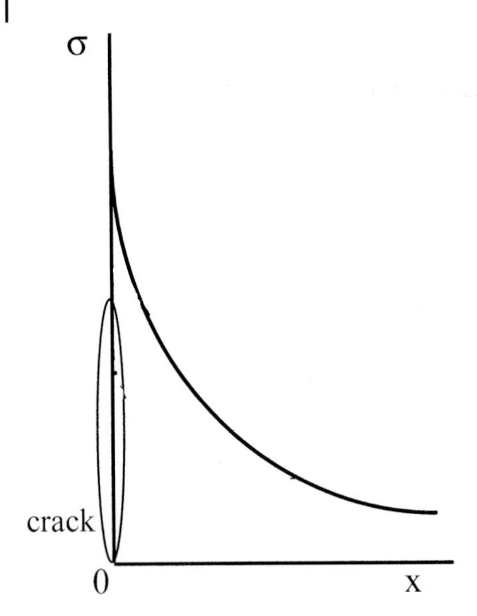


**Figure 17.10** The tensile stress near the tip of the growing crack as a function of the distance from the crack tip.

Vacancies move in the crystal lattice with a mean velocity of $V_V$. There is no need for vacancies to diffuse over relatively long distances. A relay-like motion of vacancies can occur. The growing crack absorbs the nearest vacancies as this process leads to a decrease in the free energy of the thermodynamic system the crack tip – the vacancies.

The absorption of vacancies by the crack tip results in the rupture of interatomic bondings.

The increased gradient of the vacancies leads to a displacement of the next vacancies in the direction of the crack. As a consequence, the crack that absorbs vacancies grows under the influence of alternating loading. Intersecting dislocations go on to generate new vacancies.

Thus, the displacement of vacancies to the crack tip occurs due to the gradient of tensile stresses. Moreover, the falling of vacancies into the crack produces a gradient of their concentration near the crack tip. For its part, the gradient of the vacancy concentration also facilitates the vacancy movement in the direction of the crack.

Consequently, external cycling stress creates a field of tensile stresses at the crack tip. The gradient in the tensile stress and in the vacancy concentration insure a driving force for the diffusion of vacancies in the direction of the crack tip. Vacancies are attracted to the tension area near the crack tip. The displacement of vacancies is realized in this field in the direction of the tensile stresses. If a vacancy reaches the crack, it falls down into it. The crack is a trap for the flow of vacancies. The vacancy that replaces an atom results in the breaking of interatomic bondings at the tip of the crack. This process is not reversible, of course. Figure 17.11 illustrates the process.

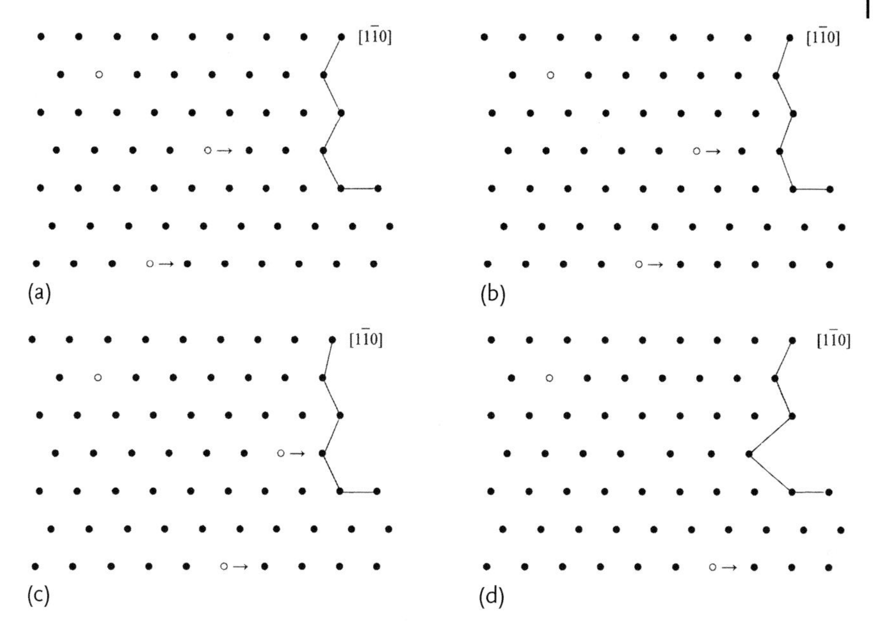


**Figure 17.11** The scheme of vacancy diffusion to the crack tip in the field of tensile stresses: (a)–(c) the sequence of vacancy displacements in the crystal lattice (shown by arrows); (d) the vacancy has fallen into the crack.

The energy of vacancy formation $E_v$ determines generally the rate of propagation of the fatigue crack. A value of $E_v$ is by-turn determined by the strength of interatomic bonding in the material.

The replacement of a bonding atom by one vacancy results in the increment of the crack length by one interatomic distance $b$, that is, by the module of the Burgers vector; $dn$ vacancies increase the length of the crack by $da$. Consequently,

$$da = b dn . \tag{17.13}$$

Differentiating (17.13) we obtain the rate of crack growth

$$\frac{da}{dt} = b\frac{dn}{dt} . \tag{17.14}$$

The number of vacancies through a cross-section $S$ during a time interval $dt$ is proportional to the gradient of stress grad$\sigma$ and values of $S$, and $dt$:

$$dn \sim \frac{d\sigma}{dx} S dt . \tag{17.15}$$

The number of vacancies, which approach the crack tip, is also directly proportional to the velocity of vacancies and inversely proportional to the activation energy of their generation.

Using the dimensional method and making reasonable physical assumptions, we arrive at the formula

$$dn = \frac{V_v}{E_v} a \frac{d\sigma}{dx} S\, dt \,, \tag{17.16}$$

where $a$ is the length of the crack, $E_v$ is the activation energy of the vacancy formation, $V_v$ is the velocity of vacancies.

Differentiating (17.16) and combining with (17.14) we obtain an expression for the rate of fatigue-crack growth in the form

$$\frac{da}{dt} = \frac{b S V_v}{E_v} a \frac{d\sigma}{dx} \,. \tag{17.17}$$

From (17.17) we obtain

$$\frac{da}{dN} = \frac{b V_v}{\nu E_v} g S a \,, \tag{17.18}$$

where $g = d\sigma/dx = \text{grad}\sigma$ and $\nu = dN/dt$ is the frequency of cycling.

Equations (17.17) and (17.18) are applicable if the crack size $a \geq a_{\text{incub}}$.

Separating the variables and integrating (17.17) we arrive at

$$a = A \exp(Gt) \,, \tag{17.19}$$

where $A$ and $G$ are constants. It is appropriate to rewrite (17.19) for calculation as

$$\ln a = \ln A + Gt \,. \tag{17.20}$$

In Figure 17.12 one can see in coordinates ln $a$–$t$ the data for non-cold-worked steel and also for plastically deformed steel. The exponential dependence between the crack length and time of fatigue test is confirmed.

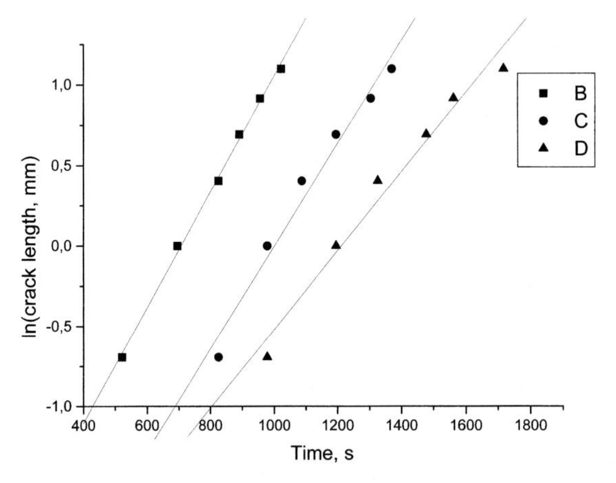


**Figure 17.12** Dependence of the logarithm of the crack length on the fatigue test time: B – non-cold-worked steel; C – preliminary stretching at 340 MPa; D – preliminary radial expansion. The frequency of tests for low-carbon steel is 23 Hz,the amplitude is 255 MPa. Test data from [90].

A cold working of the surface results in residual compressive stresses. The residual compressive stresses cause a decrease in grad$\sigma$ near the crack tip and enhance the fatigue life of the material. The residual stress at the surface reduces the applied external stress. The strengthening surface treatment retards the growth of fatigue crack considerably (compare curve B with curves C and D in Figure 17.12).

The exponent $G$ in (17.19) that strongly affects the rate of crack growth is 1.46 times less for the cold-worked specimen than for the non-cold-worked one. There is good reason to believe that it is related to a decrease in the gradient of stresses $d\sigma/dx$ near the crack.

We have discovered that the value of grad$\sigma$ at the tip of the growing crack decreases during fatigue test. From (17.18) we obtain

$$\frac{g_n}{g_1} = \frac{a_1}{a_n} \cdot \frac{da_n/dN_n}{da_1/dN_1} , \tag{17.21}$$

where subscripts 1 and $n$ denote the first and the $n$-measurements of the initial crack length, respectively. The length of fatigue cracks and the rate of crack growth have been measured [66] for steel. We have calculated relations $g_n/g_1$ according to these data. The results are illustrated in Figure 17.13.

A linear dependence is observed for the nontreated specimen as well as for the specimen with the strengthened surface. One can see from Figure 17.13 that a decrease in the stress gradient depends mainly on the length of the crack. The drop in the gradient is explained by the plastic strain that takes place in front of the growing crack.

The equation for the curve B in Figure 17.13 is

$$\frac{g_n}{g_1} = 0.832 - 0.112a . \tag{17.22}$$

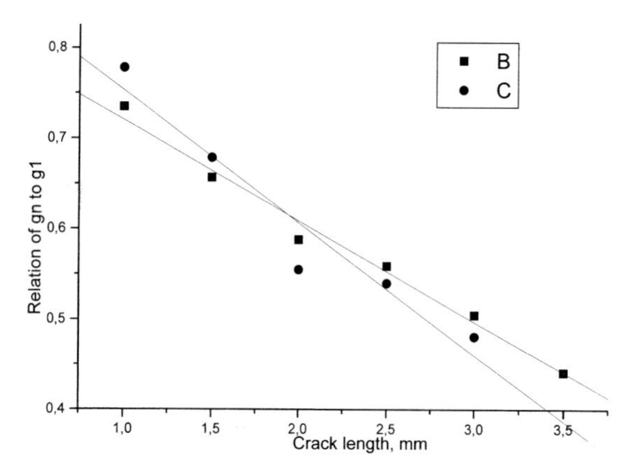


**Figure 17.13** The stress gradient at the crack tip versus the crack length: B – nontreated specimen; C – radially pre-stretched specimen.

The equation for the curve C is[2)]

$$\frac{g_n}{g_1} = 0.900 - 0.147a \,. \tag{17.23}$$

The stress gradient drops almost twice as much due to the plastic strain near the growing fatigue crack.

## 17.4 Rupture of Interatomic Bonding at the Crack Tip

Let us estimate the parameters of the crack tip proceeding from (17.17). The area of the crack tip $S$ can be expressed as

$$S = \frac{E_\mathrm{v}}{abV_\mathrm{v}}\left(\frac{da/dt}{d\sigma/dx}\right) , \tag{17.24}$$

where $E_\mathrm{v}$ is the energy of vacancy generation, $a$ is the crack length, $b$ is the Burgers vector, $V_\mathrm{v}$ is the vacancy velocity, $da/dt$ is the rate of the crack propagation, $d\sigma/dx$ is the stress gradient at the crack tip. The parameters of the crack propagation except $V_\mathrm{c}$ can be taken from experimental data, Figure 17.8. The energy of vacancy generation for $\alpha$-Fe is $E_\mathrm{v} = 2.56\times10^{-19}\,\mathrm{J\,atom^{-1}}$. $a = 0.4\times10^{-3}$ m; $b = 0.248$ nm; $da/dt = 4.60\,\mathrm{\mu m\,s^{-1}}$. The gradient of the stresses near the crack is assumed to be close to the gradient of the residual stresses near the surface. According to the data of the present author and also to the results of other authors the value of

$$\frac{|\sigma_{\mathrm{max}}| - \sigma_{\mathrm{min}}}{\Delta x}$$

varies from 3.1 to 6.3 MPa $\mathrm{\mu m^{-1}}$. We assume $d\sigma/dx = 4.7\,\mathrm{MPa\,\mu m^{-1}}$. The velocity of vacancies near the crack tip is given by

$$V_\mathrm{v} = \nu\Gamma b \,, \tag{17.25}$$

**Table 17.2** Dimension of the crack tip $\sqrt{S}$ and number of breaking interatomic bondings $N_{\mathrm{break}}$ at the crack tip per cycle of vibration.

| $\Gamma$ | $V_\mathrm{v}$ (nm s$^{-1}$) | $S$ (nm$^2$) | $\sqrt{S}$ (nm) | $N_{\mathrm{break}}$ (cycle$^{-1}$) |
|---|---|---|---|---|
| 1 | 5.7 | 442.8 | 21.0 | 85 |
| 2 | 11.4 | 221.4 | 14.9 | 60 |
| 3 | 17.1 | 147.6 | 12.1 | 49 |
| 4 | 22.8 | 110.7 | 10.5 | 42 |
| 5 | 28.5 | 88.6 | 9.4 | 38 |
| 10 | 57.0 | 44.3 | 6.6 | 27 |

2) In these empirical equations the crack length $a$ is given in mm, coefficients 0.112 and 0.147 are given in mm$^{-1}$.

where $\nu$ is the frequency vibration of loading stress, $\Gamma$ is a number of the vacancy jumps per cycle. We should make assumptions for values of $\Gamma$ because they are unknown.

Substituting the parameters in (17.17) we obtain data listed in Table 17.2.

The $\sqrt{S}$ values are of the order of the crack-tip size. It means that the needle-like tip of the crack equals 27–85 interatomic distances. Approximately this number of interatomic bondings are destroyed in front of the growing crack simultaneously per one cycle of load vibration.

# 18
# Modeling of Kinetic Processes

Processes that are dependent on time are called kinetic processes or rate processes. Creep, the crystal growth, fatigue, fracture, friction, and wear progress in time are the examples of kinetic processes in solids.

The processes that are developing in time can be described by a system of differential equations.

## 18.1
## System of Differential Equations

We should first of all ascertain which parameters determine a process under consideration. The validated approach to the problem implies that all parameters should have the well-defined physical meaning.

Next step is the development of a physical model of the process. Then we convert our physical model to a mathematical language. The mathematical formulation follows a physical model that has been worked in detail.

Let us suppose that $N$ parameters $y_1, y_2, \ldots, y_N$ affect the process under study. In a general case every parameter depends upon time and on all other parameters, that is,

$$\begin{aligned} y_1 &= \varphi_1(t, y_1, y_2, \ldots, y_N)\ ; \\ y_2 &= \varphi_2(t, y_1, y_2, \ldots, y_N)\ ; \\ &\vdots \\ y_N &= \varphi_N(t, y_1, y_2, \ldots, y_N)\ , \end{aligned} \tag{18.1}$$

where the functions $\varphi_i$ are derived on the base of our physical model.

*Interatomic Bonding in Solids: Fundamentals,Simulation,andApplications*, First Edition. Valim Levitin.

Differentiating these equations and separating the variables we arrive at a system of differential equations,

$$\begin{aligned}\frac{dy_1}{dt} &= f_1(t, y_1, y_2, \ldots, y_N)\ ;\\ \frac{dy_2}{dt} &= f_2(t, y_1, y_2, \ldots, y_N)\ ;\\ &\vdots\\ \frac{dy_N}{dt} &= f_N(t, y_1, y_2, \ldots, y_N)\ ,\end{aligned} \tag{18.2}$$

where the functions $f_1, f_2, \ldots, f_N$ on the right-hand side of equations are known. On the left-hand side are the first derivatives of the parameters with respect to time.

Thus we have obtained a system of ordinary differential equations.

We should specify values of the parameters at $t = 0$, that is the initial values of the each parameter $y_i$.

It is reasonable to use a Runge–Kutta method for integration of differential equations. The Runge–Kutta methods are known to propagate a numerical solution over an interval by combining the information from several Euler-style steps (each involving one evaluation of the right-hand side of equations) and then using the information obtained to match a Taylor series expansion up to some higher order. The step size is continually adjusted to achieve a specified precision.

Further, we consider the propagation of a fatigue crack as an illustrative example of modeling of the kinetic process.

## 18.2 Crack Propagation

The physical mechanism of growth of the fatigue crack has been considered in Chapter 17. The fatigue-crack embryo originates at the surface, then the crack embryo begins to grow. The movement of vacancies to the tip of the crack plays a dominant role in crack growth. The vacancies are generated by mobile intersecting screw components of dislocations. The external cycling stress at the crack tip creates a field of tensile stresses. The gradient in the tensile stress ensures a driving force for the relay-race diffusion of vacancies in the direction of the crack tip. Vacancies are attracted to the tension area in the tip of the crack. If a vacancy reaches a crack, it falls down inside it. The crack tip is a trap for the flow of vacancies. The replacement of an atom with a vacancy breaks the interatomic bonding in the tip of the crack.

The crack propagation, that is, an increase in the crack length, is a macroscopic process. The described atomic process is the cause of the macroscopic fracture.

Our goal here is to determine the kinetic of the crack growth. Six structure parameters of the process are of interest. The parameters of the fatigue progress with time are:

- the threshold stress for the dislocation motion $\tau_s$;
- the dislocation density $\rho$;
- the dislocation velocity $V$;
- the vacancy concentration $c_v$;
- the stress gradient near the crack tip $g$;
- the length of the fatigue crack $a$.

## 18.3 Parameters to Be Studied

If a metal specimen is subjected to alternating stresses the generation of dislocations occurs. There is a threshold stress $\tau_s$ at which this process begins [81]

$$\tau_s = \frac{\mu_s b (n_{pu}\rho)^{\frac{1}{2}}}{2\pi} , \tag{18.3}$$

where $\mu_s$ is the shear modulus; $b$ is the Burgers vector; $n_{pu}$ is the number of dislocations in the pile-up; $\rho$ is the density of dislocations. The dislocation density increases during the deformation.

Differentiating (18.3) with respect to time $t$ we arrive at

$$\frac{d\tau_s}{dt} = \frac{\mu_s b}{4\pi} \cdot \left(\frac{n_{pu}}{\rho}\right)^{\frac{1}{2}} \cdot \frac{d\rho}{dt} . \tag{18.4}$$

The number of newly generated dislocation loops is directly proportional to the mobile dislocation density and also to the dislocation velocity. Hence the multiplication rate of mobile dislocations is given by

$$\frac{d\rho}{dt} = \delta \rho V , \tag{18.5}$$

where $\delta$ is a coefficient of multiplication of mobile dislocations. The coefficient $\delta$ has the unit of inverse length.

Substituting (18.5) to (18.4) we obtain the first differential equation of the system:

$$\frac{d\tau_s}{dt} = \frac{\delta \mu_s b V (n_{pu}\rho)^{\frac{1}{2}}}{4\pi} . \tag{18.6}$$

Since $\rho$ increases during deformation, the threshold stress increases also with time.

Equation (18.5) is the second equation of the desired system.

The velocity of dislocations may be expressed as the velocity of a thermally activated process:

$$V = V_0 \exp\left[-\frac{U_0 - \gamma(\tau - \tau_s)}{k_B T}\right] , \tag{18.7}$$

where $V_0$ is a pre-exponential multiplier; $U_0$ is the activation energy for the dislocation motion; $\gamma$ is the activation volume; $\tau$ is the applied shear stress; $k_B$ is the

Boltzmann constant and $T$ is the temperature. The pre-exponential factor, $V_0$, is estimated as [81]

$$V_0 = \frac{\nu_D b^2}{l} , \tag{18.8}$$

where $\nu_D = 10^{13}\,s^{-1}$ is the Debye frequency, $l = 1/\sqrt{\rho} \approx 10^{-6}\,m$ is the mean length of the dislocation segments. The activation volume $\gamma = b^3$.

The alternating stress is dependent on time:

$$\tau = \tau_m |\sin \omega t| , \tag{18.9}$$

where $\tau_m$ is the amplitude of the applied stress; $\omega$ is the angular frequency of the stress.

Taking the derivative of (18.7) and also taking into account (18.9) and (18.6) we obtain

$$\frac{dV}{dt} = \frac{V_0 \gamma}{kT} \exp\left[-\frac{U_0 - \gamma(\tau_m|\sin \omega t| - \tau_s)}{kT}\right] \times \left[\omega \tau_m |\cos \omega t| + \frac{\delta \mu_s b V (n_{pu}\rho)^{\frac{1}{2}}}{4\pi}\right] . \tag{18.10}$$

The increase in the relative vacancy concentration in the crystal lattice is proportional to the density of both intersecting dislocation systems, and the dislocation velocity multiplied by the volume $Q$. Thus we can write

$$\frac{dc_v}{dt} = \frac{1}{4}\rho^2 V Q . \tag{18.11}$$

We find the equation for a changing of the stress gradient near the crack tip by differentiating (17.22),

$$\frac{dg}{dt} = -0.112 \times 10^{-3} g_1 \frac{da}{dt} . \tag{18.12}$$

The rate of the crack growth is given by (17.17),

$$\frac{da}{dt} = \frac{bSV_v}{E_v} ag . \tag{18.13}$$

Consequently, we have the system of six differential equations (18.6), (18.5), (18.10), (18.11), (18.12), (18.13) with six unknown quantities that have been enumerated above.

## 18.4 Results

We assume the following real and reasonable values of the physical parameters for $\alpha$-Fe:

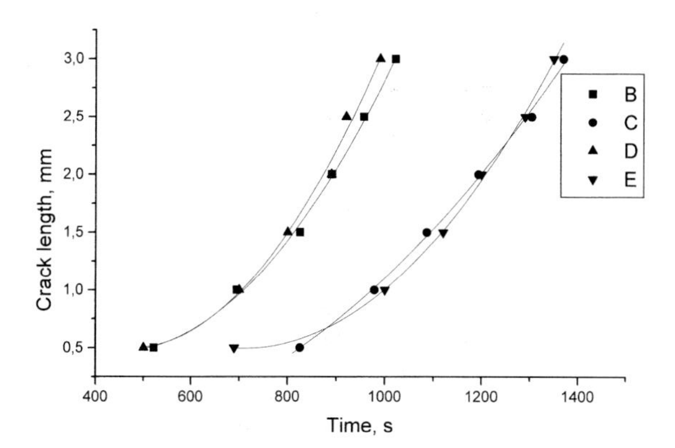


**Figure 18.1** The results of modeling of the crack growth in comparison to the experimental data. B – the fatigue test for a low-carbon steel, a stress amplitude of 170 MPa, a frequency of 23 Hz [90]; D – data for solution of system differential equations for the same steel. C – low-carbon steel after a stretching of 340 MPa; E – data for solution of system differential equations for pre-stretched steel.

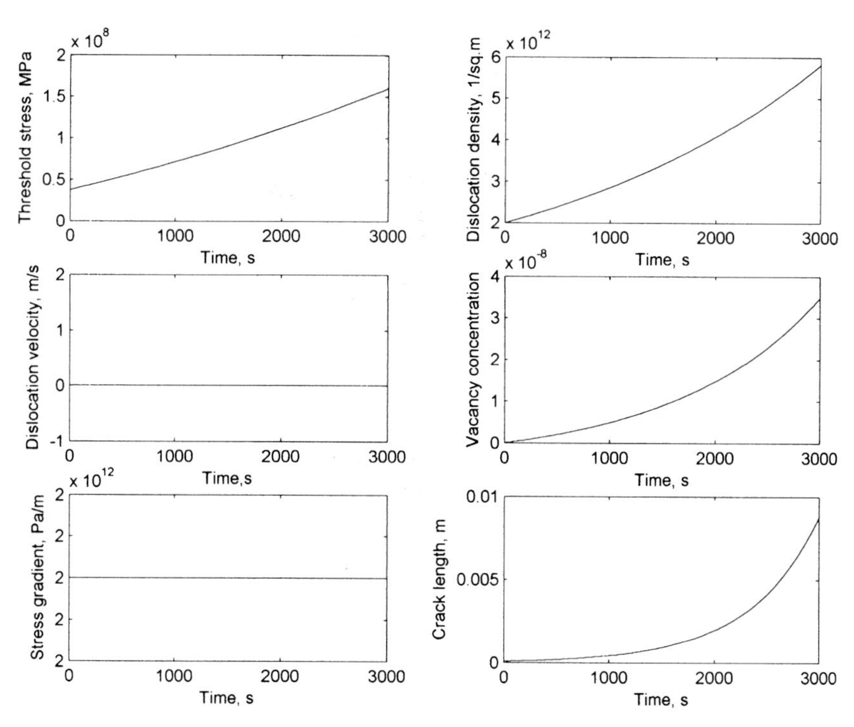


**Figure 18.2** (a–f) Variation of parameters for fatigued aluminum: description of processes by a system of six differential equations. The stress amplitude is 46 MPa, the frequency is 936 Hz, the number of cycles is $2.8 \times 10^6$.

The stress gradient at the crack tip is equal to $2 \times 10^{12}$ Pa $m^{-1}$, the initial dislocation density is $2 \times 10^{12}$ $m^{-2}$. The crack grows from 0.1 to 10 mm, see graph to the right below.

$b = 2.48 \times 10^{-10}$ m; $\mu_s = 8.3 \times 10^{10}$ Pa; $T = 293$ K; $\tau_m = 68.9 \times 10^6$ Pa; $\omega = 144.5$ rad s$^{-1}$; $\delta = 2 \times 10^4$ m$^{-1}$; $\Gamma S = 1.87 \times 10^{-18}$ m$^2$; $Q = 1.87 \times 10^{-28}$ m$^3$; $E_v = 2.56 \times 10^{-19}$ J atom$^{-1}$; $V_0 = 0.818$ m s$^{-1}$; $U_0 = 2.56 \times 10^{-20}$ J atom$^{-1}$.

Initial values are assumed to be equal to $a = 0.5$ mm; $\rho = 1 \times 10^{12}$ m$^{-2}$; $V = 1.78 \times 10^{-8}$ ms$^{-1}$; $\tau_s = 37.6$ MPa; $c_v = 10^{-11}$; grad$\sigma = 4.5 \times 10^{12}$ Pa m$^{-1}$.

Data on the comparison of the test results and solutions of system of differential equations for fatigue-crack growth in low-carbon steel are shown in Figure 18.1. It can easily be seen that the calculated curves fit the experimental data well.

The results for fatigued aluminum ($b = 2.86 \times 10^{-10}$ m, $\mu_s = 2.70 \times 10^{10}$ Pa) are presented in Figure 18.2. Following initial values of the physical parameters have been assumed for the calculations. $\rho = 2 \times 10^{12}$ m$^{-2}$; grad$\sigma = 2 \times 10^{12}$ Pa m$^{-1}$; $V = 1.78 \times 10^{-8}$ m s$^{-1}$; $c_v = 10^{-11}$; $a = 10^{-4}$ m $= 0.1$ mm.

We can deduce the following conclusions from the data obtained.

The model adequately represents physical processes of the fatigue crack propagation. During the cycling of metal the dislocation density increases by a factor of 3. The vacancy concentration increases by three orders of magnitude.

It has been found that the stress gradient near the crack tip considerably affects the rate of fatigue-crack growth. An increase of double in the stress gradient leads to a growth in the crack length of one order of magnitude.

# Appendix A
# Table of Symbols

## Roman Symbols

| Symbol | Unit | Quantity |
|---|---|---|
| $a_0$ | [m] | Bohr radius |
| $\boldsymbol{b}$ | [m] | Burgers vector |
| $\boldsymbol{a}_1, \boldsymbol{a}_2, \boldsymbol{a}_3$ | [m] | crystal lattice parameters |
| $\boldsymbol{b}_1, \boldsymbol{b}_2, \boldsymbol{b}_3$ | $[2\pi\ \mathrm{m}^{-1}]$ | reciprocal lattice parameters |
| $B$ | [Pa] | bulk modulus |
| $C_{ik}$ | [Pa] | elastic constants |
| $E$ | [eV] | energy, eigenvalue |
| $E_{\mathrm{coh}}$ | $[\mathrm{eV\ atom}^{-1}]$ | cohesive energy |
| $E_{\mathrm{corr}}$ | [Ry, eV] | correlation energy |
| $E_{\mathrm{di}}$ | $[\mathrm{eV\ molecule}^{-1}]$ | energy interaction of dipole molecules |
| $E_{\mathrm{D}}$ | $[\mathrm{eV\ molecule}^{-1}]$ | energy of dimerization |
| $E_{\mathrm{e-e}}$ | [Ry, eV] | energy of electron–electron interaction |
| $E_{\mathrm{ex}}$ | [Ry, eV] | exchange energy |
| $E_{\mathrm{F}}$ | [Ry, eV] | Fermi energy |
| $E_{\mathrm{HK}}$ | [Ry, eV] | Honnenberg–Kohn energy functional |
| $E_{\mathrm{g}}$ | [eV] | energy gap width |
| $E_{\mathrm{i-e}}$ | [Ry, eV] | energy of electron–ion interaction |
| $E_{\mathrm{ion}}$ | [Ry, eV] | ionization potential |
| $E_{\mathrm{kin}}$ | [Ry, eV] | kinetic energy per electron |
| $E_{\mathrm{KS}}$ | [Ry, eV] | Kohn–Sham energy functional |
| $E_n$ | [eV] | energy eigenvalue of $n$ band |
| $E_{\mathrm{Ry}}$ | [Ry, eV] | 1 Rydberg = 13.605 eV |
| $E_{\mathrm{T}}$ | [Ry, eV] | total energy |
| $E_{\mathrm{v}}$ | $[\mathrm{eV\ atom}^{-1}, \mathrm{J\ atom}^{-1}]$ | energy of vacancy generation |
| $E_{\mathrm{xc}}$ | [Ry, eV] | exchange-correlation energy |

*Interatomic Bonding in Solids: Fundamentals,Simulation,andApplications,* First Edition. Valim Levitin.

| | | |
|---|---|---|
| $F_p$ | [N m$^{-1}$] | interplanar force constant |
| $\boldsymbol{g}$ | [2π m$^{-1}$] | vector of reciprocal lattice |
| $\mathbf{G}$ | [–] | set of points in reciprocal lattice |
| $H$ | [Ry, eV] | Hamiltonian |
| $H_{at}$ | [Ry, eV] | Hamiltonian of a single atom |
| $k_B$ | [J K$^{-1}$] | Boltzmann constant |
| $k_F$ | [m$^{-1}$] | Fermi momentum |
| $\boldsymbol{k}$ | [2π m$^{-1}$] | wave vector |
| $m_e$ | [kg] | mass of electron |
| $n$ | [–] | energy band index |
| $n(\boldsymbol{r})$ | [m$^{-3}$] | electron density |
| $n, l, m_l, s$ | [–] | quantum numbers |
| $N$ | [–] | number of electrons |
| $N_k$ | [m$^3$] | density of levels in $\boldsymbol{k}$-space |
| $p$ | [C m] | dipole moment |
| $\boldsymbol{r}$ | [m] | crystal lattice vector |
| $r$ | [m] | intermolecular distance |
| $r_{ij}$ | [m] | interatomic distance |
| $\boldsymbol{R}$ | [m] | lattice translation vector |
| $r_s$ | [atom unit] | radius of sphere which encloses one electron |
| $\overline{u^2}$ | [m$^2$] | mean square amplitude |
| $r_{WS}$ | [m] | Wigner–Seitz radius |
| $T(n)$ | [Ry, eV] | kinetic energy of electrons |
| $T_F$ | [K] | Fermi temperature |
| $U_v$ | [eV atom$^{-1}$, J atom$^{-1}$] | energy of vacancy diffusion |
| $v_g$ | m s$^{-1}$ | group velocity |
| $V$ | [m$^3$] | volume |
| $V(\boldsymbol{r})$ | [eV] | potential of crystal lattice |
| $V^{\sigma}_{eff}(\boldsymbol{r})$ | [eV] | effective potential acting on an electron of spin $\sigma$ |
| $V_{ext}(\boldsymbol{r})$ | [eV] | external potential |

## Greek Symbols

| Symbol | Unit | Quantity |
|---|---|---|
| $\alpha$ | [–] | Madelung constant |
| $\alpha$ | [C m$^2$ V$^{-1}$] | polarizability |
| $\gamma$ | [J m$^{-2}$] | surface energy |
| $\varepsilon_{kl}$ | [–] | strain tensor |
| $\zeta = q/q_m$ | [–] | reduced wave vector |
| $\lambda$ | [m] | wavelength |

| | | |
|---|---|---|
| $\nu$ | $[s^{-1}]$ | frequency |
| $\rho$ | $[m^{-2}]$ | dislocation density |
| $\rho(E)$ | $[J^{-1}]$ | density of states |
| $\sigma_{ij}$ | [Pa] | stress tensor |
| $\Psi$ | $[m^{-3/2}]$ | wave function |
| $\psi$ | $[m^{-3/2}]$ | eigenfunction |
| $\psi_i^\sigma(\mathbf{r})$ | $[m^{-3/2}]$ | eigenfunction of an electron of spin $\sigma$ |
| $\omega$ | $[rad\,s^{-1}]$ | angular frequency |
| $\Omega$ | $[m^{-3}]$ | volume in $k$-space |
| $\Phi$ | [J] | potential energy of oscillating crystal lattice |
| $\Phi_{\alpha\beta}(lk;l'k')$ | $[N\,m^{-1}]$ | Born–von Karman force constant |

## Abbreviations

| Abbreviation | Abbreviation Expansion |
|---|---|
| AFM | atomic force microscope |
| APW | augmented plane-wave method |
| bcc | body-centered cubic structure |
| BZ | Brillouin zone |
| DFT | density functional theory |
| DOS | density of states |
| fcc | face-centered cubic structure |
| GGA | generalized gradient approximation |
| hcp | hexagonal close-packed structure |
| LAPW | linearized augumented planewave method |
| LCAO | linear combination of atomic orbitals |
| LDA | local density approximation |
| LDOS | local density of states |
| LMTO | linearized muffintin orbital method |
| MBPT | many-body perturbation theory |
| MO | molecular orbital |
| NEMS | nanoelectromechanical systems |
| NFE | near-free electron model |
| SCF | self consistent field method |
| SWNT | single-walled carbon nanotube |
| TEM | transmission electron microscopy |
| WDA | weighted density approximation |

# Appendix B
# Wave Packet and the Group and Phase Velocity

Following the de Broglie equation we should write down the relation between the properties of an electron and its wave properties. Energy $E$ of a nonrelativistic electron as energy of a classical particle is given by

$$E = \frac{1}{2} m_e v^2 = \frac{p^2}{2m_e} , \tag{B1}$$

where $m_e$ is the mass of the electron, $v$ is its velocity, $p = m_e v$ is its momentum. The velocity of the electron can be expressed as the derivative of energy with respect to momentum,

$$\frac{dE}{dp} = \frac{p}{m_e} = v . \tag{B2}$$

On the other hand, energy of a quantum particle can be expressed as

$$E = h\nu ; \quad \Rightarrow \nu = \frac{E}{h} , \tag{B3}$$

where $h$ is the Plank constant, $\nu$ is frequency. Momentum of the quantum particle equals to

$$p = \frac{h}{\lambda} ; \quad \Rightarrow \lambda = \frac{h}{p} = \frac{h}{m_e v} . \tag{B4}$$

Velocity of the wave is given by

$$v = \lambda \nu . \tag{B5}$$

Combining (B5), (B3) and (B4) we arrive at

$$\lambda \nu = \left( \frac{h}{m_e v} \right) \cdot \left( \frac{E}{h} \right) = \frac{1}{2} v . \tag{B6}$$

Thus, the velocity of the wave seems to be equal to half of the velocity of the electron.

*Interatomic Bonding in Solids: Fundamentals,Simulation,andApplications*, First Edition. Valim Levitin.

From the point of view of the particle–wave dualism, the electron that has the rest mass and charge can move in a certain direction with a certain velocity. A plane wave $\psi(x, t) = \sin(kx - \omega t)$ extends to infinity in both spatial directions, so it cannot represent a particle whose wave function is non-zero in a limited region of space.

Therefore, in order to represent a localized particle, we must superpose waves having different wavelengths. The principle is illustrated by superposing two waves with slightly different wavelengths, and using the trigonometric addition formula,

$$\begin{aligned}&\sin[(k - \Delta k)x - (\omega - \Delta\omega)t] + \sin[(k + \Delta k)x - (\omega + \Delta\omega)t] = \\ &2\sin(kx - \omega t)\cos[(\Delta k)x - (\Delta\omega)t] \, .\end{aligned} \quad \text{(B7)}$$

This formula represents the phenomenon of beats between waves close in frequency. The first term, $\sin(kx - \omega t)$, oscillates at the average of the two frequencies $\omega$. It is modulated by the slowly varying second term, which oscillates once over a spatial extent of order $\pi/\Delta k$. This is the distance over which waves initially in phase at the origin become completely out of phase. Going a further distance of order $\pi/\Delta k$, the waves will become synchronized again.

Establishing that an electron moving through space must be represented by a wave packet also resolves the paradox that the velocity of the waves seems to be different from the velocity of the electron. The point is that the electron waves have differing phase and group velocities.

The waves described by the term $\sin(kx - \omega t)$ have velocity $\frac{1}{2}v$, as previously derived. But the envelope, the shape of the wave packet, has velocity $\Delta\omega/\Delta k$ rather than $\omega/k$. These velocities would be the same if $\omega$ were linear in $k$, as it is for ordinary electromagnetic waves. But the $\omega - k$ relationship follows from the energy-momentum relationship for the nonrelativistic electron, $E = \frac{1}{2}m_e v^2 = p^2/2m_e$.

So $dE/dp = p/m_e = v$. But $E = h\nu = h\omega/2\pi$, and $p = h/\lambda = hk/2\pi$.

Therefore, $\Delta\omega/\Delta k = dE/dp = v$. So the wave packet (Figure 2.2) travels at the speed we know the electron must travel at, even though the wave peaks within the wave packet travel at one-half the speed.

# Appendix C
# Solution of Equations of the Kronig–Penney Model

Here we follow the author of [3]. Inside the first square well ($b < x < a$) we have obtained the solution of the Schrödinger equation (Section 6.2)

$$\psi_k(x) = A\exp(iKx) + B\exp(-iKx)\,, \tag{C1}$$

where

$$K = \left(\frac{2m_e E}{\hbar^2}\right)^{\frac{1}{2}}\,. \tag{C2}$$

We note that the energy of the electron $E$ is directly proportional to $K^2$. Under the potential barrier $V = V_0$ for an interval $0 < x < b$ the solution can be written as a linear combination of exponential functions, namely

$$\psi_k(x) = C\exp(\gamma x) + D\exp(-\gamma x), \tag{C3}$$

where

$$\gamma = \left[\frac{2m_e(V_0 - E)}{\hbar^2}\right]^{\frac{1}{2}}\,. \tag{C4}$$

The function $\psi_k(x)$ and its first derivative $d\psi_k(x)/dx$ must be matching across the boundaries at $x = b$ and $x = a$, respectively.

Matching at $x = b$ we have

$$A\exp(iKb) + B\exp(-iKb) = C\exp(\gamma b) + D\exp(-\gamma b) \tag{C5}$$

and

$$iK[A\exp(iKb) - B\exp(-iKb)] = \gamma[C\exp(\gamma b) - D\exp(-\gamma b)]\,. \tag{C6}$$

Matching at $x = a$ using the Bloch theorem to provide the solution for $a < x < a + b$ in terms of the solution (C3) for $0 < x < b$ we have

$$A\exp(iKa) + B\exp(-iKa) = (C + D)\exp(ika) \tag{C7}$$

*Interatomic Bonding in Solids: Fundamentals,Simulation,andApplications,* First Edition. Valim Levitin.

and

$$iK[A\exp(iKa) - B\exp(-iKa)] = (C - D)\gamma\exp(ika)\,. \tag{C8}$$

These four equations (C5)–(C8) have a solution only if the determinant of the coefficients of $A$, $B$, $C$, $D$ vanishes. After nontrivial determinantal manipulation we find

$$\left(\frac{\gamma^2 - K^2}{2\gamma K}\right)\sinh(\gamma b)\sin[K(a-b)] + \cosh(\gamma b)\cos[K(a-b)] = \cos(ka)\,. \tag{C9}$$

This equation may be simplified by considering the limit in which the barrier thickness becomes increasingly thin (that is $b \to 0$) but the barrier height becomes increasingly high (that is $V_0 \to \infty$) in such a way that the area under the barrier remains constant, that is

$$V_0 b = \frac{\hbar^2}{m_e a}\mu = \text{const}\,. \tag{C10}$$

The parameter $\mu$ measures the strength of the Kronig–Penney barrier between neighboring square wells. In this limit $\gamma \sim (2m_e V_0/\hbar^2)^{1/2} \to \infty$, whereas $\gamma b \sim (2m_e V_0 b^2/\hbar^2)^{1/2} \to 0$. Thus, substituting into (C9) and using the small argument Maclaurin expansions, $\sin\gamma b \approx \gamma b$ and $\cos\gamma b \approx 1$ we have

$$\mu\frac{\sin Ka}{Ka} + \cos Ka = \cos ka \tag{C11}$$

which links the energy of the electron $E = (\hbar^2/2m_e)K^2$ to the Bloch vector $k$. See an analysis of this equation in Section 6.2.

# Appendix D
# Calculation of the Elastic Moduli

For a cubic crystal the bulk modulus $B = (C_{11} + 2C_{12})/3$. Bulk modulus is determined from the curvature of the $E(V)$ curve (see Section 10.3). The shear moduli $C_{11} - C_{12}$ and $C_{44}$ require knowledge of the energy as a function of a lattice strain. In the case of the cubic lattice, it is possible to choose this strain so that volume of the unit cell is preserved [92].

One can use for calculation of the modulus $C_{11} - C_{12}$ the volume-conserving orthorhombic strain tensor,

$$\varepsilon = \begin{pmatrix} \delta & 0 & 0 \\ 0 & -\delta & 0 \\ 0 & 0 & \frac{\delta^2}{1-\delta^2} \end{pmatrix} . \tag{D1}$$

For the elastic modulus $C_{44}$ one uses the volume-conserving monoclinic strain tensor,

$$\varepsilon = \begin{pmatrix} 0 & \frac{1}{2}\delta & 0 \\ \frac{1}{2}\delta & 0 & 0 \\ 0 & 0 & \frac{\delta^2}{4-\delta^2} \end{pmatrix} . \tag{D2}$$

The technique of calculation of elastic constants consists of the following. A value of bulk modulus is found as has been described in Section 10.3.

According to (D1) we compute the parameters $a, b, c$ of a volume-conserving orthorhombic prism. Its edges are given by

$$a = a_0(1 + \delta) ; \tag{D3}$$

$$b = a_0(1 - \delta) ; \tag{D4}$$

$$c = a_0\left(1 + \frac{\delta^2}{1 - \delta^2}\right) , \tag{D5}$$

where $a_0$ is the edge of a basic cubic cell, $\delta$ is the strain. $\alpha = \beta = \gamma = 90°$. It is easy to see that the volume of the prism $V = a_0^3 = abc$ is changeless at any strain $\delta$.

The changes of the total energy from its unstrained value can be expressed as

$$E(\delta) = E(-\delta) = E(0) + (C_{11} - C_{12})V\delta^2 + O[\delta^4]\,, \tag{D6}$$

where $V$ is the volume of the unit cell and $E(0)$ is the energy of the unstrained lattice at volume $V$.

For the elastic constant $C_{44}$ we compute parameters $a = b, c$ and $\gamma$ according to (D2),

$$a = b = \sqrt{a_0^2 + a_0^2 \frac{\delta^2}{4}}\,; \tag{D7}$$

$$c = a_0 \left(1 + \frac{\delta^2}{4 - \delta^2}\right)\,; \tag{D8}$$

$$\gamma = \arccos\left(\frac{4\delta}{4 + \delta^2}\right)\,. \tag{D9}$$

The angles $\alpha = \beta = 90°$.

The total energy changes in this case as

$$E(\delta) = E(-\delta) = E(0) + \frac{1}{2} C_{44} V \delta^2 + O[\delta^4]\,. \tag{D10}$$

Thus, one should act in order to determine $C_{11} - C_{12}$ and $C_{44}$ as follows.

- To determine bulk modulus as it was described in Section 10.3; the volume that corresponds to the minimum of energy is an initial volume.
- To construct a structure of a volume-conserving orthorhombic prism according to formulas (D3)–(D5). To find a minimum energy of structure at strain $\delta = 0, 1, 2, 3, 4\%$.
- Plot the energy as a function of strain squared as it is presented in Figure 10.9.
- Calculate the difference $C_{11} - C_{12}$ by the angle of inclination of the straight line, (D6).

A value of $C_{44}$ is found analogically using (D7)–(D9) and graph $E - \delta^2$, (D10).

# Appendix E
# Vibrations of One-Dimensional Atomic Chain

Let an infinitely long straight-line chain consisting of identical atoms of mass $m$ be given, in which the atoms are equally spaced at intervals $a$ (Figure E.1). We take an arbitrary particle as the zero one, the next is the first one, the next is the second and so on. The atoms are bonded with each other by forces of interatomic interaction. In the equilibrium state, the kinetic energy of the atomic chain is equal to zero, and its potential energy has a minimum. Accordingly, the total force acting on each atom is equal to zero. Thus, the numbers of the atoms are 0; 1; 2; ...; $n$; $n + 1$; $n + 2$; $n + 3$; ...

Strictly speaking, when considering an infinitely long chain, some physical parameters become indefinite (for instance, the total energy of vibrations and also the length). For a chain of a finite length, the boundary conditions should be taken into account by introducing additional forces acting on the boundary atoms in order to preserve the structure of the one-dimensional lattice. However, it is easier to assume that the length of the chain is essentially greater than the interatomic distance and the forces of interatomic interaction decrease rather quickly in the distance.

Let us consider the case of longitudinal vibrations. Let the wave perturbation deviate the atoms from the equilibrium position (Figure E.1). The atoms are deviated by the length of intervals denoted by $u$: the $n$th atom is displaced an amount $u_n$; the $n + 1$th, $u_{n+1}$ and so on.

One can easily obtain an expression for the kinetic energy of a vibrating crystal. For this purpose, one should sum up the kinetic energies of all atoms:

$$T = \frac{1}{2}\sum_{n=0}^{N} m\dot{u}_n^2 \tag{E1}$$

where $\dot{u}$ is the velocity of an $n$th atom.

Contrary to the kinetic energy, the potential one is a function of the mutual location of the atoms. The total potential energy depends only on the distance between the interacting atoms. For instance, consider two atoms: the atom number $i$ and the atom number $i + p$, where $p$ is a positive integer. The coordinate of the second atom is $x = pa$. The potential energy of interaction of these two atoms can be

*Interatomic Bonding in Solids: Fundamentals,Simulation,andApplications*, First Edition. Valim Levitin.

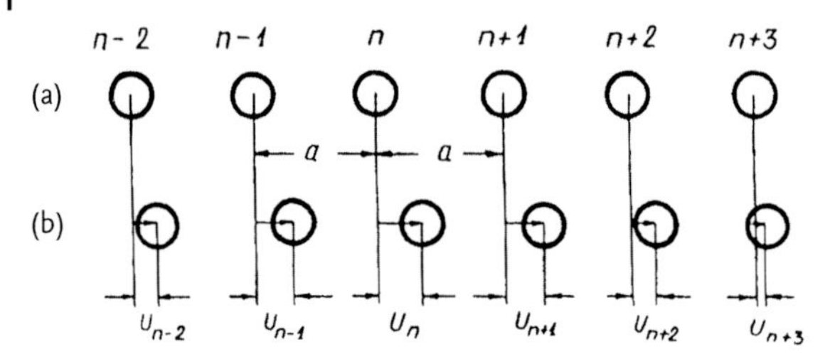


**Figure E.1** Vibrations in the infinitive one-dimensional atomic chain: (a) the atoms are in the equilibrium positions; (b) the atoms are displaced from the equilibrium positions due to passing a longitudinal wave. The vectors of the displacements $u_{n-2}, u_{n-1}, u_n, u_{n+1}, \ldots$ are shown.

written as

$$\Phi(r_{i,i+p}) = \Phi(x_{i+p} - x_i) \tag{E2}$$

where the parenthesis denotes a functional dependence. To obtain the value of the total potential energy of an one-dimensional chain, one has to sum up the interaction energy for all pairs of atoms:

$$\Phi = \sum_{i=0}^{n-1} \sum_{p=1}^{n-i} \Phi(x_{i+p} - x_i) \, . \tag{E3}$$

The number $p$ in the internal sum varies from 1 to $n - i$ and the value $i$ in the external sum varies from 0 to $n - 1$. This enables us to account the interaction of all atomic pairs once.

One can easily see that the double sum (E3) contains $n(n + 1)/2$ terms, because each of $n + 1$ atoms interacts with all the others.

Let the relative displacements $(u_{i+p} - u_i)$ be small compared with the interatomic distance. Expand the function into the Taylor series near the point $x = pa$ in powers of atomic displacements, restricting ourselves to the three terms of the series:

$$\begin{aligned}\Phi(x_{i+p} - x_i) = \Phi(pa) &+ \left(\frac{d\Phi}{du}\right)_{x=pa} (u_{i+p} - u_i) \\ &+ \frac{1}{2}\left(\frac{d^2\Phi}{du^2}\right)_{x=pa} (u_{i+p} - u_i)^2 \\ &+ \ldots\end{aligned} \tag{E4}$$

The derivatives of potential energy with respect to displacement are calculated at the point $x = pa$. Denote these derivatives $\Phi'(pa), \Phi''(pa), \ldots$

Substituting expression (E4) in (E3) we obtain, omitting the powers higher than square,

$$\Phi = \sum_{i=0}^{n-1} \sum_{p=1}^{n-i} \left[\Phi(pa) + \Phi'(pa)(u_{i+p} - u_i) + \frac{1}{2}\Phi''(u_{i+p} - u_i)^2\right] . \tag{E5}$$

Thus, we have obtained the equation for the total energy of the one-dimensional atomic chain. The force, acting on an atom, say number $r$, is equal to minus gradient of the potential energy:

$$F_r = -\frac{d\Phi}{dr} . \tag{E6}$$

One has to differentiate the double sum (E5) with respect to the displacement of the $r$th atom. Only the terms with $i = r$ or $i + p = r$ make contribution to the sum; all other terms do not depend on $u_r$. The summation over $i$ is not needed now.

The force is given by

$$F_r = \frac{\partial}{\partial u_r} \sum_{i=0}^{n-1} \sum_{p=1}^{n-i} \left[ \Phi(pa) + \Phi'(pa)(u_{i+p} - u_i) + \frac{1}{2}\Phi''(pa)(u_{i+p} - u_i)^2 \right] . \tag{E7}$$

Regrouping the terms, assuming $n = 2r$ we obtain for a force acting on the chosen atom $r$ by all other atoms as a result of their displacements from the equilibrium positions:

$$F_r = \sum_{p=1}^{n-r} \Phi''(pa)(u_{r+p} + u_{r-p} - 2u_r) . \tag{E8}$$

The equation of motion for the $r$th atom can be obtained using (E8) by Newton's Second law,

$$m\frac{\partial^2 u_r}{\partial t^2} = \sum_{p=1}^{n-r} \Phi''(pa)(u_{r+p} + u_{r-p} - 2u_r) . \tag{E9}$$

The linear homogenous differential equation of second order (E9) has a solution of the form,

$$u_r = Ae^{i\left(\frac{2\pi}{\lambda} ra - \omega t\right)} . \tag{E10}$$

This is the wave equation, see (2.12).

Substituting (E10) into (E9) and using the Euler formula we find,

$$m\omega^2 = 2\left[\sum_{p=1}^{n-r} \left(\Phi''(pa) - \Phi''(pa)\cos\frac{2\pi}{\lambda} pa\right)\right] . \tag{E11}$$

Denoting $\Phi''(pa) = F_{\rm p}$, $n - r = N$, $a = b$ and taking into account that $q = 2\pi/\lambda$, $\zeta = q/q_{\rm m}$, $q_{\rm m} = 2\pi/2b$ we arrive at

$$\omega^2(\zeta) = \frac{2}{m}\left[\sum_{p=1}^{N} F_{\rm p} - \sum_{p=1}^{N} F_{\rm p}\cos\pi p\zeta\right] . \tag{E12}$$

The equation (E12) allows one to calculate the $F_{\rm p}$ coefficients if there is an experimental dispersion curve $\nu(\zeta)$.

# References


1 Martin, R.M. (2008) *Electronic Structure: Basic Theory and Practical Methods*, Cambridge University Press, Cambridge.

2 Eisberg, R. and Resnick, R. (1985) *Quantum Physics of Atoms, Molecules, Solids, Nuclei, and Particles*, 2nd edn, John Wiley & Sons, NY, Chichester, Brisbane, Singapore.

3 Pettifor, D.G. (1996) *Bonding and Structure of Molecules and Solids*, Clarendon Press, Oxford.

4 Singleton, J. (2009) *Band Theory and Electronic Properties of Solids*, Oxford University Press, Oxford.

5 Drude, P. (1900) Zur Elektronentheorie der Metalle. *Ann. Phys.*, I, **306**, 566–613; II, **308**, 369–402, (1902); III, **312**, 687–692.

6 Mahan, G.D. (1981) *Many-Particle Physics*, Plenum Press, NY, Ltd.

7 Ashcroft, N.W. and Mermin, N.D. (1976) *Solid State Physics*, Thomson Learning, Inc., Australia et al.

8 Yin, M.T. and Cohen, M.I. (1982) Theory of static structural properties, crystal stability, and phase transformations: Application to Si and Ge. *Phys. Rev. B*, **26**, 5668–5687.

9 Landolt, H. and Börstein, R. (1981) *Metals: Phonon states, Electron states, and Fermi Surfaces*, Numerical Data and Functional relationships in Science and Technology, New Series, Group III, vol. 13, subvolume A. Springer, Berlin.

10 James, R.W. (1950) *The Optical Principles of the Diffraction of X-Rays*, London.

11 Landolt, H. and Börstein, R. (1991) *Zahlenwerte und Funktionen aus Naturwissenschaften und Technik: Neue Serie, Gruppe 3*. Kristall- und Festkörperphysik, Bd.25. Atomare Fehlstellen in Metallen, Springer, Berlin.

12 Hohenberg, P. and Kohn, W. (1964) Inhomogeneous electron gas. *Phys. Rev.*, **136**, B864–B871.

13 Kohn, W. and Sham, L.J. (1965) Self-consistent equations including exchange and correlation effects. *Phys. Rev.*, **140**, A1133–A1148.

14 Singh, D.J. and Nordström, L. (2006) *Planewaves, Pseudopotentials and the LAPW Method*, 2nd edn, Springer, USA.

15 Perdew, J.P. and Wang, Y. (1992) Accurate and simple analytic representation of the electron-gas correlation energy. *Phys. Rev. B*, **45**, 13244–13249.

16 Perdew, J.P., Burke, K. and Ernzerhof, M. (1996) Generalized gradient approximation made simple. *Phys. Rev. Lett.*, **77**, 3865–3868.

17 Yip, S., Editor (2005) *Handbook of Materials Modeling. Part A. Methods*, Springer, The Netherlands.

18 Payne, M.C., Teter, M.P., Allan, D.C., Arrias, T.A. and Joannopoulos, J.D. (1992) Iterative miniminization techniques for ab initio total energy calculations: Molecular dynamics and conjugate gradients. *Rev. Mod. Phys.*, **64**, 1045–1097.

19 Blaha, P., Schwarz, K., Madsen, G., Kvasnicka, D., and Luitz, J. (2001) WIEN2k, An Augmented Plane Wave Plus Local Orbitals Program for Calculating Crystal Properties. (Karlheinz Schwarz, Techn. Universität Wien, Austria), ISBN 3-9501031-1-2.

20 Williams, A.R., Kübler, J., and Gelatt, C.D.Jr. (1979) Cohesive proper-


*Interatomic Bonding in Solids: Fundamentals,Simulation,andApplications*, First Edition. Valim Levitin.

ties of metallic compounds: Augmented-spherical-wave calculations. *Phys. Rev. B*, **19**, 6094–6118.

**21** Bylander, D.M. and Kleinman, L. (1983) Self-consistent semirelativistic pseudopotential calculation of the energy bands, cohesive energy and bulk modulus of W. *Phys. Rev.*, **27**, 3152–3159.

**22** Philipsen, P.H.T. and Baerends, E.J. (2000) Relativistic calculations to assess the ability the generalized gradient approximation to reproduce trends in cohesive properties of solids. *Phys. Rev. B*, **61**, 1773–1778.

**23** Cortona, P. and Monteleone, A.V. (1996) *Ab initio* calculations of cohesive and structural properties of the alkali-earth oxides. *J. Phys. Condens. Matter*, **8**, 8983–8994.

**24** Mehl, M.J., Osburn, J.E., Papaconstantopoulas, D.A. and Klein, B.M. (1990) Structural properties of ordered high-melting-temperature intermetallic alloys from first-principles total-energy calculations. *Phys. Rev. B*, **41**, 10311–10323.

**25** Korhonen, T., Puska, M.J. and Nieminen, R.M. (1995) Vacancy-formation energies for fcc and bcc transition metals. *Phys. Rev. B*, **51**, 9526–9532.

**26** Sholl, D.S. and Steckel, J.A. (2009) *Density Functional Theory. A Practical Introduction.* John Wiley & Sons, Inc., Hoboken.

**27** Jahnátek, M., Krajčí, M. and, Hafner, J. (2007) Interatomic bonds and the tensile anisotropy of trialuminides in the elastic limit: A density functional study for $Al_3$(Sc, Ti, V, Cr). *Philos. Mag.*, **87**, 1769–1794.

**28** Choi, H.J., Roundy, R., Sun, H., Cohen, M. and Louie, S.G. (2002) The origin and the anomalous superconducting properties of $MgB_2$. *Nature*, **418**, 758–760.

**29** An, J.M. and Picket, W.E. (2001) Superconductivity of $MgB_2$: Covalent bonds driven metallic. *Phys. Rev. Lett.*, **86**, 4366–4369.

**30** Schweinfest, R., Paxton, A.T. and Finnis, M.W. (2004) Bismuth embrittlement of copper is an atomic size effect. *Nature*, **432**, 1008–1011.

**31** Levitin, V.V. (2004) *Atom Vibrations in Solids: Amplitudes and Frequencies.* Phys. Reviews, Volume 21, Part 2, Cambridge Scientific Publishers.

**32** Mattheiss, L.F. (1964) Energy bands for the iron transition series. *Phys. Rev.*, **134**, A970–A973.

**33** Mehl, M.J. and Papaconstantopoulos, D.A. (1996) Applications of a tight-binding total-energy method for transition and noble metals: Elastic constants, vacancies, and surfaces of monatomic metals. *Phys. Rev. B*, **54**, 4519–4530.

**34** Paxton, A.D. (2009) An introduction to tight-binding approximation – Implementation by diagonalisation. *Multiscale Simulation Methods in Molecular Sciences, Institute for Advanced Simulation, Forschungszentrum Jülich*, **42**, 145–176.

**35** Wang, C.Z., Pan, B.C. and Ho, K.M. (1999) An environment-dependent tight-binding potential for Si. *J. Phys. Condens. Matter*, **11**, 2043–2049.

**36** Tang, M.S., Wang, C.Z., Chan, C.T., Ho, K.M. (1996) Environment-dependent tight-binding potential model. *Phys. Rev. B*, **53**, 979–982.

**37** Wang, C.Z. and Ho, K.M. (2005) Environment-dependent tight-binding potential models. *Handbook of Materials Modeling*, Part A. Methods, Springer, The Netherlands, 307–347.

**38** Daw, M.S. and Baskes, M.I. (1983) Semiempirical, quantum mechanical calculation of hydrogen embrittlement in metals. *Phys. Rev. Lett.*, **50**, 1285–1288.

**39** Daw, M.S. and Baskes, M.I. (1984) Embedded-atom method: Derivation and application to impurities, surfaces and other defects in metals. *Phys. Rev. B*, **29**, 6643–6653.

**40** Baskes, M.I. (1992) Modified embedded-atom potentials for cubic materials and impurities. *Phys. Rev. B*, **46**, 2727–2742.

**41** Mishin, Y. (2005) Interatomic potentials for metals. *Handbook of Materials Modeling*, Part A. Methods, Springer, The Netherlands, 459–478.

**42** Mitev, P., Evangelakis, G.A., and Efthimios Kaxiras (2006) Embedded atom method potentials employing a faithful density representation. *Modelling and Simulation in Mater. Science and Engineering*, **14**, 721–731.

**43** Williams, P.L., Mishin, Y., and Hamilton, J.C. (2006) An embedded-atom potential for the Cu-Ag system. *Modelling and Simulation in Mater. Science and Engineering*, **14**, 817–833.

**44** Mishin, Y. (2004) Atomistic modeling of the $\gamma$ and $\gamma'$-phases of the Ni-Al system. *Acta Mater.* **52**, 1451–1467.

**45** Mishin, Y., Mehl, M.J., Papaconstantoupolos, D.A., Voter, A.F., and Kress, J.D. (2001) Structural stability and lattice defects in copper: *Ab initio*, tight-binding, and embedded-atom calculations. *Phys. Rev. B*, **63**, 224106-1–224106-16.

**46** Girifalco, L.A. and Weizer, V.G. (1959) Application of the Morse potential function to cubic metals. *Phys. Rev.*, **114**, 687–690.

**47** Born, M. and Huang, K. (1985) *Dynamical Theory of Crystal Lattice*. Oxford.

**48** Foreman, E.A.J. and Lomer, W.M. (1957) Lattice vibration and harmonic forces in solids. *Proc. Phys. Soc. B*, **70**, 1143–1150.

**49** Watson, R.E. and Weinert, M. (2001) Transition-metals and their alloys, in *Solid State Physics*, vol. 56, (eds H. Ehrenreich and F. Spaepen), Academic Press, San Diego, pp. 1–112.

**50** Ho, K.-M., Fu, C.L., and Harmon, B.N. (1984) Vibrational frequencies via total-energy calculations. Application to transition metals. *Phys. Rev.*, **29**, 1575–1587.

**51** Moruzzi, V.L., Williams, A.R., and Janak, J.F. (1977) Local density theory of metallic cohesion. *Phys. Rev. B*, **15**, 2854–2857.

**52** Gelatt, C.D.Jr., Ehrenreich, H., and Watson, R.E. (1977) Renormalized atoms: Cohesion in transition metals *Phys. Rev. B*, **15**, 1613–1628.

**53** Watson, R.E., Weinert, M., Davenport, J.W., and Fernando, G.W. (1989) Energetics of transition-metal alloy formation: Ti, Zr, and Hf alloyed with the heavier 4d and 5d elements. *Phys. Rev. B*, **39**, 10761–10789.

**54** Ho, K.M., Louie, S.G., Chelikowsky, J.R., Cohen, M.L. (1977) Self-consistent pseudopotential calculation of the electronic structure of Nb. *Phys. Rev.*, **15**, 1755–1759.

**55** Grimwall, G. and Ebbsjö, I. (1975) Polymorphism in metals, *Phys. Scr.*, **12**, 168–172.

**56** Grimwall, G. (1975) Polymorphism in metals. *Phys. Scr.*, **12**, 173–176.

**57** *Handbook Series on Semiconductor Parameters*, vol. 1,2 (eds Levinstein, M., Rumyantsev, S. and Shur, M.), World Scientific, London, 1996, 1999.

**58** Batterman, B.W. and Chipman, D.R. (1962) Vibrational amplitudes in germanium and silicon. *Phys. Rev.*, **127**, 690–693.

**59** Kyutt, R.N. (1978) Mean-square displacements of atoms and the Debye temperatures of crystals $A^{III}B^{V}$. *Solid State Phys.*, **20**, 395–398.

**60** Sirota, N.N. and Sidorov, A.A. (1985) Temperature dependence of the intensity of diffraction maxima on X-ray pictures of the semiconductor compounds GaAs, InAs, InP in the temperature range 7–310 K. *Rep. Acad. Sci.*, **280**, 352–356.

**61** Bernstein, N. (2005) Atomistic simulation of fracture in semiconductors. *Handbook of Materials Modeling*, Part A. Methods, Springer, The Netherlands, 855–873.

**62** Pérez, R. and Gumbsch, P. (2000) An *ab initio* study of the cleavage anisotropy in silicon. *Acta Mater.*, **48**, 4517–4530.

**63** Editorial. Graphene calling. *Nat. Mater.*, **6**, 169, 2007.

**64** Geim, A.K. and Novoselov, K.S. (2007) The rise of graphene. *Nat. Mater.*, **6**, 183–191.

**65** Novoselov, K.S., Jiang, D., Schedin, F., Booth, T.J., Knotkevich, V.V., Morozov, S.V., and Geim, A.K. (2005) Two-dimensional atomic crystals. *Proc. Natl. Acad. Sci. USA*, 102, **30**, 10451–10453.

**66** Frank, I.W., Tanenbaum, D.M., van der Zande, A.M., and McEulen, P.L. (2007) Mechanical properties of suspended graphene sheets. *J. Vac. Sci. Technol. B*, **25**(6): 2558–2561.

**67** Lee, C., Wei, X., Kysar, J.W., and Hone, J. (2008) Measurement of the elastic properties and intrinsic strength of monolayer graphene. *Science*, **321**, 385–388.

**68** Tsoukleri, G., Parthenios, J., Papagelis, K., Jalil, R., Ferrari, A.C., Geim, A.K.,

Novoselov, K.S., and Galiotis, C. (2009) Subjecting a graphene monolayer to tension and compression. *Small*, 5, **21**, 2397–2402.

**69** Salvetat, J.P., Bonard, J.M., Thomson, N.H., Kulik, A.J., Forró, L., Benoit, W., Zuperolli, L. (1999) Mechanical properties of carbon nanotubes. *Appl. Phys. A*, **69**, 255–260.

**70** Pauling, L. (1960) *The Nature of the Chemical Bond*, Cornell University Press, Ithaca, New York.

**71** Jeffrey, G.A. (1997) *An Introduction to Hydrogen Bonding*, Oxford University Press, Oxford.

**72** Petrenko, V.F. and Whitworth, R.W. (2002) *Physics of Ice*, Oxford University Press, Oxford.

**73** Allen, L. (1975) A simple model of hydrogen bonding. *J. Am. Chem. Soc.*, **97**, 6921–6940.

**74** Schulson, E.M. (1999) Structure and mechanical behavior of ice. *J. Miner., Met. Mater. Soc.*, **51**, 21–27.

**75** Harl, J. and Kresse, G. (2008) Cohesive energie curves for noble gas solids calculated by adiabatic connection fluctuation-dissipation theory. *Phys. Rev. B*, **77**, 045136-1–045136-8.

**76** Wright, J.D. (1995) *Molecular Crystals*. Cambridge University Press, Cambridge.

**77** Öhrström, L. and Larsson, K. (2005) *Molecule-Based Materials. The Structural Network Approach*. Elsevier, Amsterdam.

**78** Kittel, C. (1996) *Introduction to Solid State Physics*. 7 edn, John Wiley & Sons, Ltd, New York.

**79** Levitin, V. (2006) *High Temperature Strain of Metals and Alloys. Physical Fundamentals*. Wiley-VCH Verlag GmbH, Weinheim.

**80** Roytburd, A.L. and Zilberman, L.A. (1966) The motion of rectilinear dislocations with jogs. *Phys. Met. Metallogr.*, **21**, 647–656.

**81** Friedel, J. (1964) *Dislocations*, Pergamon Press, Oxford-London.

**82** Mrowec, S. (1980) *Defects and Diffusion in Solids*, Elsevier, Amsterdam.

**83** Levitin, V. and Loskutov, S. (2009) *Strained Metallic Surfaces: Theory, Nanostructuring and Fatigue Strength*, Wiley-VCH Verlag GmbH, Weinheim.

**84** Mura, T. (1994) A theory of fatigue crack initiation. *Mater. Sci. Eng.*, **A176**, 61–70.

**85** Wagner, L. (1999) Mechanical surface treatments on titanum, aluminum and magnesium alloys. *Mater. Sci. Eng.*, **A263**, 210–216.

**86** Roland, T., Retraint, D., Lu, K., Lu, J. (2006) Fatigue life improvement through surface nanostructuring of stainless steel by means of surface mechanical attrition treatment. *Scr. Mater.*, **54**, 1949–1954.

**87** Nalla, R.K., Altenberger, I., Noster, U., Liu, G.Y., Scholtes, B., Ritchie, R.O. (2003) On the influence of mechanical surface treatments – deep rolling and laser shock peening – on the fatigue behaviour of Ti-6Al-4V at ambient and elevated temperatures. *Mater. Sci. Eng.*, **A355**, 216–230.

**88** François Cardarelli (2007) *Materials Handbook*, Springer, London.

**89** Missol, W. (1973) Calculation of the surface energy of solid metals from work function values and electron configuration data. *Phys. Status Solidi (b)*, **58**, 767–773.

**90** Toparli, M., Özel, A. and Aksoy, T. (1997) Effect of the residual stresses on the fatigue crack growth behavior at fastener holes. *Mater. Sci. Eng.*, **A225**, 196–203.

**91** Almer, J.D., Cohen, J.B., and Moran, B. (2000) The effects of residual macrostresses on fatigue crack initiation. *Mater. Sci. Eng.*, **A284**, 268–279.

**92** Mehl, M.J. (1993) Pressure dependence of the elastic moduli in aluminum-rich Al-Li compounds. *Phys. Rev. B*, **47**, 2493–2500.

# Index



*Interatomic Bonding in Solids: Fundamentals,Simulation,andApplications*, First Edition. Valim Levitin.